PRINCIPLES OF
TESTING ELECTRONIC
SYSTEMS

12/07

PRINCIPLES OF TESTING ELECTRONIC SYSTEMS

SAMIHA MOURAD
Santa Clara University

YERVANT ZORIAN
Logic Vision

A Wiley-Interscience Publication
JOHN WILEY & SONS, INC.
New York • Chichester • Weinheim • Brisbane • Singapore • Toronto

Copyright © 2000 by John Wiley & Sons, Inc. All rights reserved.

Published simultaneously in Canada.

For ordering and customer service, call 1-800-CALL-WILEY.

Library of Congress Cataloging-in-Publication Data:

Mourad, Samiha.
    Principles of testing electronic systems/Samiha Mourad, Yervant Zorian.
        p. cm.
    "A Wiley-Interscience publication."
    Includes bibliographical references and index.
    ISBN 0-471-31931-7
    1. Electronic circuits—Testing. I. Zorian, Yervant. II. Title.
    TK7867 .M697 2000
    621.3815′48–dc21   99-052179

Printed in the United States of America

10 9 8 7 6 5 4 3 2 1

CONTENTS

2 Defects, Failures, and Faults   27

3 Design Representation   57

10 Boundary-Scan Testing   241

PREFACE

The challenge of testing electronic systems has grown rapidly over the last decade. This is due to the complexity of designing easily testable circuits. As the technology feature size decreases and designers keep increasing chip size, testing encounters difficult challenges. Along with this evolution in electronic systems, there is an evolution in design automation tools. Practically every phase of the design process is presently automated, including high-level synthesis. Earlier testability used to be considered only at the gate level. However, the complexity of present circuits and the advent of systems on a chip (SOCs) makes it mandatory to start considering testing early in the design process. All these factors pose many challenges to the design and test engineers. These challenges include, but are not limited to, development of more accurate failure modeling, examining testability on a higher level of design representation, and embedding more effective test constructs prior or during synthesis. Another major challenge to VLSI testing is the large volume of testing data that has to go through the IC.

The rapid change in technology and the growing need for efficient solutions to the problems encountered in design and test are still not matched with an increase in the number of qualified engineers to handle these tasks. Many practicing engineers are involved in testing and design for test, but they rarely have the opportunity to study the field in a systematic fashion. All their learning is mainly on the job. Although learning by doing is an acceptable mode of acquiring knowledge, it is much more productive if before being immersed in testing, these engineers are knowledgeable of its foundation. It is for these engineers that the book is mainly written. The major features of this book include:

- Relevant discussion throughout the book of where the test is situated in the design process
- Detailed discussion of issues related to scan path and ordering of scan chain registers
- Self-test methodologies for random logic and memories
- Test methodologies for RAM-based FPGAs
- Chapter on testing a system on chip

To have a good foundation in testing, one must understand five basic disciplines: (1) physical defects and their manifestation on the circuit level, (2) test pattern generation for detecting faults, (3) the relationship of testing to the design cycle, (4) design for test practice, and (5) testing the manufactured chip.

The first four disciplines are detailed in the book. The approach followed is not theoretical. The intent here is to sacrifice theoretical elegance in favor of a phenomenological understanding of the topic. For example, it is possible to use solely a mathematical basis to explain the D-algorithm for test pattern generation. However, a flowchart explanation of the algorithm is sufficient and more appropriate for its implementation in software tools. This pragmatic approach does not exclude giving the readers the opportunity to explore the theoretical foundation in depth. For this, the book includes a bibliography of specialized testing books, a list of magazines and archival journals, and a list of Web sites.

The book is organized into five main parts. Part I consists of four chapters. It is the road map for the book. From the onset it relates design and testing for obtaining reliable electronic products. Chapter 1 is an introduction to the objectives of the book in which we distinguish design verification from testing. Whereas design verification checks the compliance of the design to specifications, the intent of testing is to show that the design implementation on silicon is error-free. In this chapter we also briefly describe the preparation needed to test the final product and the processes and tools used in this preparation: test pattern generation and design to facilitate testability of the product.

Before detailing the topics presented in Chapter 1, it is important to know the causes of product failures. In Chapter 2 we provide a thorough description of the possible defects encountered in electronic products and explain the need for mapping these defects to faults on the structural and behavioral levels of the design. Most of the failures of an electronic product are due to manufacturing defects, but they may also be due to noise induced by operation of the product. Since the overwhelming majority of VLSI systems utilize CMOS, the emphasis will be on MOS technology.

Preparation for testing the product starts from its inception as a design idea. It is thus very important to learn how the design is represented at different stages of its life cycle. Toward this end, we present in Chapter 3 a taxonomy of design that is used to explain the various activities of the design cycle in Chapter 4. The automated design processes are known as synthesis and follow algorithms that operate primarily on graphs. The knowledge gained in this part has a twofold benefit. First, it will facilitate the understanding of the following parts of the book. Second, it will mold a mind-set on why design and test can no longer be independent topics.

Part II comprises three chapters. Although much effort at present is put to use formal verification techniques in CAD tools, simulation is still the main vehicle for design verification. It plays an indispensable role in product design and testing. In addition, fault simulation helps in assessing the quality of the test pattern generated to test the products. In Chapter 5 we show how fault simulation evolved to facilitate the verification of today's very dense and large circuits.

In Chapter 6 we address what is known as deterministic test pattern generation. In general, test pattern generation relied on fault detection by measuring the voltage signal at the circuit outputs. Today, in addition to voltage measurement, current measurement is also used. This approach, which is known as I_{DDQ} testing, is possible only for CMOS circuits that constitute most present-day electronic circuits. In Chapter 7 we present current testing, which is very effective at uncovering many defects that voltage testing cannot detect. A new paradigm of *defect detection* is becoming very relevant and is utilized in addition to traditional fault detection. Moreover, current testing facilitates *defect diagnosis.*

In Part III we utilize the knowledge presented in the preceding seven chapters and pave the way to designs that facilitate testing. This part of the book consists of four chapters. In Chapter 8 we outline common sense approaches to DFT. As simple as they are, these ad hoc techniques can be very powerful. For example, a divide-and-conquer approach may make quite a difference in testing a product. The next three chapters cover structured techniques. In Chapter 9 we describe a way that helps reduce the complexity of mainly synchronous sequential circuits. Scan-path design is very widely used and is becoming a standard feature in electronic circuits. It has also been practiced on asynchronous circuits. Chapter 10 is on boundary-scan design, a technique initially thought to be used for printed circuit boards, which, however, is also becoming very useful for integrated circuits. This DFT technique follows IEEE Standard 1149.1 and is used for debugging and diagnosis. Chapter 11 is the last chapter in this part. Built-in self-test (BIST) technique is effectively applied to random logic. It is also used for memory testing, microprocessors, and FPGAs. After its principles are explained, its use in conjunction with scan design for random logic is also described.

In Part IV, which comprises two chapters, we investigate how testing is performed or applied to the ubiquitous RAMs, FPGAs, and microprocessors. Each of the three structures presented makes use of a functional fault model, since the use of structural fault models would make their testing untraceable. Testing based on the functional model also detects failures on the associated circuitry: decoding logic and the attached registers.

In Chapter 12 we investigate RAMs that have regular structures but are very dense and cause a whole set of problems in their testing. These problems are compounded when the RAM is embedded in logic chip or in a SOC. In Chapter 13 we address another two commonly used circuits. FPGAs come in different varieties, but the main interest here is in RAM-based ones. They also have regular array structures and are much less dense than RAMs. Their regular structure is exploited to make them C-testable. Microprocessors are some of the most used designs, but their testability is not explored sufficiently. Few models are used for testing microprocessors. Most embedded testability constructs, such as scan path and BIST, are widely used in microprocessor testing.

Part V is a synthesis of all the main topics covered in earlier chapters. As we advocated there, the design cycle should incorporate test preparation and design for testability constructs and be thought of as the design for a testability cycle. The two-pass approach that first completes the design then adds on these constructs and usually yields a lower-than-desirable circuit performance. Instead, a one-pass approach that includes the testability issues from the onset of the design cycle results in a circuit that is optimized for area, performance, and testability. In Chapter 14 we illustrate how constraints are placed on the synthesis process to achieve this one-pass approach. All knowledge acquired so far is utilized in Chapter 15. However, now additional problems are encountered, which are caused by embedding a variety of pre-designed blocks on the same IC. The volume of testing data increases but the throughput is still low. Testing SOCs is presenting a challenge to the testing community.

The presentation in the book combines the long teaching experience of one of the authors and the intensive industrial experience of the other author. In selecting and presenting the topics, we strove for a balance between the educational merits and the practical reality of testing.

ACKNOWLEDGMENTS

This book has been motivated and shaped by several friends, colleagues, and students. We are grateful to all of them and, in particular, Sunil Das, University of Toronto, Toronto, Canada; Yacoub M. EIZiq, SUN Microsystems, Menlo Park, California; F. Joel Ferguson, University of California at Santa Cruz, California; Suresh Gopalakrishnan, Cabeltron, Santa Clara, California; Daryl Hoot, Marvell, Sunnyvale, California; London Jin, Toshiba, Sunnyvale, California; Samy Makar, Transmeta, Santa Clara, California; Rafic Makki, University of North Carolina at Charlotte, North Carolina; Takashi Nanya, University of Tokyo, Tokyo, Japan; Kenneth P. Parker, Hewlett Packard, Fort Collins, Colorado; Prachi Sathe, Santa Clara University, Santa Clara, California; and A. J. van de Goor, Delft University of Technology, The Netherlands.

We also thank George Telecki and Cassie Craig for their helpful suggestions and their patience with our several delays in submitting the manuscript. The preparation of the manuscript would not have been possible without the help of

Martha Giannini, Mike Daly, and Amit Hakoo. The authors also acknowledge the benefits received from the comments of many Santa Clara students who used the first draft of this book. We are extremely grateful to everyone who contributed to our endeavor.

SAMIHA MOURAD
YERVANT ZORIAN

Palo Alto, California
Santa Clara, California

PART I

DESIGN AND TEST

1

OVERVIEW OF TESTING

1.1 RELIABILITY AND TESTING

The reliability of electronic systems is no longer a concern limited to the military, aerospace, and banking industries, where failure consequences may have catastrophic impact. Reliability and testing techniques have become of increasing interest to all other applications, such as computers, telecommunications, consumer products, and automotive industry, because of the following factors: (1) at present, electronic systems are becoming ubiquitous in the workplace and are now being used in harsher environments; (2) because of the proliferation of their use, users of electronic systems are not necessarily experienced people and may misuse the machines inadvertently; and (3) the continuous decrease in technology feature size, accompanied by an increase in systems complexity and speed, has resulted in newer failure modes.

A key requirement for obtaining reliable electronic systems is the ability to determine that the systems are error-free [Breuer 1976]. Electronic systems consist of hardware and software. In this book we investigate only hardware testing. The majority of the hardware used today consists of digital circuits, and this book concentrates on *digital testing*. A circuit must be tested to guarantee that it is working and continues to work according to specifications. Such testing detects failures due to manufacturing defects. It can also detect many field failures due to aging, environmental changes, power supply fluctuations, and so on. Test pattern generation is a complex problem. Often, simulation patterns developed for design verification are augmented with patterns that are generated manually or by an automatic test pattern generator (ATPG) to obtain a complete test set, capable of detecting all faults in the circuit. This test also verifies the functionality of its logic [Abadir 1989]. The patterns are then applied to the circuit using automatic test equipment (ATE). Because of the complexity of testing processes, a design approach aimed at making digital circuits more easily testable has been formulated. This approach to design, known as *design for test* (DFT) or *design for testability*, is discussed in subsequent chapters. Its aim is to make these circuits more controllable and observable by embedding test constructs into the design.

Another aspect of reliability is the system's ability to run dependently on demand. This requires that the system be fault tolerant. Fault tolerance is a vast and rich field that involves digital testing, but it is not possible in this book to give it the justice it deserves. To explore this topic further, it is recommended that you consult such references as [Siewiorick 1982] and [Johnson 1989].

In this chapter we first give a brief description of the digital design process to better understand the role and scope of verification and testing and the relationship between the two processes. In the balance of the chapter we give a synopsis of what is discussed in the remainder of the book.

1.2 DESIGN PROCESS

As a result of the continuous decrease in the minimum feature size of transistors, both device density and design complexity have steadily increased. Densities of hundreds of million transistors are now a reality, and to manage such complexity, it is only natural to design on a hierarchical basis. In addition to their complexity, digital systems have a life cycle that sometimes becomes shorter than their design cycle. To stay competitive in the electronics field, vendors need to increase designers' efficiency and to reduce time-to-market.

The design process has been changing continually to reflect the status of the technology and of the computer-aided design (CAD) tools available to designers. We can view the process as divided into three main phases: system design, logic design, and physical design. The time taken to accomplish any of the three phases has changed due primarily to the availability of automation. This is illustrated in Fig. 1.1. Each of the design phases comprises a means to enter the design and verify it, then transform it to the next phase. Usually, simulation is used for verification, although more recently, formal verification has been gaining in importance.

The shift toward submicron technologies has allowed integrated-circuit (IC) designers to increase the complexity of their designs to the extent that an entire system can now be implemented on a chip. This new paradigm of *system on a chip* (SOC) has changed the approach to design and testing. To increase the design productivity, and hence to decrease time-to-market, the reuse of previously designed modules is becoming common practice in SOC design. This is known as *core-based design*. The reuse approach is not limited to in-house designs, but is extended to modules that have been designed by others. Such modules are referred to as *embedded cores*.

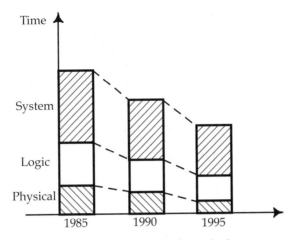

Figure 1.1 Design time for each phase.

An example of an SOC is shown in Fig. 1.2. It consists of different cores (rectangular blocks) and user-defined logic (UDL). The cores may be processors, DSP, RAMs, and so on. The UDL components are "gluing" the various cores for the intended system. The cores may have been designed previously. They may also be described in different levels, from specifications to hardware description language to layout. Because of the complexity of designing an SOC, it is likely that the system design phase will be longer than the other design phases, as illustrated in Fig. 1.1. We defer the discussion of the design and testing of SOCs to Chapter 15. In the rest of the section, we discuss how a core or a UDL component is designed from specifications.

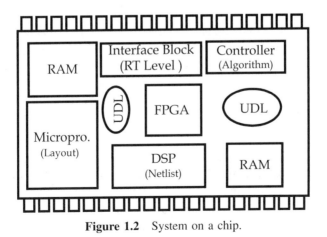

Figure 1.2 System on a chip.

Whereas in the past, hardware design was done primarily on paper, it is now described in hardware description language (HDL). Expressing the design in HDL has several advantages. The most important advantages are management of complexity and shortening the design cycle. At present, the design may be entered on a behavioral level or immediately in register transfer level (RTL), then mapped into logic level as illustrated in Fig. 1.3. Finally, the design is transformed into layout masks. These automated transformations are known as *design synthesis*. Both synthesis and simulation tools utilize libraries of components. For example, consider the logic description of a design. Assume that it consists of basic logic gates such as NAND, NOR, and so on. These components form the library on the logic level. Every logic gate has a layout view that is process dependent. The layouts of the various gates constitute the physical library. Designers have the option of using a standard cell library available from vendors. They may augment these libraries or even develop their own libraries. The transformation from the gate level to the physical level is known as *technology mapping*.

In stepping from one level to another, simulation is repeated at each level until the design is satisfactory. Upon completion of the layout, it is possible to extract parameters such as the load resistance and capacitance. With this information, it is thus possible to verify timing of the circuit. Parameter extraction is becoming significantly more important in present deep submicron technology.

The artwork, the layout masks, are then used to fabricate the design in the form of a die on a wafer. After fabrication, the wafer is tested and defective dies are marked. This process is known as *wafer sort*. The good dies are then packaged and tested again using ATE equipment. Because of the close relationship of the design process to testing, two chapters are devoted to this topic and to design representation.

1.3 VERIFICATION

Testing a circuit prior to its implementation is known as *design verification*. No engineer would take a design to foundry prior to verifying it. The question is not *whether one should verify* but *how well to verify* in order to have confidence that the device will comply with its specifications. During its design cycle, a circuit has several descriptions, as illustrated in Fig. 1.3. The four representations of the design—behavioral, RTL, gate level, and layout—are different perspectives of the same circuit. In mapping the design from one phase to another, it is likely that some errors are produced. Errors in mapping may be due to the CAD tools or to human mishandling of the tools. At each stage, the design is verified to assert that it is the same design from the previous phase as well as that it adheres to the specifications. At present, simulation is the most *popular* tool for hardware verification. Although to date, extensive research has been done in formal verification, there are few practical tools to prove correctness of design based on the formal verification. Exhaustive simulation is equivalent

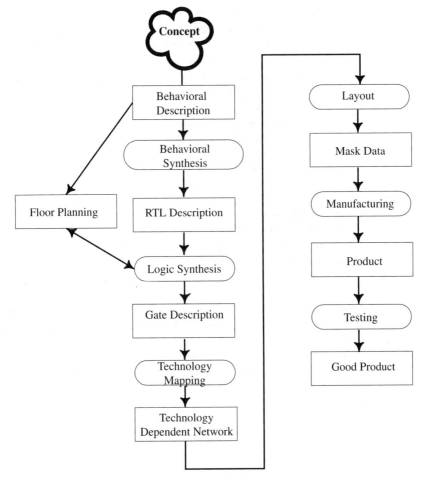

Figure 1.3   Design cycle.

to formal verification. However, such a simulation is not feasible for present complex circuits. Two types of simulations are used to verify the design: *functional simulation* and *timing simulation*.

1.3.1 Functional Simulation

In functional simulation, the designer may verify the correctness of the design but not necessarily at operational speed; that is, no delays (or, at most, constant delay) of the functional units are included. This is usually known as *zero* (or *unit*) *delay*. The primary concerns are (1) to check if each block performs its intended function and (2) to modify the design to evaluate alternatives before finalizing the design.

The simulator translates the circuit description into an appropriate internal data structure that facilitates its interpretation. For example, a schematic is mapped into a *netlist*, which is a data structure listing of all the gates and their connectivities. In Chapter 3 we describe the various circuit representations and their influence on the performance of CAD tools. The designer prepares a set of inputs, verification patterns, applies them to the circuit, and then examines the output for correctness. Then, if the designer is satisfied with the outcome, the translation should be performed. It is a good practice to use the same set of test vectors on the actual device to confirm that the circuit is a correct implementation of the design entered into the simulator.

1.3.2   Timing Simulation

Simulation results build sufficient confidence in the design to invest in the production of a prototype. Actual verification of the prototype gives more assurance, since it embodies the process-dependent parameters. Fortunately, present CAD tools include parameter extraction that permits simulation under actual process conditions. Timing verification can be performed in conjunction with functional verification or independently. The delays associated with the various gates are assigned. Usually, nominal delays are assigned to the gates (or functional units in the case of an HDL model). The delays are part of the library used in the design, or they may be modified. In this type of simulation, the net delays are not included, and thus the timing verification is not complete. It is only when the design is placed and routed that actual delays can be assigned to the gates and to their interconnects. Also, loading can be evaluated and assigned to the output pads.

1.4   TESTING

Once the design is implemented on silicon, it can be verified by applying the appropriate stimuli and checking the responses. However, testing is not design verification. It is meant to verify the manufacturing correctness. There are two principal categories of testing: *parametric testing* and *functional testing*. The first category is concerned with the parameters of the circuit, such as current and voltage measurements. Unless the nominal voltages and currents are assured, no further testing of the circuits is really needed. Functional testing is the subject of this book.

As we mentioned earlier, the purpose of testing is to demonstrate that the manufactured IC is error-free. One needs first to define an error. Early techniques of testing digital circuits were concerned primarily with functional verification. To switch to a testing method that takes into consideration the structure of the circuit was suggested in a paper presented by R. Eldred at the ACM meeting in August 1959 [Eldred 1959]. The opening sentence of this paper is: "In order for the successful operation of a test routine to guarantee that a computing system has no faulty components, the test conditions imposed by the routine should be devised at the level of the components themselves, rather than at the level of programmed orders."

To make testing more manageable, it is thus important to characterize the defects in the circuit as logical or electrical value on the nodes connecting the various components of the circuit, that is, to represent the failure modes by a logical value. This amounts to representing physical defects by models on the logic level. If this mapping is one-to-one, there will be a large number of models to represent. As a model, the fault does not have to be an exact representation of the defects, but rather, be useful in detecting the defects. For example, the most common fault model assumes *single stuck-at* (SSA) *lines* even though it is very clear that this model does not accurately represent all actual physical failures. However, despite its popularity, the stuck-at fault is no longer sufficient for present circuits and technologies. With the advent of MOS technology, it has become evident that additional fault models are needed to represent more comprehensively the failure modes in this technology [El-Ziq 1981]. Fault modeling is the topic of Chapter 2.

To this point we have asserted that digital testing is performed on the actual IC using test patterns that are generated to demonstrate that the product is fault-free. This view of testing implies that test generation is done at the gate-level design. This used to be the case. Nowadays, there is a testability cycle that parallels the design cycle, as illustrated in Fig. 1.4. It is realistic, therefore,

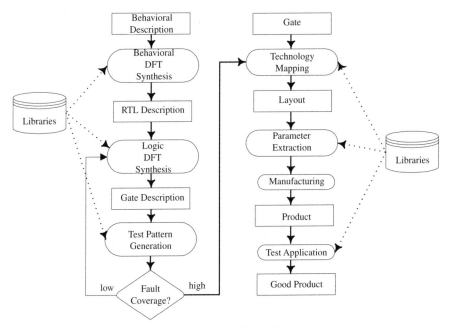

Figure 1.4   DFT cycle.

to associate with every design verification step a counterpart in the testability cycle. For example, at the logic level, test pattern generation is performed to provide the patterns and they are verified using *fault simulators*. As the trend is moving toward designing on higher levels of abstraction, testability is also being performed on the same higher levels. Attempts for test pattern generation on the RTL and behavior levels have been reported. However, at these levels of design abstraction, there are other means of assessing the testability of the circuit without actually developing the patterns. Usually known as *testability measures*, they are described in a later section and discussed in more detail in Chapter 8.

Another fundamental part of testing that is more intimately related to the design is the practice of design for testability (DFT). This approach advocates introducing embedded test constructs into the circuit to make testing easier and more efficient. Ease and efficiency here refer to test pattern generation and test application processes. Testing is also becoming a factor in design optimization. Designers customarily strive for an optimal design: a high-speed, low-power design occupying the smallest possible area. However, depending on the use of the product, the designers often optimize one of the three attributes: speed (or delay), area, and power. At present, a fourth attribute is considered: testability.

1.5 FAULTS AND THEIR DETECTION

In this section we describe what a fault is and how a test pattern detects it. We use the single stuck-at fault model as an example to illustrate the concept, but any other model would work as well. A *single stuck-at* (SSA) *fault* represents a line in the circuit that is fixed to logic value 0 or 1. One may think of it as representing a short between the faulty line and the ground or V_{dd} rail. Examples of failure modes that manifest themselves as stuck-at faults are shown in Fig. 1.5. The first example shows a short in a bipolar inverter that forced the input to be held to ground, thus causing a stuck-at-0 fault. The other example of a stuck-at-0 (SA0) fault on the input of a CMOS inverter is due to oxide breakdown between the gate and the source of the pull-down transistor. In these two examples, the SA0 models represent the failure mechanism. The short in the resistance, R2, actually resulted in a short between the base of the bipolar transistor, Q1, and ground. Whatever signal is imposed on the input of the XOR gate will be sensed by this gate as a zero logic and will change the functionality of the gate from an XOR, $Z = A \oplus B$, to an inverter, $Z = B'$. For the CMOS inverter, the input being stuck at 0 is equivalent to holding the output at logic 1. Similarly, had the input been stuck at 1 (SA1), the output would appear as stuck at 0. Thus the SA faults (SSA) on the inputs on an inverter are equivalent to those at its output.

Next we examine the behavior, on the logic level, of a two-input AND gate when stuck-at faults are injected, one at a time, on all input and output leads. This is illustrated in Fig. 1.6. All input combinations are given in the first col-

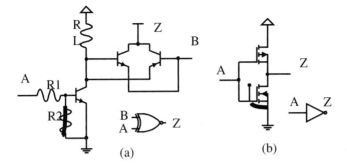

Figure 1.5 Stuck-at faults: (*a*) bipolar XNOR gate; (*b*) CMOS inverter.

umn of the table. The fault-free and faulty circuit's responses, R and R_f, respectively, are listed in the other columns of the table for each stuck-at fault. The fault is detected whenever there is an input combination such that $R \oplus R_f = 1$. Stuck-at faults on line A are indicated by $A/0$ for stuck-at 0 and $A/1$ for stuck-at 1. Similar notations are used for the other lines. A careful observation of the table indicates that a faulty response of the circuit is not always observable. For example, with $A/0$, it is expected that the output will always be zero irrespective of the input combination. Thus the faulty response differs from the fault-free response only when the input combination $AB = 11$ is applied on the circuit. This combination is considered as a *test pattern* that detects the fault $A/0$. In a similar fashion, we can determine that this pattern also detects $B/0$ and $Z/0$. This pattern detects any of the faults but does not help diagnose which fault actually occurred. We notice also that a fault may be detected by more than one pattern. This is the case with $Z/1$. Any of the three test patterns 10, 01, and 00 should be sufficient to detect the fault. The latter pattern (00) is the only one that determines that the failure is due to $Z/1$. If the main aim is to detect the failures rather than diagnose them, only three patterns are necessary to accomplish the task. They are 01, 10, and 11 and they form a test set of length 3.

Summing up the SSA faults for the two-input AND gate, we have shown

Inputs	FF	Faulty Response					
AB	Response	A/0	B/0	Z/0	A/1	B/1	Z/1
00	0	0	0	0	0	0	1
01	0	0	0	0	1	0	1
10	0	0	0	0	0	1	1
11	1	0	0	0	1	1	1

Figure 1.6 Stuck-at faults on a two-input AND gate and their detection.

that (1) three patterns are sufficient to detect all faults; (2) the three faults $A/0$, $B/0$, and $Z/0$ are equivalent; and (3) detecting stuck-at-1 faults on the inputs guarantees detection of the same fault on the output.

1.6 TEST PATTERN GENERATION

The process followed in generating test patterns for the two-input AND gate is relatively straightforward. Conceptually, it can be repeated for any circuit. However, this is an oversimplification of the process. For any meaningful circuit size where there are several hundreds of thousands of gates and lines, the process becomes tedious and time consuming. It has actually been proven mathematically that the process is NP-complete [Ibarra 1975]. Informally speaking, this means that no solution can be obtained within a time that can be expressed as a polynomial in n, where n is the size of the circuit. Therefore, heuristics are used in commercial ATPGs. A *heuristic* is a commonsense rule (or set of rules) intended to increase the probability of solving some problem without guaranteeing an optimal solution. Algorithms and heuristics are discussed in more detail in Chapter 4. The oldest technique is the D-algorithm [Roth 1966]. Other algorithms include PODEM [Goel 1981] and SOCRATES [Schultz 1988]. These algorithms are discussed at greater length in Chapter 6. Developing test patterns for sequential circuits is even more complex [Miczo 1983], while for a combinational circuit, one test pattern is sufficient to detect a fault; however, a sequence of patterns is needed for a sequential circuit. It is necessary to put the circuit in a known state before sensitizing the fault to the output.

We illustrate this complexity with a very simple example. Consider the *SR* latch shown in Fig. 1.7 and assume that we would like to detect the stuck-at-0 fault on line A. To provoke (or excite) the fault, we need to control A to logic 1 ($A = 1$). This implies that both inputs of gate 2 have to be low, $SQ = (00)$. But for $Q = 0$, $R = 1$. Thus $SR = (01)$ is needed to control A to 1. To sensitize the fault to the primary output Q through gate 1, we must have $R = 0$. To detect the fault $A/0$, it is necessary first to apply the pattern $SR = (01)$, followed by the pattern $SR = (00)$. The patterns have to be in this sequence. A sequential circuit needs to be placed in a certain state before sensitizing (propagating) the fault

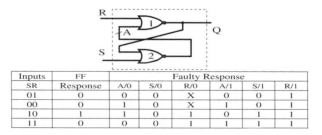

Inputs	FF	Faulty Response					
SR	Response	A/0	S/0	R/0	A/1	S/1	R/1
01	0	0	0	X	0	0	1
00	0	1	0	X	1	0	1
10	1	1	0	1	0	1	1
11	0	0	0	1	1	1	1

Figure 1.7   Testing a sequential circuit: SR latch.

to the output. In most cases more than two patterns are needed. For example, if we need to detect a SA0 fault on the ripple carry-out of a 32-bit counter, it is necessary to apply 2^{31} clock cycles!

1.7   FAULT COVERAGE

The effectiveness of the test sets is usually measured by the *fault coverage*. This is the percentage of detectable faults in the circuit under test (CUT) that are detected by the test set. The set is complete if its fault coverage is 100%. This level of fault coverage is desirable but rarely attainable in most practical circuits. Moreover, 100% fault coverage does not guarantee that the circuit is fault-free. The test checks only for failures that can be represented by the model used, such as a stuck-at-fault model. Other failures are not necessarily detected. The fault coverage is calculated using a *fault simulator*. This is a logic simulator in which faults are injected at the appropriate nets of the circuits, usually one at a time. The response of the circuits to test pattern applications is compared with the good response of the circuit. The fault is considered detected if at least one of the test patterns has a response different from the good circuit response. Fault simulation is a topic of Chapter 5.

1.8   TYPES OF TESTS

In this section we distinguish between various types of tests according to the test generation method. Tests are categorized as off-chip or on-chip and can also be categorized according to the test application method, discussed in Section 1.9.

1.8.1   Exhaustive Tests

Since a test pattern is a combination of the values applied on the primary inputs of a circuit under test, it is conceivable to use all possible combinations (an exhaustive test set) and apply them to the circuit. This exhaustive approach to testing has the advantage of being easy to generate and of yielding 100% fault coverage. However, such a testing method is efficient only for purely combinational small circuits. Applying an exhaustive test to a 20-primary inputs circuit using a 1-MHz tester would take an entire second [Fujiwara 1986]. It should be clear from the preceding section why exhaustive testing is not applicable for sequential circuits since in these circuits the sequence in which patterns are applied is crucial for fault detection.

1.8.2   Pseudoexhaustive Tests

An alternative to exhaustive testing was proposed [McCluskey 1981]. The approach is to test the components of a circuit exhaustively without having to apply an exhaustive test on the entire circuit. We illustrate the approach with the circuit in Fig. 1.8. The circuit has eight primary inputs. The length of an

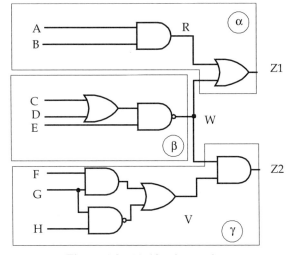

Figure 1.8   Verification testing.

exhaustive test is 256 patterns. Instead, the circuit is partitioned into three sub-circuits: α, β, and γ. It is possible to test the first component exhaustively using four patterns and observing through the primary output, Z_1. For this, the other input to the OR gate should be controlled to zero. The other two components each depend on three of the inputs. They can be tested exhaustively with eight patterns each. The total length of the test is the sum of the three individual tests, 20 patterns instead of 256 patterns. To test subcircuit γ, we need to sensitize the response to the test through Z_2. This implies keeping $W = 1$. As for component β, the response may be sensitized through Z_1 or Z_2, and hence we need to keep $R = 0$ or $V = 1$. This type of testing requires an efficient way of partitioning the circuit. It is considered as an approach to design in order to ease testability and will be revisited in Chapter 9.

Another special case of exhaustive testing is known as *verification testing* [McCluskey 1984]. It is applicable to circuits where each primary output is a function of only a subset of primary inputs. The circuit we used for pseudoex-haustive testing has two primary outputs, Z_1 and Z_2. Each output is dependent on a subset of the primary inputs, 5 and 6, respectively, as can be determined from Fig. 1.8. The corresponding exhaustive test sets are of lengths 32 and 64. The length of the test for the circuit is then the sum of both test sets and is equal to 96 patterns. This is definitely a much shorter test set than the exhaustive test of length 256 patterns.

1.8.3   Pseudorandom Tests

Test patterns may also be generated in random order. The cost of generating the test is minimal. A fault simulator is needed to grade the test and assess the fault coverage. The advantage of random testing is that it has been shown to

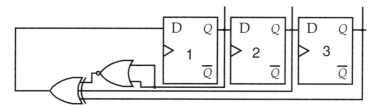

Figure 1.9   Pseudorandom testing using a 3-bit linear feedback shift register.

detect a large percentage (possibly 85%) of stuck-at faults. Consequently, many commercial ATPGs use random testing as a first stage of the test pattern generation and then apply heuristics to deal with the still undetected faults, which are called *random pattern-resistant* (RPR) *faults*. In purely random testing, a test pattern may be generated more than once. However, pseudorandom (PR) test generation is more appropriate to ensure that there is no repetition of patterns. PR test sets may be generated by a software program or by a linear feedback shift register (LFSR) that is built on the chip. This is called *on-chip* test pattern generation. Figure 1.9 shows a three-stage LFSR that can generate all 3-bit combinations. The number of feedback leads and their positions determine the length of the test. We discuss this topic at greater length in conjunction with logic built-in self-test (BIST) in Chapter 11 and memory BIST in Chapter 12.

1.8.4   Deterministic Tests

Deterministic tests are fault-oriented tests. In this case, patterns are generated targeting a specific fault model, as we described in Section 1.6. Generating these patterns is an NP-complete problem and requires heuristics to speed up the process. Deterministic test pattern generation is the topic of Chapter 6. In contrast to pseudorandom testing, deterministic tests have to be generated *off-chip*.

1.9   TEST APPLICATION

The value of a test set is in detecting faults when it is applied to a manufactured IC. In addition, a test set serves in defect analysis during diagnosis. Testing may also be used for debugging, repair of subassemblies, and reliability purposes while the circuit is in normal operation. Test applications may be performed on the wafer before it is diced and on the packaged ICs. In this book we concentrate on testing for the purpose of *screening ICs*. However, we first distinguish between on-line and off-line testing. Then we describe the use of automatic test equipment (ATE). Finally, we examine on-chip versus off-chip testing.

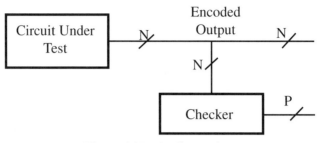

Figure 1.10   On-line testing.

1.9.1   On-Line Versus Off-Line Tests

Depending on the time at which test sets are applied, testing is categorized as off-line or on-line. *Off-line testing* is performed when the circuit is not in use. On the other hand, *on-line testing* can be performed while the circuit is working in normal mode. This testing requires special codes and checkers and duplication techniques. It is illustrated symbolically in Fig. 1.10. One of the main advantages of on-line testing is the capability to detect faults that are transient or intermittent. These types of faults are described in Chapter 2. In this book, however, we use the term *testing* in the sense of off-line only. Test application methods are different for patterns generated off-chip and patterns generated on-chip. A test manager controls test generation and application for on-chip testing. This is described in Chapters 11, 12, and 13 for logic BIST, memory BIST, and FPGA BIST, respectively.

1.9.2   Automatic Test Equipment

For off-chip testing, automatic test equipment (ATE) is required to apply the test. It consists of a fixture, hardware, and software. The fixture is used to hold the IC under test. Each IC pin is supported by pin electronics that either drives it, measures its signal, and/or acts as a load. The pin electronics are connected to the measurement instrumentation by means of a complex cabling system. The hardware includes a computer system with sufficient storage for test patterns and the expected responses. According to the function of the parts that are being tested, testers are categorized as digital, memory, or analog. Logic and memory digital ICs are handled differently. Memory ICs are generally tested in parallel with several parts being given the same input and their outputs being ORed together, while logic parts have to be tested one at a time. Even the generation of input data is different. The data for a memory IC can be produced more easily by an algorithmic pattern generator because of the regularity of these devices' architecture, as we describe in Chapter 12. However, memory ICs are much denser than logic ICs, and for this reason, they have unique problems, such as crosstalk. *Crosstalk* is a noise created by coupling between signals on the chip that may cause the memory element to change values unexpectedly. Digital testers consist, then, of memory testers and logic testers.

Although in this book we concentrate on digital circuits, we compare digital testers to analog circuit testers. The reason is that most ICs in the near future will include both digital and analog parts. This type of ICs is known as a *mixed-signal ICs*. The biggest difference between analog and digital testers is manifested in the configuration of the individual pins. Whereas every digital pin is interchangeable with any other and is therefore supported by the same circuitry, analog pins usually have a large variation in their purpose, so the support circuitry is more unique. Combining digital and analog circuitry on the same chip requires a great deal of care to coordinate and properly isolate the two types of signals and grounds. This requirement, in addition to the various pin types, makes mixed-signal testers much more expensive per pin than other testers, which are already capital-intensive.

The cost of digital testers increases with the number of pins. The high cost per pin is primarily because they are testing leading-edge technology using existing technology—a paradox? Scheduling their use efficiently can offset this high cost. The tester is more economical when its use is amortized over a large production quantity. That is, cost is kept low by making testing time minimal. Therefore, it is important to keep the test set as short as possible. Another way of minimizing test application time and cost is to combine off-chip with on-chip tests.

1.9.3 On-Chip Versus Off-Chip Testing

Shortening test application time is difficult for several reasons. First, as the device density increases, the volume of test data is becoming extremely large. Second, the growing disparity between the internal clock frequencies and the output capability of the I/Os makes at-speed testing of ICs extremely difficult, if not impossible. Third, the number of transistors per IC pin keeps increasing. All these factors make the *external bandwidth* much lower than the internal bandwidth. The external bandwidth is defined here as the product of the number of I/Os by the switching speed of the I/Os, while the internal bandwidth is the product of the number of switching transistors by the internal frequency.

To minimize interaction with the external world while keeping the on-chip overhead reasonable, it is possible to partition the test function into on-chip and off-chip resources. In this fashion, testing can be achieved at an optimal rate. This approach will be particularly beneficial for present SOCs that include analog components and RAMs in addition to the digital cores, as illustrated in Fig. 1.11. From the presentation on ATE above, it is evident that off-chip testing will require insertion of the IC in three different testers: one for logic, one for memory, and one for analog test. Embedding test pattern generation on-chip as shown to the right in each part of the figure makes it possible to use a single low-bandwidth external tester [Zorian 1999].

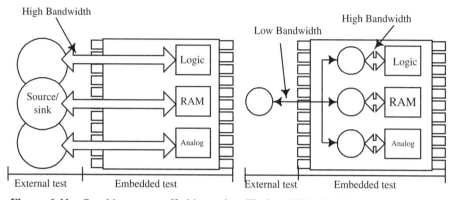

Figure 1.11 On-chip versus off-chip testing [Zorian 1999 © IEEE. Reprinted with permission].

1.10   DESIGN FOR TEST

The complexity of test pattern generation motivated the development of design approaches that facilitate testing. The design for test (DFT) discipline started formally only in the mid-1970s after publication of the scan-path technique [Williams 1973] and its adoption by IBM [Eichelberger 1977]. Kobayashi developed the same technique earlier in Japan [Kobayashi 1968]. Also, this design style was practiced informally due to its serviceability and maintainability at IBM [Carter 1964].

Ad hoc approaches to design for testability as well as other techniques, such as *scan path* design, built-in self-test (BIST), and boundary scan (JTAG) are detailed in subsequent chapters. They are all approaches to enhance *testability*. There is no formal definition for testability. An interesting attempt was given as: "A digital IC is testable if test patterns can be generated, applied, and evaluated in such a way as to satisfy pre-defined levels of performance (e.g., detection, location, application) within a pre-defined cost budget and time scale" [Bennetts 1984]. This definition is still vague and it is interpreted differently by IC designers. One of the key words is "cost." It is probably the cost of testing that deters the semiconductor manufacturers from doing as much testing as is really needed for reliability. An attempt to quantify testability, in the sense of fault detection, was proposed by [Chang 1974]. Two testability measures (TMs) were then defined as controllability and observability.

1.10.1   Controllability

Controllability reports the cost of placing a node in the circuit at a predetermined logic value. Placing this value on a primary input is "free." However, it is possible to assign a minimal cost, say, 1, to any primary input. The cost increases as the node depth in the circuit increases. This cost will also depend

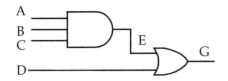

Figure 1.12 Example to illustrate TM calculations.

on the type of gate, the logic value to be imposed on the line, and whether the circuit is combinational or sequential. For example, controlling the output of a multi-input AND gate to 1 requires the control of all of its inputs to 1, while for an OR, it is sufficient to control only one of the inputs to 1. Also, controlling a node at a second level, such as node E in Fig. 1.12, requires controlling the input at level 1 (E) as well as those at level 0 (the primary inputs, A, B, and C). There is a cost, then, for stepping from one level to another.

1.10.2 Observability

The *observability* of a node indicates the effort needed to observe the logic value on the node at a primary output. Again, there is no cost associated with observing a primary output. To observe an internal node, for example, node E of the circuit in Fig. 1.12, it necessary to control D to 0 in order to sensitize through the OR gate to the primary output G. Similarly, the observability of D at G requires that E be controllable to 0. The other primary inputs, A, B, and C, have equal observability. The most popular testability measure (TM) program is SCOAP [Goldstein 1980]. These measures are quick estimates of the degree of difficulty generating test patterns without actually generating them. Calculating TM is not as time consuming as test pattern generation. The complexity is only $O(n)$, where n is the circuit size (number of lines). There are several variations of testability measures. Also, TMs have been generalized to design described on RTL and functional levels. They have not been really successful in making a design easier to test, but they still serve as an aid in test pattern generation and in decisions about selecting points of observation for nodes with low observability. Examples of TM calculations are given in Chapter 8.

1.11 TESTING ECONOMICS

We started this chapter discussing testing and reliability, indicating that the question was not whether to test, but rather, how thorough the test must be to ensure a highly reliable product. Achieving high quality requires a large investment in time and money. As product lifetime is becoming shorter than its development time, it is difficult to justify testing if it delays introduction of the product on time. The loss due to delay in reaching the market at the appro-

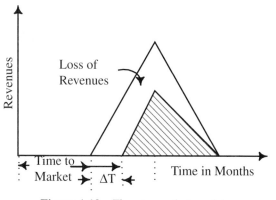

Figure 1.13   Time-to-market model.

priate moment results in a loss of market share as indicated in Fig. 1.13. The solid line indicates that revenues increase to a peak and then decrease to the end of the product cycle. A delay Δt in arrival of the product to market with respect to its original introduction results in a shorter peak and lower revenue. The hashed area shows this loss of revenues. Thus, if testing causes delays and also incurs cost, it is often hard to justify it on a purely financial basis using this time-to-market as a basis. However, this model does not mention the loss of revenues due to rejects because of problems with the product. It is thus important to show how a slight delay may be compensated for a better quality. Also, as DFT constructs are introduced from the onset in the design, product introduction would not be delayed. For this we first define two major quality controls, yield and defect level.

1.11.1   Yield and Defect Level

The *yield Y* of IC manufacturing is defined as the fraction of the parts that are defect-free. Thus it is determined by $Y = G/(G + B)$, where G is the number of good parts that pass all tests and B is the number of parts that failed some tests. It is difficult to find the exact value of this fraction, for three good reasons. First, there is a lack of data for parts once they are sold. Second, it is not possible to test all chips. Only sampling is used and hence a projected yield is calculated. Third, testing may result in passing, as good, bad chips. Usually, a test will target fault models, and thus they may miss a defect that is not represented by the fault model used.

Many factors affect the yield, including the die area of the wafer, process maturity, and number of process steps. Efforts to understand the defects and their causes—fault analysis—has become an important factor for manufacturers. This is witnessed by a preponderance of papers on the topic at testing conferences [ITC 1997]. There are several mathematical models to determine the yield. The first model, developed by [Murphy 1964], is given as $Y =$

$[(1 - e^{-AD})/AD]^2$, where A and D are the area and defect density, respectively. The wafer area is increasing constantly. To improve the yield, fabricators are interested in understanding the types of defects in order to minimize them.

The good dies are packaged and tested before placing them on the market. Of those that passed the test, some are actually bad. The *defect level* DL, the fraction of these bad chips that pass the test, is usually measured in defects per million (DPM). Thus a defect level of 0.1% is equivalent to 1000 DPM.

1.11.2   Fault Coverage and Defect Level

Among the several theoretical formulas relating DL to Y is the following expression due to [Williams 1981]:

$$DL = 1 - Y(1 - T)$$

where T is the fault coverage of the functional test used. For small values of DL, say, less than 1000 DPM [McCluskey 1989], this expression can be written as

$$DL = TT[-\ln(Y)]$$

where $TT = 1 - T$ is the *testing transparency*, which is the fraction of defects that are not detected. It is often approximated by $1 - C$, where C is the single-stuck-at-fault coverage. But actually, $TT \geq 1 - C$. Accordingly, TT can be considered as the percentage of ICs that did not pass the test.

The relationship between DL and T is shown in Fig. 1.14 for two values of

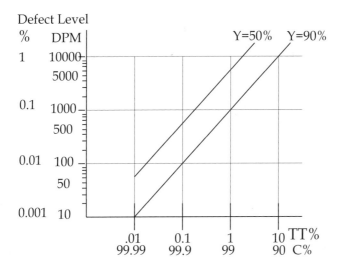

Figure 1.14   Relationship of defect level versus fault coverage for given yield values [McCluskey 1989 © IEEE. Reprinted with permission].

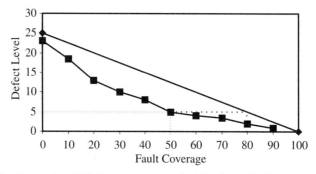

Figure 1.15 Defect level, fault coverage, and yield theoretical versus experimental results.

the yield, 50% and 90%. Inspection of the graphs indicates that high fault coverage is required to ensure low defect levels, 1 to 100 DPM. Interpretation of experimental data by [Maxwell 1991] indicates that Williams's formula is rather pessimistic and that less fault coverage is needed for low values of DL. The dashed line in Fig. 1.15 represents the experimental result using functional testing, and the linear relation is a plot of Williams's expression. The relationship between the two graphs can be interpreted as follows: Although the theoretical 90% fault coverage is required to guarantee a DPM level of 5, it is actually sufficient to reach 50% fault coverage.

Initial publications on testing cost are used to concentrate on the automatic test equipment and their deployment in manufacturing testing. Pioneer papers in DFT cost appeared in the 1980s [Pittman 1984; Varma 1984]. Only recently has interest in this area peaked, and culminating in a special workshop in 1994 on economics of design, test, and manufacturing [D&T 1997].

1.12 TO EXPLORE FURTHER

For those who wish to investigate in depth some of the topics covered in this chapter, the best approach to further exploration is finding resources to aid in your study of testing. For example, surveying testing publications listed in Appendix A and exploring the Web sites of the Test Technology Technical Council (TTTC) can be helpful.

REFERENCES

Abadir, M., et al. (1989), Logic design verification via test pattern generation, *IEEE Trans. Comput.-Aided Des. Integrated Circuits Syst.*, Vol. CAD-7, No. 1, pp. 138–149.

Bennetts, R. G. (1984), *Design of Testable Logic Circuits*, Addison-Wesley, Reading, MA, p. 164.

Breuer, M. A., and A. D. Friedman (1976), *Diagnostics and Reliable Design of Digital Systems*, Computer Science Press, New York.

Carter, W. C., et al. (1964), Design of serviceability features of the IBM system/360, *IBM J. Res. Dev.*, Vol. 8, No. 4, pp. 115–126.

Chang, H. Y., et al. (1974), *Fault Diagnosis of Digital Systems*, Krieger Publishing, Melbourne, FL.

D&T (1997), *IEEE Des. Test Comput.*, Vol. 14, No. 4.

Eichelberger, E. B., and T. W. Williams (1977), A logic design structure for LSI testability, *Proc. Design Automation Conference*, pp. 462–468.

Eldred, R. D. (1959), Test routines based on symbolic logic systems, *J. ACM*, Vol. 6, No. 1, pp. 33–36.

El-Ziq, Y. M., and R. J. Cloutier (1981), Functional-level test generation for stuck-open faults in CMOS VLSI, *Proc. IEEE International Test Conference*, pp. 536–546.

Fujiwara, H. (1986), *Logic Testing and Design for Testability*, MIT Press, Cambridge, MA.

Goel, P. (1981), An implicit enumeration algorithm tests for combinational circuits, *IEEE Trans. Comput.*, Vol. C-30, No. 3, pp. 215–222.

Goldstein, L. H., and E. L. Thigpen (1980), SCOAP: Sandia Controllability/Observability Analysis Program, *Proc. Design Automation Conference*, pp. 190–194.

Ibarra, G. H., and S. K. Sahni (1975), Polynomially complete fault detection problems, *IEEE Trans. Comput.*, Vol. C-24, No. 3, pp. 242–249.

ITC (1997), *Proc. IEEE International Test Conf.*, pp. 633–653.

Johnson, B. W. (1989), *Design and Analysis of Fault Tolerant Digital Systems*, Addison-Wesley, Reading, MA.

Kobayashi, A., et al. (1968), A flip-flop circuit suitable for FLT (in Japanese), *Annual Meeting of the Institute of Electronics, Information and Communications Engineers*, Manuscript 892, p. 962.

Maxwell, P. C., et al. (1991), The effect of different test sets on quality level prediction: when is 80% better than 90%? *Proc. IEEE International Test Conference*, pp. 358–364.

McCluskey, E. J., and S. Bozorgui-Nesbat (1981), Design for autonomous test, *IEEE Trans. Comput.*, Vol. C-30, No. 11, pp. 866–875.

McCluskey, E. J. (1984), Verification testing: a pseudo-exhaustive test technique, *IEEE Trans. Comput.*, Vol. C-33, No. 6, pp. 541–546.

McCluskey, E. J., and F. Buelow (1989), IC quality and test transparency, *IEEE Trans. Ind. Electron.*, Vol. IE-36, No. 2, pp. 197–202.

Miczo, A. (1983), The sequential ATPG: a theoretical limit, *Proc. IEEE International Test Conference*, pp. 143–147.

Murphy, B. T. (1964), Cost-size optima of monolithic integrated circuits, *IEEE Proc.*, Vol. 52, No. 12, pp. 1537–1545.

Pittman, J. S., and W. C. Bruce (1984), Test logic economic considerations in a commercial VLSI chip environment, *Proc. IEEE International Test Conference*.

Roth, J. P. (1966), Diagnosis of automata failures: a calculus and a method, *IBM J. Res. Dev.*, Vol. 10, No. 7, pp. 278–291.

Schultz, M. H., et al. (1988), SOCRATES: a highly efficient automatic test pattern generation system, *IEEE Trans. Comput.-Aided Des.*, Vol. CAD-8, No. 1, pp. 126–137.

Siewiorick, D. P., and R. S. Swarz (1982), *The Theory and Practice of Reliable System Design*, Digital Equipment Corporation Press, Bedford, MA.

Varma, P., et al. (1984), An analysis of the economics of self-test, *Proc. IEEE International Test Conference*, pp. 20–30.

Williams, M. J. Y., and J. B. Angel (1973), Enhancing testability of large scale integrated circuits via test points and additional logic, *IEEE Trans. Comput.*, Vol. C-22, No. 1, pp. 46–60.

Williams, T. W., and N. C. Brown (1981), Defect level as a function of fault coverage, *IEEE Trans. Comput.*, Vol. C-30, No. 12, pp. 987–988.

Zorian, Y. (1999), Testing the monster chip, *IEEE Spectrum*, Vol. 36, No. 7, pp. 54–60.

PROBLEMS

1.1. Use the method shown in Section 1.5 to find test patterns to detect stuck-at faults in (**a**) two-input OR gate, (**b**) two-input XOR gate, and (**c**) three-input NAND gate.

1.2. For each of the circuits shown in Fig. P1.2:

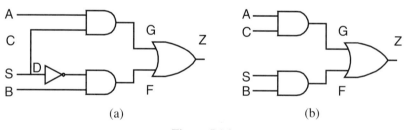

(a)   (b)

Figure P1.2

    (**a**) Determine the total number of stuck-at faults.

    (**b**) Develop a test to detect all stuck-at faults on the primary inputs, then find the corresponding fault coverage.

1.3. Plot the *detectability profile* of the circuit in Fig. P1.2a. This profile is defined as the frequency of faults detected by 1, 2, 3, ... patterns. (*Hint*: First develop the exhaustive test set, and then determine the number of the patterns that detect each fault.)

1.4. For the test set developed in Problem 1.3, determine the number of faults detected by each pattern, then plot the cumulative frequency of faults detected.

1.5. How many test patterns detect the fault stuck-at 0 on line E in Fig. P1.5?

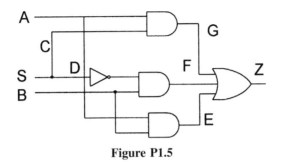

Figure P1.5

1.6. For the circuit in Fig. P1.6, develop a pseudoexhaustive test and compare its length to that of the exhaustive test.

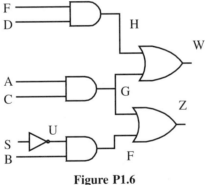

Figure P1.6

1.7. If the yield of a process is 90% and you wish to decrease the DPM from 1000 to 100, by how much do you need to increase the fault coverage? Use the graph in Fig. 1.14.

1.8. Verify the authenticity of the following statement: A complete test set for a NAND/NAND implementation of the circuit detects all SSFs of the NOR/NOR implementation of the same circuit.

1.9. Assume that the relative cost of repairing defects, C_d, expressed as a function of the percentage t of faults tested, is $C_d = (100 - 0.7t)m$, where m is the number of units to be manufactured. Further, assume that the cost C_p of achieving a particular test percentage t is $C_p = t/(100 - t)$. What value of t will minimize the total cost? [Miczo 1986].

2

DEFECTS, FAILURES, AND FAULTS

2.1 INTRODUCTION

Failures in integrated circuits can be characterized according to their duration, permanent or temporary, and by their mode, parameter degradation or incorrect design. *Permanent failures*, also called *hard failures*, are usually caused by breaks due to mechanical rupture, some wear-out phenomenon, or an incorrect manufacturing procedure. They occur less frequently than *temporary failures*, which are failures that cannot be replicated. Temporary failures, also called soft *failures*, are categorized as transient or intermittent. *Transient failures* are induced by external perturbation such as power supply fluctuations or radiation. *Intermittent failures* are usually due to some degradation of the component parameters. Figure 2.1 shows classifications of IC failures.

In this chapter we examine the physical defects and their manifestation on the electrical level as *failure modes*. Failure modes are then modeled as *faults* at the logic and behavioral levels. The mapping from the physical domain to the logic and behavioral domains facilitates the detection process of failures. Figure 2.2*a* shows a representation of the scanning electron microscope (SEM) image of an embedded particle in an IC that produced a short [Henderson 1997]. It is clear that the short possibly spans several metal lines. We assume here, for simplicity, that it affects only two lines, L1 and L2, as shown in Fig. 2.2*b*. On the logic level, the failure mode may be interpreted in different ways, depending on the technology used and the relationship of the two metal lines. Four different possibilities are illustrated in Fig. 2.2*c* and *d*. One of the metal lines may be power or ground, and thus the output of the inverter is held high or low. This results in *stuck-at-1* and *stuck-at-0 faults*, respectively. The second

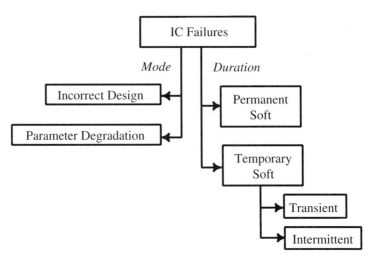

Figure 2.1   Classifications of IC failures.

line may, instead, be the output of the same gate, and the short results in a *feed-back bridging fault*. Finally, if the second line is the output of another gate, a *simple bridging fault* results.

We first describe the main causes of physical failures in MOS technology.

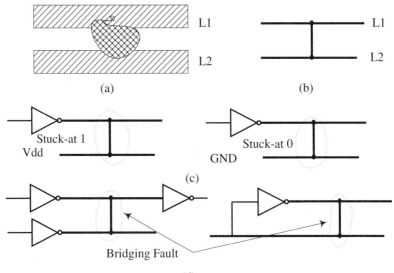

Figure 2.2   Mapping physical defects into faults: (*a*) metal mask with dust causing extra metal; (*b*) failure mode: a short; (*c*) faults on the logic level: stuck-at faults; (*d*) bridging faults.

The manifestation of these defects as two main failure modes, short and open, is then explained. The remainder of the chapter is devoted to fault modeling. Notice that the importance of the fault is to facilitate detection of the defect even if it is not a faithful representation of this defect. Thus a fault model may represent different failures [Sachdev 1998].

2.2 PHYSICAL DEFECTS

Physical defects are due to different *failure mechanisms*, which are largely dependent on the technology and layout of the circuit. A list of the principal failure mechanisms is given below.

- Surface and bulk effect
  - Passivation pits and cracks
  - Gate oxide breakdown
    - Pinholes or thin spots in oxide
    - Electrical overstress
  - Surface potential instability

- Metallization and metal semiconductor
  - Open metal at oxide steps
  - Wire bonding failure
  - Intermetallic compound formation

- Electromigration
- Package-related failures
  - Mass transport of metal atoms
  - Momentum exchange with electrons

Reference is made to the various steps used to fabricate an IC, which are several repetitions of depositing, conducting, and insulating material, oxidation, photolithography, and etching [Sze 1983]. Transistor formation is assumed to occur where the gate (poly) and active layers (diffusion) intersect. Other reasons for physical defects are the instabilities in the process conditions. These include (1) random fluctuation in the actual environment: for example, turbulent flow of gases used for diffusion and oxidation; (2) inaccuracies in the control or furnace; and (3) variation in the physical and chemical parameters of the material, such as fluctuation in the density and viscosity of the photoresist, and water and gas contaminants. Physical defects may often occur because of human errors such as mishandling wafers or processing equipment. Next we examine three more frequently occurring types of defects: (1) extra and missing material, (2) oxide breakdown, and (3) electromigration.

2.2.1   Extra and Missing Material

Extra and missing material defects may be caused by dust particles on the mask or wafer surface, or in the processing chemicals. During the photolithography steps, these particles lead to unexposed photoresist areas or resist pinholes, thus causing unwanted material or unwanted etching of the material on a layer [Strapper 1980]. For this reason, these defects are sometimes referred to as *photo* or *lithography defects* [Strapper 1976]. These defects cause extra and missing polysilicon, active and metal. An extra metal may cause shorts such as those described in the preceding section. An extra poly over an active area will form an additional unwanted transistor.

2.2.2   Oxide Breakdown

The primary cause of poor yield in gate oxide is the formation of pinhole defects due to insufficient oxygen at the interface of silicon (Si) and silicon dioxide (SiO_2), chemical contamination, nitride cracking during field oxidation, and crystal defects [Schugraf 1994]. Oxide breakdown may also be due to the operational conditions. For example, as the oxide thickness decreases without reducing the power supply voltage, the electric field across the oxide is increased and causes local breakdown. A large discharge through the oxide may cause the same damage. The charge can be transferred from other objects or from people, causing what is commonly known as electrostatic discharge (ESD). The damage may be permanent if the energy was enough to cause material to flow, for example, between the gate and the channel of a transistor [Bakoglu 1990]. Also, hot electrons cause the trapping of a charge in the gate oxide and eventually break it down [Schugraf 1994].

2.2.3   Electromigration

Electromigration is one of the major failure mechanisms in IC interconnects [Black 1969, Ho 1982, Merchant 1982]. It is caused by the transport of metal atoms when a current flows through the wire. Because of a low melting point, aluminum has large self-diffusion properties, which increase its electromigration liability. As atoms are displaced, three main phenomena may happen as illustrated in Fig. 2.3. The metal line breaks and causes an open in the interconnect. The displaced atoms may accumulate and spread in very thin lines (whiskers) to another metal line in the vicinity, or they may migrate through the silicon dioxide insulation to another metal line on an adjacent layer (hillock) [Bakoglu 1990].

Electromigration is related to stresses applied to the IC, such as increased current densities or temperature. The mean time to failure (MTTF) is proportional to the width and thickness of the metal lines [Woods 1984] and inversely proportional to the current density. This cause of IC failure is becoming more prevalent now that the trend is toward very deep submicron tech-

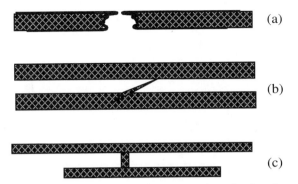

Figure 2.3 Electromigration's effects: (*a*) open in a line; (*b*) short between two lines on the same layer (whisker); (*c*) short between lines on different layers (hillhop) [Bakoglu 1990 © Addison-Wesley. Reprinted with permission].

nology (VDSM): $2\lambda \leq 0.5$ μm. Complete characterization of physical defects is difficult. However, simulation studies using a statistical approach based on data from experimental studies have proved to be very effective in developing a systematic classification of physical defects in MOS [Maly 1984, Shen 1985, Walker 1987, Ferguson 1988].

In this section we highlighted the main failure mechanisms that are current in present technologies. Could we have skipped this information without affecting the knowledge to be gained from this book? The answer is, probably, yes. However, to increase the IC yield, it is becoming more important to understand the failure mechanisms. Interest in this area is evident by the large number of fault analysis articles recently published [ITC 1997, D&T 1998].

2.3 FAILURE MODES

Most manufacturing failure mechanisms are manifested on the circuit level as *failure modes*. The most common failure modes are open and short interconnections or parameter degradation.

2.3.1 Opens

Opens may be formed when missing material spans a line or via on an interacting layer, splitting a net into unconnected branches. If the failure occurs at a fan-out, several gates will have an open gate. For example, a missing metal defect can break a metal line, or a missing first-level via defect can cause an open via. Opens may also be caused by extra material if, for example, this material completely covers a via so that the lower layer is no longer connected to the upper layer.

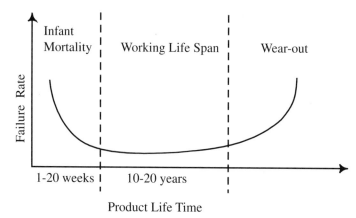

Figure 2.4 Failure rate versus product lifetime.

2.3.2 Shorts

Some extra material defects may result in connecting a metal line to another on the same layer, or to one on a layer below it through incomplete vias, or to a layer above it by blocking a via. Gate oxide breakdown causes *resistive* shorts between the gate and the source, drain, or channel. A resistive short is the type of failure that connects the gate to another terminal through a finite, but not necessarily zero, resistance [Hawkins 1986].

An IC can fail at different stages of its lifetime. The failure rate over the lifetime is given by the well-known bathtub curve shown in Fig. 2.4. Defects that escape visual and optical scanning may cause many chips to fail within 20 weeks of their operation (infant mortality period). At the end of this period, the failure rate tends to stabilize for 10 to 20 years (normal lifetime period). Eventually, due to excessive use of the components, there is an exponential increase in the failure rate (wear-out period). In present ICs, a major factor causing wear-out failure is excessive heat dissipation. This is due to high device density and increased circuit activities. Other failure modes that do not have roots in physical defects result from noise induced within the circuit, such as crosstalk and ground bounce. These faults are discussed in Section 2.13.

2.4 FAULTS

Failure modes are manifested on the logical level as incorrect signal values. A *fault* is a model that represents the effect of a failure by means of the change that is produced in the system signal. Several defects are usually mapped to one fault model. It is a many-to-one mapping. But some defects may also be represented by more than one fault model. Table 2.1 lists the common fault models. Fault models have the advantage of being a more tractable representation than

TABLE 2.1   Most Commonly Used Fault Models

Fault Model	Description
Single stuck-at faults (SSAs)	One line takes the value 0 or 1.
Multiple stuck-at faults	Two or more lines have fixed values, not necessarily the same.
Bridging faults	Two or more lines that are normally independent become electrically connected.
Stuck-open faults (SOPs)	A failure in a pull-up or pull-down transistor in a CMOS logic device causes it to behave like a memory element.
Stuck-on faults (SONs)	A transistor is always conducting.
Delay faults	A fault is caused by delays in one or more paths in the circuit.
Intermittent faults	Caused by internal parameter degradation. Incorrect signal values occur for some but not all states of the circuit. Degradation is progressive until permanent failure occurs.
Transient faults	Incorrect signal values caused by coupled disturbances. Coupling may be via power bus capacitive or inductive coupling. Includes internal and external sources as well as particle irradiation.

physical failure modes. As a model, the fault does not have to be an exact representation of the defects, but rather, to be useful in detecting the defects. For example, the most common fault model assumes *single stuck-at* (SSA) lines even though it is clear that this model does not accurately represent all actual physical failures. The rationale for continuing to use the stuck-at-fault model is the fact that it has been satisfactory in the past. Also, test sets that have been generated for this fault type have been effective in detecting other types of faults. However, as with any model, a fault cannot represent all failures, and a stuck-at fault is no longer sufficient for present circuits and technologies. With the advent of MOS technology, it has become evident that other fault models are needed to represent more accurately the failure modes in this technology.

2.5   STUCK-AT FAULTS

2.5.1   Single Stuck-at Faults

A single stuck-at (SSA) fault represents a line in the circuit that is fixed to logic value 0 or 1. One may think of it as representing a short between the faulty line and the ground or V_{dd} rail. Examples of failure modes that manifest themselves as stuck-at faults are shown in Fig. 1.5. The first example shows how a defective resistance caused the input of the bipolar XOR gate to short to ground. This is represented by a stuck-at-0 fault. Similarly, in the CMOS

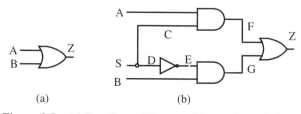

Figure 2.5   (a) Two-input OR gate; (b) 2-to-1 multiplexer.

gate, the breakdown in gate oxide causing a resistive short between gate and source results in a stuck-at-0 fault on the input. Another example is shown in Fig. 2.2c, which resulted from extra metal shorting the output of a gate to the power or ground rail.

Independent of how accurately the stuck-at fault represents the physical defect, we next continue investigating how to generate patterns that detect these faults. In Chapter 1 we developed a truth table listing the good and faulty responses of a simple AND gate to stuck-at faults on the input–output leads. Here we first consider the dual function, OR gate, and use the concepts of controllability and observability to determine the test patterns. Consider a stuck-at 0 on input A of the OR gate in Fig. 2.5a. To determine whether or not A is stuck at 0, we apply logic 1 to it and sensitize the result to the output of the gate, Z. To observe the outcome on the output, we need to hold the other input, B, low. If the output is low, this confirms that A actually is SA0 and it is detected; otherwise, it is not stuck at 0. Therefore, the pattern $AB = (10)$ will detect the $A/0$. Due to the symmetry of the circuit, the pattern $AB = (01)$ will detect $B/0$. Using the same reasoning, we should be able to generate the test pattern for $A/1$, which is $AB = (00)$. Again, because of symmetry, the same pattern detects $B/1$. To detect SA faults on the output, we need only control this node to the complement logic value to which it is stuck. Thus controlling Z to 1 detects $Z/0$. This can then be accomplished by either of the patterns that detect $A/0$ or $B/0$. Similarly, to detect $Z/1$, we force it to 0, which can be done with the pattern $AB = (00)$. Summing up the SSA faults for the two-input OR gate, we showed that (1) three patterns are sufficient to detect all faults (10, 01, 00); (2) the three faults $A/1$, $B/1$, and $Z/1$ are detected by the same pattern (00); and (3) detecting stuck at 0 on the inputs guarantees detection of the same fault on the output.

Let us consider next a more complex logic gate, a 2-to-1 multiplexer (MUX), as shown in Fig. 2.5b. This circuit helps illustrate the following important features: fan-out and reconverging fan-out. The input S has two fan-out lines, C and D. Usually, S is called the *stem* and the other lines the *branches*. It is obvious that lines C and D are electrically identical to the primary input, S. They might be considered indistinguishable from S, as the three lines are all stuck at 1 or stuck at 0 simultaneously. However, there are cases when a failure on one of the branches does not affect the other branch or the stem. For example, due to electromigration, a branch may be detached from the stem. If the input to the inverter is open,

TABLE 2.2   Stuck-at Faults for a 2-to-1 MUX

Input SAB	ff	$S/0$	$S/1$	$C/0$	$C/1$	$D/0$	$D/1$
				Response			
000	0	0	0	0	0	0	0
001	1	1	**0**	1	1	1	**0**
010	0	0	**1**	0	**1**	0	0
011	1	1	1	1	1	1	**0**
100	0	0	0	0	0	0	0
101	0	**1**	0	0	0	**1**	0
110	1	**0**	1	**0**	1	1	1
111	1	1	1	**0**	1	1	1

branch D is no longer electrically connected to the stem S or to the other branch, C. With this explanation, we find that the MUX has nine lines in total. However, here we concentrate on the stuck-at faults on three lines, S and its branches, C and D. A table similar to that constructed for the AND gate in Chapter 1 is developed for the MUX. In Table 2.2 the first column lists all possible test patterns, a combination of the three primary inputs. The second column lists the fault-free (ff) responses to these patterns. The other columns give the faulty response for each of the stuck-at faults listed in the first row. Whenever the response diverges from the good circuit response, column ff, the corresponding input combination is a test set that detects the fault. There is only one test pattern, $SAB = (010)$, that detects $C/1$. This pattern also detects $S/1$, a fault that is also detected by $SAB = (001)$. Detecting $C/1$ implies the detection of $S/1$. We observe also that there is a common pattern, $SAB = (110)$, that detects both $C/0$ and $S/0$, although each of these faults is also detected by other patterns: $SAB = (101)$ and (111), respectively. A similar relationship exists between SA faults on D and S. The pattern $SAB = (101)$ is the only pattern that detects $D/0$, and it also detects $S/0$. In addition, the test pattern $SAB = (001)$ detects $D/1$ and $S/1$. We then conclude that detecting the SA faults on branches C and D leads to the detection of the SA faults on the stem, S.

Constructing similar tables for circuits other than elementary gates is not a feasible solution to determine the test patterns. The higher the number of lines in the circuit, the more complex the problem. As we will discover in Chapter 6, test pattern generation is a complex proposition, whenever possible, it is advantageous to minimize the fault list. For this we generalize the observations made on the simple gates by defining more formally in subsequent sections of this chapter the concepts of *equivalence* and *dominance*.

2.5.2   Multiple Stuck-at Faults

A defect may cause multiple stuck-at faults. That is, more than one line may be stuck-at high or low simultaneously. With decreased device geometry and increased gate density on the chip, the likelihood is greater that more than one

TABLE 2.3   Number of Multiple Stuck-at Faults in an *N*-Line Circuit

Number of Nodes	Number of Faults		
	Single	Double	Triple
N	$2N$	$4C(N, 2)$	$8C(N, 3)$
10	20	180	960
100	200	19,800	1.3×10^6
1,000	2,000	1,998,000	$>10^9$
10,000	20,000	199,980,000	$>10^{12}$

SSA fault can occur simultaneously. It has been recommended to check *m*-way stuck-at faults up to $m = 6$ [Goldstein 1970]. This is particularly true with present technology circuits because of the high device density. The number of faults increases exponentially with *m* as indicated in Table 2.3. A set of *m* lines has 2^m combinations of SA faults. Since the total number of *m*-sets of lines in an *N*-line circuit is

$$C(N, m) = \frac{N!}{m!(N - m)!}$$

the total number of *m*-way faults is $2^m C(N, m)$. Thus the total number of multiple faults is $\sum_{m=1}^{N} 2^m C(N, m) = 2^N - 1$.

In detecting multiple stuck-at faults, it is always possible to use exhaustive and pseudoexhaustive testing. However, this is not practical for large circuits. It has been shown that using a SSA fault test yields high fault coverage in detecting multiple stuck-at faults [Hughes 1984]. The most important factors that affect the detectability of multiple stuck-at faults are the number of primary outputs and the reconverging fanouts [Schertz 1971, Hughes 1984]. Comparing a multiple-output circuit such as an ALU to a parity tree, the multiple fault coverage of SAF test decreased from 99.9% to 83.33% [Mourad 1986]. The high reconverging-path nature of a parity tree causes fault masking, but the fault coverage increased to 96% when the test was increased.

2.6   FAULT LISTS

To generate tests for digital circuits, the test tools are provided with a circuit description, a *netlist*. The tool then creates a list of all faults to be detected. For large circuits, the list can be quite long. It is thus beneficial to minimize the list whenever possible. As we have seen previously, some SA faults may be detected by the same test patterns. Therefore, only one of these faults needs to be included in the fault list. In the following sections we define and use the concepts of *equivalence* and *dominance* to collapse the faults and reduce the fault list.

2.6.1 Equivalence Relation

Definition 1. Two faults are called *equivalent* if every pattern that detects one of the faults also detects the other. That is, their test sets, T_1 and T_2, are identical: $T_2 \subseteq T_1$ and $T_1 \subseteq T_2$.

As pointed out for the two-input AND gate, SA0 faults on the inputs and output are equivalent. All three are detected by the same test pattern. Any one of them can represent the three faults. Accordingly, the fault list is reduced from six to four faults. Similarly, SA1 for faults on the I/O of an OR gate are equivalent. For an inverter, the SA0 (SA1) on the input is equivalent to the SA1 (SA0) on the output. We can then use this knowledge about the three elementary gates, AND, OR and NOT, to develop equivalence for a circuit built of such gates.

2.6.2 Dominance Relation

Definition 2. A fault, f_1, dominates another fault, f_2, if the test set of the latter, T_2, is a subset of the test set of the former, T_1; that is, $T_2 \subseteq T_1$. Any test pattern that detects f_2 will also detect f_1. Therefore, f_2 implies f_1 and it is sufficient to include f_2 in the fault list.

Going back to the case of the two-input AND gate, we have found that a test pattern for the SA1 fault on any of the inputs detects SA1 on the output. $Z/1$ can be dropped from the fault list, which is then reduced to three faults: $\{A/0, A/1, B/1\}$. In general, an N-input AND gate has only $N + 1$ unique faults. In other words, the gate has $N + 1$ equivalence classes. This is also true for other simple logic functions, such as OR, NAND, NOR, and XOR.

2.6.3 Fault Collapsing

Using equivalence and dominance, we have shown how to reduce the fault list of any simple N-input gate. Next we generalize the concept to other circuits. Consider the fan-out-free circuit shown in Fig. 2.6. It has 10 lines and the fault list must have 20 faults. Applying the equivalence relation to the AND and OR gates, we get the following equivalence classes:

1. $\{A/0, B/0, H/0\}$
2. $\{C/1, D/1, F/1, G/0\}$
3. $\{E/0, G/0, V/0\}$
4. $\{H/1, V/1, Z/1\}$
5. $\{F/0, G/1\}$

Based on the transitive property of equivalence relations, and since classes 2 and 3 include the same fault, $G/0$, they can be merged. All faults in the two classes are equivalent. Next, we list the dominance relations:

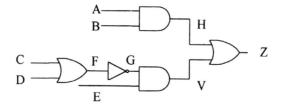

Figure 2.6   Example of fanout-free circuit.

6. $A/1 \rightarrow H/1$; thus $A/1$ can represent $H/1$ and all its equivalent faults in class 4.

7. $C/0 \rightarrow F/0$; thus $C/0$ can represent $F/0$ and all its equivalent faults in class 5.

8. $V/0 \rightarrow Z/0$, but $V/0$ belongs to equivalence class 3, which has been merged into class 2. Any fault from this class is dominated by $Z/0$.

9. $B/1 \rightarrow H/1$

10. $D/0 \rightarrow F/0$

11. $E/1 \rightarrow V/1$

From all these equivalence and dominance relations, we can reduce the fault classes from 11 to 7. An example of the shortened list will include a fault from class 1 ($A/0$), one from the merged classes 2 and 3 ($C/1$), a fault from classes 4 and 5; or faults that imply all the faults in each of these classes as shown by dominance relations 6 and 7: $A/1$ and $C/0$; and finally, faults from classes 9, 10, and 11: $B/1$, $D/0$, and $E/1$. In summary, a fault list is given by $\{A/0, A/1, B/1, C/0, C/1, D/0, E/1\}$. This list is not unique since faults $A/0$ and $C/1$ may be represented by other faults from the same equivalence class as representative of the class. Notice that the fault list contains only faults on the primary inputs. This is a convenient property of reconverging fan-out-free circuit [Breuer 1976]. However, most circuits are more complex. They include fan-out and also reconverging fan-out.

We will form the fault list for the 2-to-1 MUX of Fig. 2.5b. We have learned in Section 2.5 that the faults on S dominate those of C and D. Hence, in developing the fault list, we exclude SA faults on S since they will be covered with test patterns developed for branches C and D. Excluding S reduces the circuit to a fan-out-free circuit. It is thus sufficient to generate patterns for the primary inputs and the fan-out branches. The primary inputs and the branches of all fan-outs of the circuit are called *checkpoints*. Using checkpoints is valid only for *irredundant* circuits. An irredundant circuit is an implementation of a logic function that does not contain redundant literal or terms. Not all faults on redundant gates are detectable, as we will show next. In addition, the faults on the checkpoints as defined above do not cover all SSF in a redundant circuit. In such a case, additional patterns have to be used.

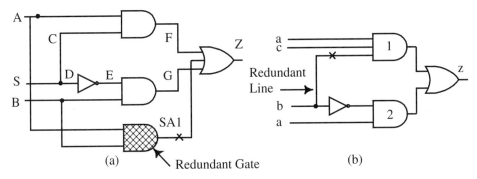

Figure 2.7 Redundant faults: (*a*) redundant product to eliminate hazard; (*b*) redundant literal.

Redundancy is sometimes included to eliminate static hazard. Other times, redundancy is introduced inadvertently as a result of revising the circuit. The 2-to-1 MUX of Fig. 2.5*b* has potential for static hazard as *S* changes from 1 to 0 while both *A* and *B* are high. This is due to the reconverging structure of the circuit: $F(A,B,S) = AS + BS'$. To remove the hazard, we add the term *AB*. $F(A,B,S) = AS + BS' + AB$ and we obtain the circuit shown in Fig. 2.7. There are no test patterns to detect the $H/0$ since this requires that $A = B = 1$; hence, either *F* or *G* will be equal to 1 and the fault on *H* cannot be sensitized to the output Z. Such a fault is called a *redundant fault*. It is simply undetectable because of redundancy in the circuit. Implementation of the function $F(a,b,c) = ab' + abc$, which is shown in Fig. 2.7*b*, also includes a redundant fault, $b/1$, as indicated in the figure. This function includes a *redundant literal*, *b* in the second product, since it may be expressed as $F(a,b,c) = ab' + ac$.

2.7 BRIDGING FAULTS

Bridging faults occur when two or more lines are shorted together and create wired logic [Mei 1974]. When the fault involves *r* lines with $r \geq 2$, it is said to be of *multiplicity r*; otherwise, it is a *simple bridging fault*. Multiple bridging faults are more likely to occur at the primary inputs of a chip. Bridging faults are becoming more predominant because the devices are becoming smaller and the gate density higher. The total number of all possible simple bridging faults in an *m*-line circuit is $C(m,2)$. However, in reality most pairs of lines are not likely to be shorted. Thus the actual number is much smaller than theoretically calculated and is layout dependent.

The behavior of the circuit in the presence of bridging faults is dependent on the technology. The short between the outputs of two inverters as shown in Fig. 2.2*d* can be modeled as a wired logic. If the implementation is in TTL

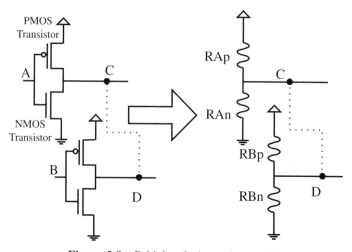

Figure 2.8 Bridging faults voting model.

technology, it is a wired AND; in the case of ECL technology, it is a wired-OR. In the case of CMOS technology, the wired logic depends on the type of gate driving the shorted lines and the input on these gates. We use the example in Fig. 2.8 to illustrate these effects. The two CMOS inverters are represented by their pull-up and pull-down resistances. Of course, the value of these resistances will depend on the inverters' input signals. For example, if $A = 0$, the NMOS transistor is off, and its corresponding resistance, RA_n, is infinite while the PMOS is on and RA_p assumes the value of the on resistance. The circuit may then be represented as a voltage divider, and the value of the output will depend on the on resistance of the various transistors, as listed in Table 2.4.

If the input signals are the same for both gates, the output will be the same for both gates and the fault will not be detected. Now if $A = 0$ and $B = 1$, the output will depend on RA_p and RB_n. If $RA_p > RB_n$, the output is 0, as if the two outputs, C and D, are wired ANDed. Following the same reasoning, one can understand the remainder of the entries in Table 2.4.

TABLE 2.4   Bridging Faults Models for the Circuit in Fig. 2.8

Input Condition	Relative Drive	Output Values	Wired Logic
$A = B$	Any ratio	$C = D = A' = B'$	AND, OR
$A = 0, B = 1$	$RA_p > RB_n$	$C = D = 0$	AND
	$RA_p < RB_n$	$C = D = 1$	OR
$A = 1, B = 0$	$RA_n > RB_p$	$C = D = 1$	OR
	$RA_n < RB_p$	$C = D = 0$	AND

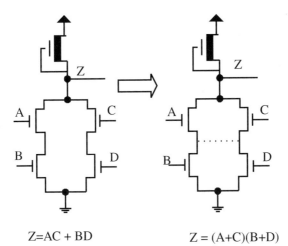

$$Z = AC + BD \qquad\qquad Z = (A+C)(B+D)$$

Figure 2.9 Change in functionality due to bridging faults.

Bridging faults may cause a change in the functionality of the circuit that cannot be represented by a known fault model. An example of this type of fault is shown in Fig. 2.9, where the function of the good NMOS circuit is $AB + CD$, while the bridging fault changes the functionality of the gate to $(A + C)(B + D)$. Also, the NOR gate has been transformed into a sequential circuit. Another important consequence of bridging faults is observed where the bridged wires are the input and the output of the same gate. Called a *feedback bridging fault*, this is illustrated on the right in Fig. 2.2d. The fault transforms a combinational circuit into a sequential circuit and increases the number of states in a sequential circuit.

SSA fault test sets have been used to detect bridging faults. They yield 100% fault detection for some special circuits. The approach is to alter the order of the patterns. It is also possible to use exhaustive test sets. Detection of bridging faults by current testing are detailed in Chapter 7 [Hawkins 1994].

2.8 SHORTS AND OPENS FAULTS

Some defects in CMOS technology defy representation by stuck-at faults [Wadsack 1978, El-Ziq 1981, Soden 1993]. The main reason for this is that MOS combinational circuits do not necessarily remain combinational under all faulty conditions (other than feedback bridging faults). There are several failure modes in MOS technology: (1) Transistors shorts and opens; (2) open-on gate, drain, or source contacts; and (3) shorts between gate and drain, or source, or channel.

Defects of the third category are due primarily to gate oxide breakdown, and they have been studied thoroughly [Syrzycki 1989, Segura 1995, 1996]. Models for these defects are shown in Fig. 2.10 for MOS transistors with a gate

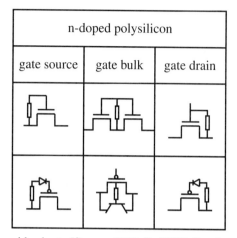

Figure 2.10   Gate oxide shorts [Segura 1996 © IEEE. Reprinted with permission].

of *n*-doped polysilicon [Segura 1995]. These *resistive shorts* are mentioned in Section 3.2. It has been proven that they can be modeled by a resistance for *n*-channel transistors and a resistance and a diode for the *p*-channel transistor. Unless the short resistance is negligible, most of these defects tend to pass undetectable; that is, the circuit behaves correctly on the functional level. However, the dynamic behavior of the circuit is affected due to the delays. These defects, however, are observable using current testing, which is the topic of Chapter 7.

In the remainder of this section we examine the effects of shorts and opens assuming zero resistance, but we show the effect of the resistance value in one case. We consider both NMOS and CMOS technology using the inverters shown in Figs. 2.11 to 2.13. In these figures, the numbers on some of the transistor terminals refer to short or open on these terminals. Then we consider the case of a short between gate and bulk.

2.8.1   NMOS Circuits

Site 1 in Fig. 2.11 is an open of the pull-down to GND, site 2 is a short between the gate and source of the NMOS driver, site 3 is a short between the source and drain of the driver transistor, and so on. The logic table next to the circuit includes one column for the input, *A*, and another for the good circuit response on the output, *Z*. The remaining columns give the outputs of the faulty circuit outputs in the presence of faults at the corresponding sites. A close observation of this table shows that for sites 1 through 4, the failure modes can easily be modeled by a stuck-at fault—*A*/0 in the first three cases and *A*/1 for the fourth case. The other sites need more careful examination to clarify the failure in MOS technology. The open at site 5 results in a floating gate. The trapped charge at the gate will keep the transistor on or off, depending on its charge

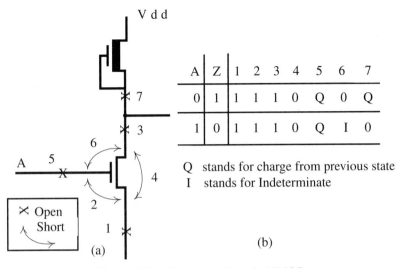

Figure 2.11   Open and shorts in NMOS.

value. However, it is probable that the charge will eventually leak and the output will become high.

The short between gate and drain at site 6 will tie the output to the input. When the input is low, so will be the output. However, when the input is high, the transistor will be on and will pull the output down, but the output will also be pulled up through the load transistor as well as through the gate driving the inverter under test. The output will thus be undetermined. Finally, the open at site 7 will cause the output to depend on the input signal. If the input is high, $A = 1$, the output will be discharged to ground ($Z = 0$). For $A = 0$, the transistor is not conducting and will retain the previous value, which is indicated by Q. This failure mode has caused the circuit to memorize the previous state, and as such, the circuit needs first to be put in a known state ($A = 1, Z = 0$), then forced to change value by making $A = 0$ and observing the output. If the output is still 0, this indicates that the fault is detected; otherwise, the circuit is not faulty. In this case, as in the case of the simple SR latch in Chapter 1, a sequence of test patterns is needed to detect the fault.

2.8.2   CMOS Circuits

2.8.2.1   Stuck-Open Faults.   In the case of CMOS circuits, open failures in pull-up and pull-down transistors cause the inverter to behave as a memory element. Detection of an open in such a circuit requires a specific test sequence, as in the case of the fault at site 7 of the NMOS inverter. For an open at sites 1 and 2 in Fig. 2.12, the sequence $A = 0$ ($Z = 1$), followed by $A = 1$, will still detect the fault if $Z = 1$; otherwise, the circuit is not faulty. Similarly, for sites

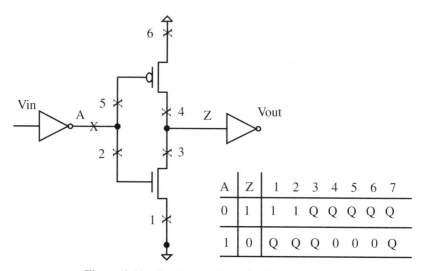

Figure 2.12   Stuck-open faults in CMOS circuits.

4, 5, and 6, the needed sequence to detect these faults is $A = 1$, $A = 0$. This type of fault, called *stuck-open*, has been investigated extensively by [Banerjee 1984]. It has also been demonstrated that such faults may sometimes be avoided by proper design practice [Koeppe 1987] that makes SOP faults appear as SA faults. Test generation algorithms for MOS circuits have been recommended and are addressed in a later chapter.

2.8.2.2   Stuck-On Faults and Shorts.   Next, we examine possible shorts as indicated in Fig. 2.13. Shorts between the source and drain of a transistor are equivalent because this transistor is on all the time. This failure mode is represented by a stuck-on fault (SON). The stuck-on fault at site 1 will keep the output pulled down to 0, and 1 can represent this failure mode as a stuck at 0 on Z. It is interesting to notice, though, that when $A = 0$, the PMOS transistor is on and there is a direct flow of current from V_{dd} to ground that will continue to flow as long as A is kept low. In such conditions, it is possible to observe the fault by monitoring the current, since under normal operating conditions, current flows only during switching and not in a static situation. This is known as current testing or I_{DDQ} testing and is discussed in detail in Chapter 7. Because of the symmetry of the CMOS transistor, the SON fault on the PMOS transistor can be observed by keeping the input $A = 1$.

If the NMOS is stuck-on (site 1) or the gate is shorted to the source in the PMOS (site 4), the circuit behaves as if the input of the inverter is stuck at 0 and stuck at 1. Similarly, for sites 2 and 3, the inverter behaves as if its input is stuck at 0 ($A/0$). Shorts at sites 5 and 6 are equivalent. They cause the circuit to behave in a fashion similar to the NMOS inverter in the presence of the short at site 7 (Fig. 2.11). The output follows the input.

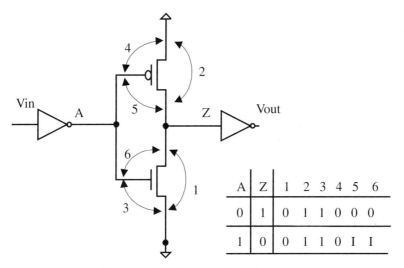

A	Z	1	2	3	4	5	6
0	1	0	1	1	0	0	0
1	0	0	1	1	0	I	I

Figure 2.13 Shorts in CMOS circuits.

So far in this section we have ignored the resistance of the shorts. However, we know from Section 2.2 that the short is resistive. The value of the resistance depends on the severity of the failure mechanism. The resistance varies from a few ohms to several thousand ohms. We reconsider the short between the gate and drains of the CMOS inverter in Fig. 2.13. The simulation results are shown in Fig. 2.14. The characteristic curves for different values of the short resistance, R_{short}, varying from 1000 kΩ to 100 Ω are shown in Fig. 2.14a, and the corresponding currents are given in Fig. 2.14b. Although for high values of the shorting resistance the output appears to be acceptable, the current is elevated and this fact will be used for current testing, which is the topic of Chapter 7.

2.9 DELAY FAULTS

It is possible for a circuit to be structurally correct but to have signal paths with delays that exceed the bounds required for correct operation. In such a case, a *delay fault* is said to have occurred. Delay faults may not be provoked if the operating frequency is low. The ultimate goal for detecting delay faults is to determine that the circuit works without malfunctions at the designed clock frequency. Thus it is appropriate to assume that the bound for correct operations should be the slack of the signal at the lead at which the fault is detected. The slack is the difference between the clock period and the longest delay path. Delay testing of a circuit determines if it contains signal paths that are too slow or too fast in propagating input transitions. Two main models are used for delay testing. The first, *gate delay fault* (GDF), is gate-oriented. The second model,

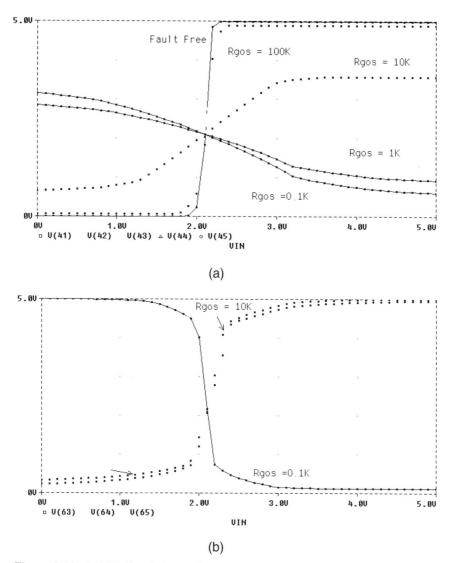

Figure 2.14 SPICE Simulation: Effect of the short resistance on (*a*) output voltage and (*b*) switching current.

path delay fault (PDF), is path-oriented [Smith 1985]. The GDF assumes that the delay faults are lumped at the faulty gate. The delay at the output of the gate will depend on whether this signal is switching from 0 to 1 (rise) or vice versa (fall). There is a disadvantage in adopting this type of fault because it does not capture the cumulative effects from other gates, and it also ignores the delays in the interconnect wires. Also, the gate delay may cause a local

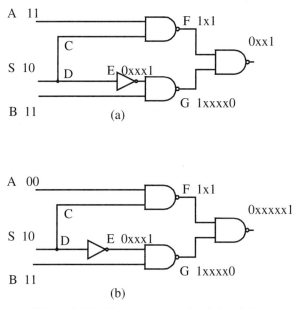

Figure 2.15 Two-vector test for delay faults.

delay at its output without affecting the delay of the circuit. For example, no delay is noticed at the output of the multiplexer in Fig. 2.15a when $A = B = 1$ and S switches from 0 to 1. This is not true, however, when the switching is from 1 to 0.

The PDF model takes into account the cumulative delay from the primary input to the output. Although this model requires the consideration of too many paths, it is more realistic, particularly for present technology circuits, where delays are due primarily to the interconnect wires. As the technology features decreased, gate delays have been reduced. Meanwhile, the resistance of the interconnect wires has increased due to the reduction of their cross-sectional area. Except for domino logic gates, delay testing of CMOS circuits consists of applying a pair of input patterns at the desired operational speed and observing the outputs for early or late transitions. Application of the pattern pairs to sensitize and propagate the fault is illustrated in Fig. 2.15b. The first test initializes the circuit, and the second pattern sensitizes the delay to the output. The circuit shows a 2-to-1 MUX that is implemented in NAND gates. The delay in the inverter is propagated to the output by the pair of patterns $(p_1, p_2) = (011, 001)$. The input signals at A and B remain the same in order to sensitize a path for any transition on S to appear on the output.

A two-pattern test is said to be a *robust delay test* for transition on a path, L, if and only if when L is faulty and a test pair (p_1, p_2) is applied, the circuit output is different from the expected state at sampling time independently of the delays along the gate [Lin 1987, Lam 1997]. Deterministic test generation tech-

TABLE 2.5   Temporary Faults

Type	Causes
Transient	Power supply disturbances
	Electromigration interference
	Charged particles
	Atmospheric discharges
	Electrostatic discharges
Intermittent	Parameter degradation

niques for delay faults are proposed in [Hsieh 1977, Shedletskey 1978, Malaiya 1984]. In addition to algorithmic delay test generation, pseudorandom testing [Wagner 1985] is also used. The latter is more suitable for a BIST environment. Simulation issues in delay testing are discussed in [Koeppe 1986]. Delay testing is becoming a very important aspect of digital testing.

2.10   TEMPORARY FAILURES

Hard failures result in a need to change a component or repair it, causing a long mean time to repair; soft or temporary failures are, however, more frequent [Ball 1967, Siewiorek 1978]. Temporary failures are much harder to track because it is usually not possible to reproduce the fault when a component, chip, or board is tested. Hence they are not studied as thoroughly as are hard failures. They are encountered in different digital components, but are particularly important in memory chips and microprocessors. Memory defects and testing are investigated in more detail in Chapter 12.

There are two main types of temporary failures: *transient* and *intermittent* (recurring). The first type is usually due to some temporary external conditions, while the second is due to varying hardware states: parameter degradation or improper timing. Table 2.5 lists the various types of temporary failures and some of their main causes.

2.10.1   Transient Faults

A transient fault occurs when a logic signal has its value temporarily altered by noise signals, and the resulting signal may be interpreted incorrectly by the rest of the circuit [McCluskey 1986]. Such a fault is difficult to diagnose and correct. It is thus important to minimize the noise in the circuit and increase the circuit's noise immunity. Transient faults may be caused by fluctuations in the power supply, metastability, or cosmic radiation.

2.10.1.1   Power Supply Disturbances. Power supply disturbances are known to cause errors in the operation of digital systems. Experiments were run to measure circuit susceptibility to power supply disturbances by measuring the change in the outputs of the circuit whose inputs are kept at constant signals. These experiments related the disturbances to the noise immunity of the circuits. Later experimentation was carried out on circuits in different technologies: CMOS gate array, CMOS breadboard, and bipolar breadboard. Here the experiment was carried out under more realistic assumptions: that the logic signal changes with time. It was found that the susceptibility of the circuits to power supply voltage disturbance is related to the operating frequencies. Errors are more likely to occur as the operation frequency increases. Also, propagation delay variation is the dominant effect, and noise immunity plays a smaller role in error occurrences. Failures due to power supply disturbance can be modeled by delay faults.

2.10.1.2 Metastability. Another form of a transient fault is caused by metastability in latches and flip-flops. Metastability occurs when a latch is given only enough energy to switch its state halfway to another stable state, and when the latch enters a critical stable, or metastable state, which exists somewhere between the two stable states. The latch will remain in this metastable state for some indeterminate amount of time but will eventually leave it for one of the stable states. No method of eliminating all metastability is known. It is important, however, to determine to a certain degree the probability of metastability occurrences. For this, metastability sensors have been developed [Stucki 1979]. The mean time between metastability (MTBM) failures can be predicted in terms of latch parameters [Chaney 1979].

2.10.1.3 Radiation-Induced Failures. Depending on the sensitive area in an electronic device, high-energy particles such as alpha particles or cosmic rays may have different effects [Sharma 1997]:

- Single-event upset (SEU) is the ionizing radiation-induced logic change in the cell of a storage device such as a RAM cell. This may cause hard or soft failure [May 1979, Berger 1985]. That is, the logic reversal may be permanent or temporary.

- Transient errors are dependent on the ionizing dose rate rather than on the total amount of radiation delivered.

- Single-event latch-up is caused by a latched low-impedance state by turning on the parasitic *pnp* and *npn* in the bulk of the CMOS circuit.

There is a general consensus that radiation hardening and proper packaging of integrated circuits are sufficient to decrease the occurrence of radiation-induced transient faults [Dieh 1982, Messenger 1992]. However, this approach is not effective in presenting the effect of alpha particles emanating from the impurities in the metal used in the fabrication.

2.10.2 Intermittent Faults

Intermittent failures are recognized to be an important cause of field failures in computer systems. Very little is known about the failure mechanisms because spontaneous intermittent failures are difficult to observe and control. However, artificially induced intermittent failures can easily be produced, controlled, and observed. Several papers [Koren 1977, Varshney 1979, Savir 1980] have addressed the problem of testing for intermittent failures. The intermittent fault models presented in these papers assume signal-independent faults. This assumption has turned out to be inappropriate after experimental evidence of pattern-sensitive intermittent failures.

Pattern sensitivity was first encountered in memory testing [Hayes 1980a,b] and microprocessors [Hackmeister 1975]. It is possible to explain pattern sensitivity in a microprocessor as being due to the embedded large RAMs. These failures are better described by delay faults than by pattern sensitivity. It is reasonable to conjecture that different instruction streams exercise different portions of the chip, and failures are caused by delay faults due to supply voltage reductions as described in [Cortes 1986].

An attempt to collect data on intermittent failures on Sperry–Univac computers was reported by [O'Neill 1980]. Hard failures that were believed to have appeared earlier as intermittent failures were analyzed. Instead, two classes of failure mechanisms were responsible for this type of failure: (1) metal-related open and short circuits, and (2) violations of operating margins. The latter class of failures has been investigated further by [Cortes 1986]. In his study, catalog parts are forced into intermittent faulty behavior by stressing supply voltage, temperature, and loading. The temperature stress changes the voltage transfer characteristics. The voltage stress affects the noise immunity. The loading stress reduces the driving capability. All stresses used in the experiments have some impact on the logic interfacing between two gates, thereby causing the circuit under stress to exhibit behavior similar to that of a marginal circuit under normal operating conditions [McCluskey 1987].

2.11 NOISE FAILURES

IC failures so far have been attributed mainly to process and manufacturing defects. Except for single-event transient errors, it has been implied that the circuits are noise immune. Also, attention was given only to the gates forming the circuit independently of their interconnects. Changes in technology have resulted in digital ICs with the following characteristics:

- The smaller devices are capable of switching faster, thus allowing an increase to the operating frequency. The slope of the ramp function mentioned above is not longer but is much smaller than the device dimensions.

- The higher device density enables the designer to make more complex and larger circuits.
- The power supply voltage was decreased from 5 V to 3.3 V and, very soon, to less than 2 V. Thus the circuit is becoming less immune to noise.
- The use of thinner and longer interconnect wires, resulting in a smaller cross section, has caused an increase in the resistance. Because of the fine pitch, fringing capacitances have been added to capacitance to ground. Also, mutual capacitance between interconnects on the same layer and on adjacent layers cannot be neglected.

As a result of these changes, the interconnect can no longer be modeled as a lumped model but as a distributed network or, not in the far future, as a transmission line. Also, as the number of switching devices has increased, the switching current has exaggerated the resistance of the substrate and the inductance of the ground and power pins. These new physical characteristics of very deep submicron technology together with the increased operating frequencies have resulted in compromising signal integrity. The most important types of failures that are due to noise are *crosstalk* and *simultaneous switching noise* (SSN), which is more commonly known as *ground bounce*. The first phenomenon is due to coupling capacitance between long wires and can be minimized by careful routing, which we discuss in Chapter 4. The other phenomenon is due to the package pin's inductance. It has usually been tamed by shorting the power and ground rails with appropriate capacitances. Efforts in characterizing the effects of these phenomena are ongoing [Haydt 1994, Breuer 1996, Rubio 1996].

REFERENCES

Bakoglu, H. B. (1990), *Circuits, Interconnections, and Packaging*, Addison-Wesley, Reading, MA.

Ball, M. O., and F. Hardie (1967), Effects and detection of intermittent failures in digital systems, *IBM*, 67-825-2137.

Banerjee, P., and J. A. Abraham (1984), Characterization and testing of physical failures in MOS logic circuits, *IEEE Des. Test Comput.*, Vol. 1, No. 8, pp. 76–86.

Berger, E. R., et al. (1985), Single event upset in microelectronics: third cosmic ray upset experiment, *IBM Tech. Direct. Fed. Syst. Div.*, Vol. 11, No. 1, pp. 33–40.

Black, J. R. (1969), Electromigration: a brief survey on some recent results, *IEEE Trans. Electron Devices*, Vol. ED-16, No. 2, p. 388.

Breuer, M. A., and A. D. Friedman (1976), *Diagnosis and Reliable Design of Digital Systems*, Computer Science Press, New York.

Breuer, M. A., and S. K. Gupta (1996), Process aggravated noise: new validation and test problems, *Proc. IEEE International Test Conference*, pp. 914–923.

Chakravarty, S., and Y. Gong (1995), Voting model based diagnosis of bridging faults in combinational circuit, *IEEE International Conference on VLSI Design*, pp. 338–342.

Chaney, T. J., and F. U. Rosenberger (1979), Characterization and scaling of MOS flip-flop performance in synchronizer applications, *Proc. CALTECH Conference on VLSI*, pp. 357–374.

Cortes, M. L., et al. (1986), Properties of transient errors due to power supply disturbances, *Proc. IEEE International Symposium on Circuits and Systems*, pp. 1046–1049.

D&T (1998), *IEEE Des. Test Comput.*, Vol. 15, No. 4.

Dieh, S. E., et al. (1982), Error analysis and prevention of cosmic ion-induced soft errors in static CMOS-RAMs, *IEEE Trans. Nucl. Sci.*, Vol. NS-30, No. 6, pp. 2032–2039.

El-Ziq, Y. M. (1981), Functional-level test generation for stuck-open faults in CMOS VLSI, *Proc. IEEE International Test Conference*, pp. 536–546.

Ferguson, F. J., and J. P. Shen (1988), Extraction and simulation of realistic CMOS faults using inductive fault analysis, *Proc. IEEE International Test Conference*, pp. 475–484.

Goldstein, H. (1970), *A Probabilistic Analysis of Multiple Faults in LSI Circuits*, Repository R-77, IEEE Computer Society, Long Beach, CA.

Hackmeister, D., and A. C. L. Chiang (1975), Microprocessor test technique reveals instruction pattern-sensitivity, *Comput. Des.*, No. 12, pp. 81–85.

Hawkins, C. F., and J. M. Soden (1986), Reliability of electrical propertices of gate oxide shorts in CMOS ICs, *Proc. IEEE International Test Conference*, pp. 443–451.

Hawkins, C. F., et al. (1994), Defect classes: an overdue paradigm for CMOS IC testing, *Proc. IEEE International Test Conference*, pp. 413–425.

Haydt, M. S., R. Owens, and S. Mourad (1994), Modeling and characterization of ground bounce, *Proc. IEEE International Test Conference*, pp. 279–285.

Hayes, J. P. (1980a), Detection of pattern-sensitive faults in random-access memories, *IEEE Trans. Comput.*, Vol. C-29, No. 3, pp. 713–719.

Hayes, J. P. (1980b), Testing memory for single-cell pattern-sensitive fault, *IEEE Trans. Comput.*, Vol. C-29, No. 3, pp. 713–719.

Henderson, C., et al. (1997), Yield and failure analysis challenges in IC manufactoring, *Future Fab Int.*, Vol. 1, No. 2, pp. 335–343.

Ho, P. S. (1982), Basic problems for electromigration in VLSI, *Proc. 20th Annual International Reliability Physics Symposium*, p. 288.

Hsieh, E. P., et al. (1977), Delay test generation, *Proc. 14th ACM-IEEE Design Automation Conference*, pp. 486–491.

Hughes, J. L. A., and E. J. McCluskey (1984), Multiple stuck-at fault coverage of single stuck-at fault test sets, *Proc. IEEE International Test Conference*, pp. 52–58.

ITC (1997), *Proc. IEEE International Test Conference.*

Koeppe, S. (1986), Modeling and simulation of delay faults in CMOS stuck-open logic circuits, *Proc. IEEE International Test Conference*, pp. 530–536.

Koeppe, S. (1987), Optimum layout to avoid CMOS stuck-open faults, *Proc. 24th ACM-IEEE Design Automation Conference*, pp. 829–835.

Koren, I., and Z. Kohavi (1977), Diagnosis of intermittent faults in combinational networks, *IEEE Trans. Comput.*, Vol. C-26, No. 11, pp. 1154–1157.

Lam, W. K., et al. (1997), Delay fault covering, *IEEE Trans. Comput.-Aided Des.*, Vol. CAD-16, No. 1, pp. 32–44.

Lin, C. J., and S. M. Reddy (1987), On delay fault testing in logic circuits, *IEEE Trans. Comput.-Aided Des.*, Vol. CAD-6, No. 9, pp. 694–703.

Malaiya, Y. K., and R. Narayanaswamy (1984), Modeling and testing for timing faults in synchronous sequential circuits, *IEEE Des. Test Comput.*, Vol. 1, No. 4, pp. 62–74.

Maly, W., F. J. Ferguson, and J. P. Shen (1984), Systematic characterization of physical defects for fault analysis of MOS IC cells, *Proc. IEEE International Test Conference*, pp. 237–245.

May, T. C., and M. H. Wood (1979), Alpha-particle-induced soft errors in dynamic memories, *IEEE Trans. Electron Devices*, Vol. ED-26, No. 1, pp. 2–7.

McCluskey, E. J. (1986), *Principles of Logic Design*, Prentice Hall, Upper Saddle River, NJ.

McCluskey, E. J., and S. Mourad (1987), Comparing causes of IC failures, in *Developments in Integrated Circuit Testing*, D. M. Miller (ed.), Academic Press, San Diego, CA, pp. 13–46.

Mei, K. (1974), Bridging and stuck-at faults, *IEEE Trans. Comput.*, Vol. C-23, No. 7, pp. 720–727.

Merchant, P. P. (1982), Electromigration: an overview (VLSI metallization), *Hewlett-Packard J.*, Vol. 33, No. 8, p. 28.

Messenger, G. C., and M. Ash (1992), *The Effects of Radiation on Electronic Systems*, Van Nostrand Reinhold, New York.

Mourad, S., et al. (1986), Effectiveness of single stuck-at fault tests to detect multiple faults in parity trees, *Comput. Math. Appl.*, Vol. 13, No. 5/6, pp. 455–459.

Mourad, S., and E. J. McCluskey (1987), Fault models, in *Testing and Diagnosis of VLSI and ULSI*, F. Lombardi and M. Sami (eds.), Kluwer Academic, Norwell, MA, pp. 49–58.

O'Neill, E. J., and J. R. Halverson (1980), *Study of Intermittent Field Hardware Failure Data in Digital Electronics*, NASA Contractor Rep. 159269.

Rubio, A., et al. (1996), An approach to the analysis and detection of crosstalk faults in digital circuits, *IEEE Trans. Comput.*, Vol. C-36, No. 3, pp. 387–394.

Sachdev, M. (1998), *Defects Oriented Testing for CMOS Analog and Digital Circuits*, Kluwer Academic, Norwell, MA.

Savir, J. (1980), Testing for single intermittent failures in combinational circuits by maximizing the probability of fault detection, *IEEE Trans. Comput.*, Vol. C-29, No. 5, pp. 410–416.

Schertz, D. R., and G. Metze (1971), On the design of multiple fault diagnosable networks, *IEEE Trans. Comput.*, Vol. C-20, No. 11, pp. 1361–1364.

Schugraf, K. F., and C. Hu (1994), Reliability of thin SiO_2, *Proc. IEEE*, Vol. 9, Sept., pp. 989–1004.

Segura, J., et al. (1995), A detailed analysis of GOS defects in MOS transistors: testing implications at circuit level, *Proc. IEEE International Test Conference*, pp. 544–551.

Segura, J., et al. (1996), A detailed analysis of electrical modeling of gate oxide shorts in MOS transistors, *J. Electron. Test. Theory Appl.*, Vol. 8, No. 3, pp. 229–240.

Sharma, A. K. (1997), *Semiconductor Memories: Technology, Testing, and Reliability*, IEEE Press, Piscataway, NJ.

Shedletskey, J. (1978), Delay testing LSI logic, *Proc. 8th International Fault Tolerance Computing Symposium*, June 1978, pp. 410–416.

Shen, J. P., W. Maly, and F. J. Ferguson (1985), Inductive fault analysis of MOS integrated circuits, *Des. Test Comput.*, Vol. 2, No. 6, pp. 13–26.

Siewiorek, D. P. (1978), A case study of C.mmp, Cm* and Cvmp," *Proc. IEEE*, Vol. 66, No. 10, pp. 1178–1220.

Smith, G. L. (1985), Models for delay faults based on path, *Proc. International Test Conference*, Philadelphia, Oct., pp. 342–349.

Soden, J. M., and C. F. Hawkins (1993), Correct CMOS defect models for quality testing, *Proc. NASA Symposium on VLSI Design*, Albuquerque, NM.

Strapper, C. (1976), LSI yield modeling and process monitoring, *IBM J. Res. Dev.*, Vol. 20, No. 3, pp. 228–234.

Strapper, C., A. McLaren, and M. Dreckmann (1980), Yield model for productivity optimization of VLSI memory chips with redundancy and partially good product, *IBM J. Res. Dev.*, Vol. 24, No. 3, pp. 398–409.

Stucki, M. J., and J. R. Cox, Jr. (1979), Synchronization strategies, *Proc. CALTECH Conference on VLSI*, pp. 375–393.

Syrzycki, M. (1989), Modeling of gate oxide shorts in MOS transistors, *IEEE Trans. Comput.-Aided Des.*, Vol. CAD-8, No. 3, pp. 193–202.

Sze, S. (1983), *VLSI Technology*, McGraw-Hill, New York.

Varshney, P. K. (1979), On analytical modeling of intermittent faults in digital systems, *IEEE Trans. Comput.*, Vol. C-28, No. 10, pp. 786–791.

Wadsack, R. L. (1978), Fault modeling and logic simulation of CMOS and MOS integrated circuits, *Bell Syst. Tech. J.*, Vol. 57, No. 5, pp. 1449–1474.

Wagner, K. D. (1985), The error latency of delay faults in combinational and sequential circuits, *Proc. IEEE International Test Conference*, pp. 334–341.

Walker, D. M. H. (1987), *Yield Simulation for Integrated Circuits*, Kluwer Academic, Norwell, MA.

Woods, M. H. (1984), The implications of scaling on VLSI reliability, *Proc. 22nd Annual International Reliability Physics Symposium*, p. 288.

PROBLEMS

2.1. The two patterns (000) and (111) toggle all nodes of a three-input NAND gate. Are they sufficient to detect all stuck-at faults in the gate? Explain why or why not.

2.2. Prove that the patterns that detect the SSF on the primary inputs will detect all SSF in a fan-out-free combinational circuit with single output.

2.3. Show that the implementation of the function $F(a, b, c) = ab' + abc$ has a redundant fault.

2.4. Determine the checkpoints of each of the two circuits in Fig. P2.4.

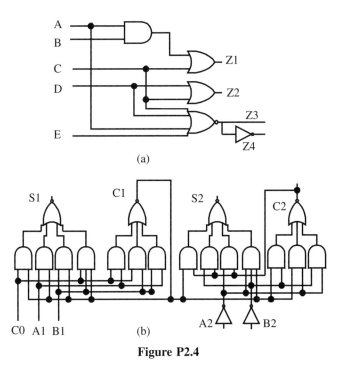

Figure P2.4

2.5. Use the equivalence and dominance properties to reduce the faults on the checkpoints determined in Problem 2.4.

2.6. For the circuit in Fig. P2.4a, generate test patterns to detect the bridging faults among the primary inputs. Model this fault by a wired AND.

2.7. Determine test sequences that detect stuck-open on the gates of all transistors of the CMOS circuits implementing **(a)** a two-input NAND gate, **(b)** a two-input NOR gate, and **(c)** the gate representing $[(ab)' + c]'$.

2.8. Consider the circuit in Fig. 2.13. Assume that there is a resistive short between the gate and drain, R_{short}. Use a SPICE-like simulator to examine the effect of the load on the driving inverter (the first inverter). That is, add a capacitive load, C_L, and plot the output voltage versus C_L.

3

DESIGN REPRESENTATION

3.1 LEVELS OF ABSTRACTION

In this chapter we examine design representations, also called *design models.* From the onset it is important to distinguish between design models and design specifications. The specifications describe the design in terms of its results, while the models describe the design's procedure. The model is used to simulate the design in a given domain and at a given level and to assert whether or not it conforms to specifications. It is also used in mapping the design to another level or domain. Use of the terms *domain* and *level* refer to the taxonomy illustrated with modifications in Fig. 3.1 [Gajski 1983, Walker 1985, Michel 1992].

For each of the domains—*behavioral, structural,* and *physical*—we distinguish several levels: system, RTL, logic, and circuit. The behavioral domain gives a functional representation, while the structural domain describes the architectural blocks; the physical domain is the actual chip. For the logic level, the three domains are illustrated in Fig. 3.2, and Fig. 3.3 shows the levels in the structural domain. Table 3.1 shows the different levels of the design for each domain. The same level of a domain may be described in different formats: equations, tabular, programming language, or hardware description languages. In this chapter we examine these formats and illustrate their use in the testing field. Familiarity with these representations is important in order to understand their effect on the efficiency of CAD tools, which we discuss in Chapters 4 and 14.

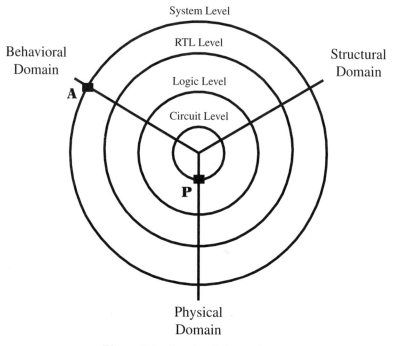

Figure 3.1   Levels of abstractions.

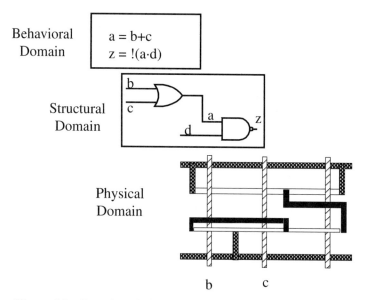

Figure 3.2   Domains of circuit representations on the logic level.

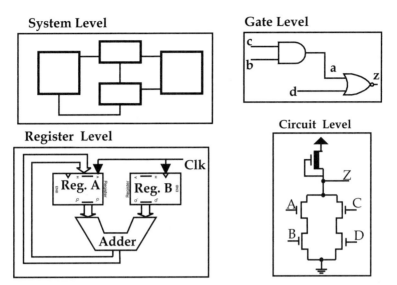

Figure 3.3   Levels of abstractions in the structural domain.

TABLE 3.1   Domains and Level of Design

	Domain		
Level	Behavioral	Structural	Physical
System	System specifications	Blocks	Chip
RTL	RTL specifications	Registers	Macro cells
Logic	Boolean functions	Logic gates	Standard cells
Circuit	Differential equations	Transistors	Masks

3.2   MATHEMATICAL EQUATIONS

Mathematical formulation is used to describe designs on the logic level as well as on the circuit level. Examples are combinational switching functions, finite state machines, and transistor-level circuits.

3.2.1   Switching Functions

The combinational functions that we customarily use in logic design are a special case of Boolean functions. They are described more accurately as switching functions. Any such function $f(x)$ has a range B and a domain B^n where $B = \{1,0\}$ and $f(x): B^n \rightarrow B$.

Definitions

- For any element $c \in B$, the constant function is $f(x_1, \ldots, x_n) = c$, where $x_i \in B^n$.
- For any $x_i \in B^n$, the projection function is $f(x_1, \ldots, x_n) = x_i$.
- The set of variables $\{x_1, x_2, \ldots, x_n\}$ is called the *support* of the function.
- If g and h are n-variable functions, the functions $g + h$, $g \cdot h$, and g' are defined by

$$(g + h)(x_1, \ldots, x_n) = g(x_1, \ldots, x_n) + h(x_1, \ldots, x_n)$$

$$g \cdot h(x_1, \ldots, x_n) = g(x_1, \ldots, x_n)h(x_1, \ldots, x_n)$$

$$g'(x_1, \ldots, x_n) = (g(x_1, \ldots, x_n))'$$

- There is only a finite set of distinct functions of the n-variable, $|B|^{2^n}$.

However, the same function may be expressed in a different form—for example, $f(x_1, x_2, x_3) = x_1 x_2' + x_2 = x_1 + x_2$—by virtue of switching algebra postulates [Kohavi 1978, McCluskey 1986].

Definition. The Shannon expansion of any switching function is given by

$$f(x_1, \ldots, x_n) = x_i' f(x_1, \ldots, x_i = 0, \ldots, x_n) + x_i f(x_1, \ldots, x_i = 1, \ldots, x_n) \quad (3.1)$$

or

$$f(x_1, \ldots, x_n) = [x_i' + f(x_1, \ldots, x_i = 0, \ldots, x_n)][x_i + f(x_1, \ldots, x_i = 1, \ldots, x_n)] \quad (3.2)$$

The cofactors of x_i are the residues of the functions for $x_i = 0$ and $x_i = 1$. The residue of the function for $x_i = 1$ is the value of this function when x_i is substituted by 1. Successive applications of the Shannon expansion would eventually result in expressing the function in terms of products of x_i. These products are called the *minterms*. The function is thus expressed as a sum of minterms. For example, the function

$$f(x_1, x_2, x_3) = x_1 x_2' + x_3 \quad (3.3)$$

may be expressed after three successive applications of Shannon expansion, in terms of x_3, x_2, and x_1, into the sum of minterms

$$f(x_1, x_2, x_3) = \begin{cases} (x_1x_2')x_3 + (x_1x_2')x_3' + x_3 \\ x_1x_2'x_3 + x_1x_2'x_3' + x_2'(x_3) + x_2(x_3) \\ x_1x_2'x_3 + x_1x_2'x_3' + x_1'(x_2'x_3) + x_1(x_2'x_3) + x_1'(x_2x_3) + x_1(x_2x_3) \end{cases}$$

and using $x + x = x$, we can eliminate the fourth term, which is identical to the first one, and obtain

$$f(x_1, x_2, x_3) = \begin{cases} x_1x_2'x_3 + x_1x_2'x_3' + x_1x_2x_3 + x_1'x_2x_3 + x_1'x_2'x_3 & (3.4) \\ \sum m(1, 3, 4, 5, 7) & (3.5) \end{cases}$$

3.2.2 Boolean Difference

Switching functions are used extensively in logic synthesis and, in particular, in minimization [Hachtel 1996]. They are used to develop test pattern generation as described below. To observe a stuck-at fault at any of the outputs of the circuit, we must have the response of the good circuit, R, different from the faulty circuit, R_f. In terms of Boolean algebra, this can be expressed as

$$R \oplus R_f = 1 \qquad (3.6)$$

If the faulty node is a primary input, x_i, then for a function $f(x_1, \ldots, x_n)$, R and R_f are the residues of the function with respect to x_i. Equation (3.6) can then be expressed as

$$f(x_1, \ldots, x_i = 0, \ldots, x_n) \oplus f(x_1, \ldots, x_i = 1, \ldots, x_n) = 1 \qquad (3.7)$$

This last expression, known as the *Boolean difference with respect to* x_i is denoted by $df(x)/dx_i$, where x is the support of the function. For simplicity, we express it as $df(x)/dx_i = f_i(0) \oplus f_i(1)$, where f_i is the residue of f with respect to x_i. The Boolean difference is then equated to 1 to determine the values of the variables that will allow the observability of the fault on node x_i at the primary output. In addition, for a SA1 or SA0 fault, we need $x_i = 0$ or $x_i = 1$. The test patterns to detect SA0 and SA1 on x_i are, respectively, given by

$$x_i \frac{df(x)}{dx_i} = 1 \quad \text{and} \quad x_i' \frac{df(x)}{dx_i} = 1 \qquad (3.8)$$

As an example, let us consider the function

$$f(x) = g(x) + x_3 \qquad \text{where } g(x) = x_1x_2 \qquad (3.9)$$

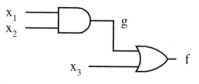

Figure 3.4 Example for test pattern generation using Boolean algebra.

which is illustrated by the gate-level description in Fig. 3.4. To determine the test patterns for the stuck-at faults on the primary input, x_2, we first form the Boolean difference and equate it to 1. Thus $df(x)/dx_2 = x_3 \oplus (x_1 + x_3) = x_3'x_1 = 1$. This implies that $x_1 = 1$ and $x_3 = 0$. For the SA1 and SA0 faults on x_2, the patterns are then $x_1x_2x_3 = (100)$ and (110), respectively. We repeat this calculation for stuck-at faults on x_3. First, we calculate the Boolean difference: $df(x)/dx_3 = g(x) \oplus 1 = x_1x_2 \oplus 1 = (x_1x_2)'$. Then we equate its products with x_3 and x_3' to 1. The patterns to detect the faults $x_3/0$ and $x_3/1$ are then $x_3(x_1x_2)' = 1$ and $x_3'(x_1x_2)' = 1$. For the first fault, we must then have $x_3x_1' + x_3x_2' = 1$. This results in three patterns: $x_1x_2x_3 = (001, 011, \text{or } 101)$, and for the other fault, we have $x_1x_2x_3 = (000, 010, \text{or } 100)$.

The fault may also be on an internal node of the circuit. In this case we include this node among the variables in expressing the function. For instance, for the internal node $g(x)$ of our example, we can express the function as $f(x_3, g)$ and the test pattern that detects $g/0$ by

$$g(x)\, \frac{df(x, g(x))}{dg(x)} = g(x)[f(x, g = 0) \oplus f(x, g = 1)] = 1 \qquad (3.10)$$

From Eq. (3.9) we have $f(x, g = 0) = x_3$ and $f(x, g = 1) = 1$. Thus the expression in Eq. (3.10) becomes $g(x)\{x_3 \oplus 1\} = g(x)x_3' = 1$. Substituting for $g(x)$, we get $x_1x_2x_3' = 110$, which is the same pattern that detects $x_2/0$. This was expected from our knowledge of fault equivalence that was explained in Section 2.6.

The Boolean difference has several properties that can be used in dealing with more complex functions [Fujiwara 1986, Miczo 1986]. However, this approach is not computationally efficient for test pattern generation.

3.2.3 Finite State Machines

A finite state machine (FSM) is formally expressed as a six-tuplet $(I,S,\delta,S_0,O,\lambda)$, where

- I is the input alphabet, that is, a finite nonempty set of inputs.
- S is the finite and nonempty set of states.
- $\delta: S \times I \rightarrow S$ is the next-state function.
- $S_0 \subseteq S$ is the set of initial states.

- O is the output alphabet.
- $\lambda: S \times I \rightarrow O$ is the output function for a Mealy machine [Mealy 1955].
- $\lambda: S \rightarrow O$ is the output function for a Moore machine [Moore 1956].

An example of an FSM, M, is given by the sets

$$I = \{0, 1\}$$
$$S = \{A, B, C, D\}, O = \{0, 1\}$$

and the next-state and the output functions

$$
\begin{array}{llll}
\delta(A, 0) = C, & \delta(A, 1) = B, & \delta(B, 0) = C, & \delta(B, 1) = B \\
\delta(C, 0) = D, & \delta(C, 1) = C, & \delta(D, 0) = A, & \delta(D, 1) = C \quad (3.11)\\
\lambda(A, 0) = 1, & \lambda(B, 0) = 0, & \lambda(C, 0) = 1, & \lambda(D, 0) = 1 \\
\lambda(A, 1) = 0, & \lambda(B, 1) = 1, & \lambda(C, 1) = 1, & \lambda(D, 1) = 0 \quad (3.12)
\end{array}
$$

When applied to a machine that is initially in state A, an input string (10110) will yield an output string (00111) and leave the machine in the final state D:

Time t	0	1	2	3	4	5
Input I	1	0	0	1	0	
State $\delta(S, 1)$	A	B	C	C	C	D
Output $\lambda(S, 1)$	—	0	0	1	1	1

In this example, all states and outputs are fully specified. M is said to be a completely specified machine. Often, not all next states or outputs are known for each input combination. In such cases, the FSM is *incompletely specified*.

The mathematical representation of an FSM is cumbersome and other means are used more typically. A graph representation, illustrated in Section 3.5, is another useful way of expressing an FSM. This diagrammatic representation enhances the visualization of state transitions. However, when the FSM is large, that is, when the cardinality of the set S is too large, the mathematical equations and the graph representation become cumbersome. It is more practical then to represent the FSM in a tabular format, as explained in Section 3.3.

3.2.4 Transistor-Level Representation

On the circuit level, however, equations ar still the most useful way for simulation and timing analysis. Before the early 1970s, circuits were analyzed almost exclusively by hand [Hodges 1983]. The first important automation software

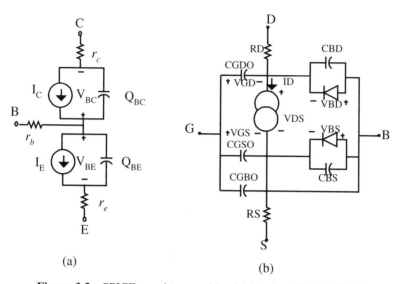

Figure 3.5   SPICE transistor models: (*a*) bipolar; (*b*) MOSFET.

was SPICE (simulation program, IC emphasis), developed at the University of California–Berkeley [Nagel 1975, Quarles 1994]. The simulation makes use of mathematical models of the transistor. Usually, the model is given as a set of equations that relates the voltages and currents through the network. The model equations are complex since the transistor exhibits nonlinear resistive and capacitive characteristics. For a bipolar transistor, the Eber–Moll and the charge-based Gummel–Poon models have been merged to represent the currently used model, which is shown in Fig. 3.5*a* [Getreu 1976]. For a MOS transistor, the models used are also those developed primarily for a SPICE simulator. There are four different models for the various levels of this simulator. They represent the effects of charge storage associated with thin oxide. Level 1 uses the Shichman–Hodges model shown in Fig. 3.5*b*, which relies on the current square-law expression in terms of the gate voltage. The other levels utilize more complex models to reflect the effect of submicron and *deep* submicron technologies. Submicron refers to feature size, $2\lambda < 1$ μm, and deep submicron is usually for $2\lambda < 0.5$ μm. For example, level 3 takes into account second-order effects such as short channel threshold voltage, subthreshold conduction, scattering-limits velocity saturation, and charge-controlled capacitances. There are several references that detail these models, among them [Antognetti 1988]. Recently developed level 4, the Berkeley short-channel IGFET model (BSIM) model, is based on different parameters that are extracted from experimental data. Currently, it is the most used in industry to model the behavior of submicron MOS transistors [Jeng 1983, Sheu 1988, Cheng 1996].

The circuit description for SPICE simulation, usually called the *deck*, consists of the topology of the circuit and specifications of the transistor model, the parameters, and the technology. An example for a typical CMOS inverter is shown below.

```
* CMOS inverter
* The circuit
Mn 3 2 0 0 CMOSN W=1.8U, L=0.5U
Mp 3 2 1 1 CMOSP W=5.4U, L=0.5U
CL 3 0 2pF
* Voltages used
V_DD 1 0 DC 5V
Vin 2 0 0 3.3 0.2
Vin 2 0 PULSE (0V 3.3V 0n 0n 0n 50n 100n)
* Transistor Models used
.MODEL CMOSN NMOS LEVEL=3 PHI=0.700000 TOX=9.6000E-09 J=0.200000U
+ TPG=1 VTO=0.6566 DELTA=6.9100E-01 LD=4.7290E-08 KP=1.9647E-04
+ UO=546.2 THETA=2.6840E-01 RSH=3.5120E+01 GAMMA=0.5976
+ NSUB=1.3920E+17 NFS=5.9090E+11 VMAX=2.0080E+05 ETA=3.7180E-02
+ KAPPA=2.8980E-02 CGDO=3.0515E-10 CGSO=3.0515E-10 CGBO=4 0239E-10
+ CJ=5.62E-04 MJ=0.559 CJSW=5.00E-11 MJSW=0.521 PB=0.99
* Weff = Wdrawn - Delta_W
* The suggested Delta_W is 4.1080E-07
.MODEL CMOSP PMOS LEVEL=3 PHI=0.700000 TOX=9.6000E-09 J=0.200000U
+ TPG=-1 VTO=-0.9213 DELTA=2.8750E-01 LD=3.5070E-08 KP=4.8740E-05
+ UO=135.5 THETA=1.8070E-01 RSH=1.1000E-01 GAMMA=0.4673
+ NSUB=8.5120E+16 NFS=6.5000E+11 VMAX=2.5420E+05 ETA=2.4500E-02
+ KAPPA=7.9580E+00 CGDO=2.3922E-10 CGSO=2.3922E-10 CGBO=3.7579E-10
+ CJ=9.35E-04 MJ=0.468 CJSW=2.89E-10 MJSW=0.505 PB=0.99
* Weff = Wdrawn - Delta_W
* The suggested Delta_W is 3.6220E-07
.TRAN 5N 1000N
.END
```

The first three lines describe the topology of the circuit with input node 2 and output node 3. Node 0 is connected to ground and node 1 is connected to V_{dd}, respectively. In this example, Mn and Mp are the n- and p-channel transistors that employ the user-defined models CMOSN and CMOSP, respectively. Their lengths and widths are indicated for the smallest devices in 0.5-μm technology. These models are described in the latter part of the SPICE deck. Only some parameters are specified; the others take default values. All such parameters are technology specific.

A set of differential equations for the voltage, current, and delays is developed and solved numerically. Many other derivatives of this program have been used in industry: HSPICE [Avanti 1997] and PSPICE [MicroSim 1994]. SPICE-like simulators are now the standard in circuit-level simulation. One of the major problems with these programs is that under some conditions, the solution does not converge. To overcome this problem, lookup tables are used in the iterative solution process.

3.3 TABULAR FORMAT

The tabular form is one of the most popular representations of digital design because of its suitability for use with computer programs, an essential computer-aided design (CAD) tool for present designs. Behavioral or functional models on the logic level are often represented in a tabular form. Most readers are familiar with truth tables and state tables for defining combinational logic functions and finite state machines from logic design courses.

3.3.1 Truth Tables

A truth table of a combinational logical function, $f(x_1, x_2, \ldots, x_n)$, consists of 2^n rows. Each row includes a minterm (a zero cube) and the corresponding value of the function, 0 or 1. Three examples of simple functions are given in Table 3.2. The first function, $f(x_1, x_2, x_3) = \sum m(1,4,6,7)$, is fully specified. For each minterm the function has one logic value, 0 or 1. The second function is *incompletely specified*; for minterms 3 and 4, the value of the function is indicated by d, which stands for "don't care." The designer may assign it a value of 1 or 0 for the benefit of simplifying the design. The third function is expressed in terms of 0-cubes as well as 1-cubes, where x implies that the function value is valid when $x = 0$ and 1.

3.3.2 State Tables

State tables easily represent finite state machines. A state table may be in one of the two formats illustrated in Table 3.3 for a machine, M, defined in Section 3.2.3. The first column in Table 3.3a contains the present state (PS) and the remaining columns list, for the various input combinations, the next state

TABLE 3.2 Combinational Functions Defined in Tabular Form

Minterm	$f_1(x_1, x_2, x_3)$	$f_2(x_1, x_2, x_3)$	Minterm	$f_3(x_1, x_2, x_3)$
000	1	1	000	1
001	1	1	0x1	1
010	0	0	010	0
011	0	d	1x0	1
100	1	d	101	0
101	0	0	111	1
110	1	1		
111	1	1		
	(a)			(b)

TABLE 3.3   State Table for FSM *M*

Present State	Next State			*I*	PS	NS	Output
	I = 0	*I* = 1					
A	C, 1	B, 0		0	A	C	1
B	C, 0	B, 1		1	A	B	0
C	D, 1	C, 1		0	B	C	0
D	A, 1	C, 0		1	B	B	1
				0	C	D	1
				1	C	C	1
				0	D	A	1
				1	D	C	0
	(a)					*(b)*	

(NS) and the output. An alternative form is shown in Table 3.3*b*. Here the first column represents the input combinations; the second through fourth columns represent the present state, the next state, and the output, respectively. This tabular description of an FSM is useful in synthesis tools as well as in some testing practices, for example, finding the *checking sequence* of the machine [Hennie 1964, Hachtel 1996]. This sequence is to an FSM what exhaustive testing is to combinational circuits and it is described in Chapter 6.

3.4   GRAPHICAL REPRESENTATION

Graphical representation of a circuit is a convenient notation for designers. We are all familiar with schematic capture on the logic and circuit levels, which are illustrated in Fig. 3.3. As the designs become very large, on the order of 1 million gates, it becomes difficult to display such designs graphically as a whole. However, it is still convenient to deal with one part of the design at a time. All CAD tools continue to display schematic capture even for those designs that have been entered in HDL. This is done for the benefit of the designers to help in visual perception of the design.

A generic model for an FSM is shown in Fig. 3.6*a*. The combinational part represents the next-state forming logic and the output forming logic. The one flip-flop shown represents all flip-flops in the circuit. Here we assume that only D flip-flops are used. This assumption is very realistic and does not limit the applicability of the scheme to other flip-flop types. An FSM may also be represented as a *flowchart*. This is used to represent a counter or any arbitrary control unit of a processor. This is suitable for an algorithmic representation of the FSM. An *algorithmic state machine* (ASM) gives a functional (behavioral) description of the FSM. Figure 3.6*b* shows the flowchart for the FSM *M* defined in Section 3.2.3. This representation can become very cumbersome for larger

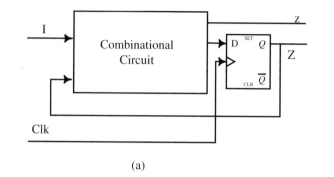

(a)

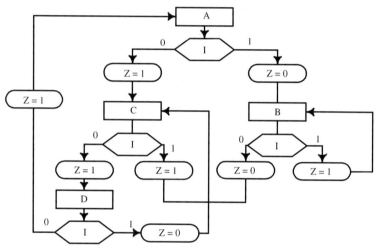

(b)

Figure 3.6   Finite-state machine representations: (*a*) Huffmann model; (*b*) flowchart model.

machines. In these cases, using hardware description language would be much more manageable, as we demonstrate later in the chapter.

3.5   GRAPHS

In this section we introduce the basic concepts of graph theory, which are used whenever needed throughout the book. A graph $G(V, E, W)$ consists of a set of *vertices V*, a set of *edges E*, and a set of *weights W*. The vertices are also called *nodes* and we use these two words interchangeably. An edge $e(u, v) \in E$, where $u \in V$ and $v \in V$ is a relation between vertices u and v. It may also be denoted

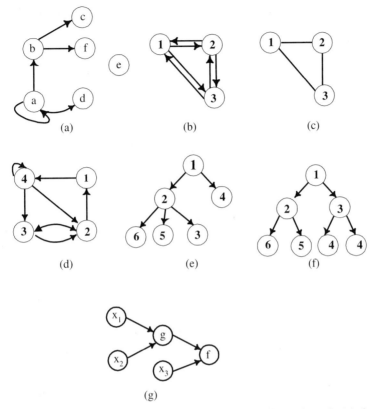

Figure 3.7 Examples of graphs: (*a*) directed graph; (*b*) undirected graph; (*c*) simplified undirected graph; (*d*) directed graph; (*e*) tree; (*f*) binary tree; (*g*) another directed graph.

as (u, v). If the relation is from u onto v, but not vice versa, the graph is called a *directed graph* (digraph). Figure 3.7*a* provides an example of a directed graph on $V = \{a, b, c, d, e, f\}$ with $E = \{(a,a), (a,b), (b,c), (d,a), (b,f)\}$. The arrow indicates the direction of the edge. The edge (a,a) is a loop. The vertex a is said to have a self-loop. Nodes related by an edge are called *adjacent*. Node e is not related to any of the other nodes; it is an isolated node. The number of edges emanating from a vertex is called the *outdegree* of the vertex, and the number of edges incoming to the vertex is its *indegree*. Vertex b has an indegree of 1 and an outdegree of 2 since it points to both nodes, c and f.

The relation between the vertices may be symmetric. In such a case, we have $e(u, v) \equiv e(v, u)$, and one edge is sufficient to represent the relation in an *undirected graph*. This is illustrated by the graphs in Fig. 3.7*b* and its simplified equivalent in Fig. 3.7*c*. In general, the graph is assumed undirected unless stated otherwise.

A *path* is a sequence of edges: (v, u), (u, w), (w, r). A graph is said to be *connected* if there is a path from any vertex to any other vertex. The graph in Fig. 3.7*a* is not connected, whereas that in Fig. 3.7*b* is connected. A *cycle* is a closed path. The path (a, a) of the graph in Fig. 3.7*a* is a cycle of length 1. The graph in Fig. 3.7*d* has cycles of length 1, 2, 3, and 4. A graph is said to be *complete* if it is such that every node has an edge to all other nodes.

A *tree*, *T*(*V*, *E*, and *W*), is an acyclic graph. An example is shown in Fig. 3.7*e*. It has a root node with indegree 0 and terminal nodes with outdegree 0. Any terminal node is called a *leaf*. A special type of tree, a *binary tree*, is shown in Fig. 3.7*f*. All of its internal nodes are such that each has an *indegree* of 1 and an *outdegree* of at most 2.

Given a graph, *G*(*V*,*E*), a subgraph of *G*, *S*(*U*,*F*) is such that $U \subset V$ and $F \subset E$. A subgraph that is complete is called a *clique*. Node *e* of the graph in Fig. 3.7*a* can be a subgraph and a clique, although it consists of only one node.

In digital design and testing, graphs are used extensively. The design at any level of representation is modeled as a graph, and operations on the design are developed for its graph model. Operations on graph models of the design include system partitioning, synthesis on logic and behavioral levels, synthesis for testability, test pattern generation, and physical design—placement and routing, back-annotation, simulation, and so on. The schematic capture discussed in the preceding section is easily represented by a graph, as illustrated in Fig. 3.7*g* for the circuit in Fig. 3.4. The vertices of the diagram are the primary inputs and the results of the operations of the various gates. The edges are the interconnections between the gates. In using this type of a graph for delay calculations, a weight is associated with each edge. To place the gate on a floor of a chip, the weight assigned may be the lengths of the wires connecting the gates. Another important graph representation is that of finite state machines. In this case the graph is called a *state transition graph* (STG). The vertices of the graph are the states and the edges are the transitions between states. The STG of machine *M* defined in Section 3.2.3 is shown in Fig. 3.7*d*. It is an isomorphism of the state transition table given in Table 3.3.

3.6   BINARY DECISION DIAGRAMS

Binary decision diagrams (BDDs) are directed acyclic graphs (DAGs). They were proposed over 40 years ago [Lee 1959] and were developed further in the digital testing field [Akers 1978] and in digital design synthesis [Bryant 1986]. A BDD is a rooted acyclic directed graph that represents a Boolean function. The set of vertices is partitioned into three sets: internal nodes (with indegree = 1 and outdegree = 2), the leafs (with indegree = 2 and outdegree = 0), and the root node with (indegree = 0 and outdegree = 2). The values of the leaf vertices are 0 and 1, the logic value of the function. The other vertices represent each one of the variables. Thus the labels of the internal nodes are $l(v) \in S_f$, where S_f is the support of the function.

The development of BDDs is based on the recursive use of the Shannon

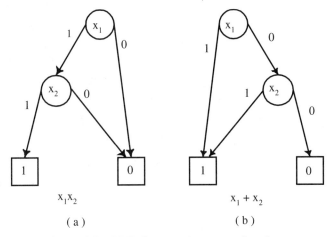

$x_1 x_2$

$x_1 + x_2$

(a)

(b)

Figure 3.8 BDD for two elementary functions.

expansion given by Eq. (3.1). We illustrate the formation of BDDs using an elementary function: $f(x_1, x_2) = x_1 x_2$. Since this function is symmetrical, we can start with either variable. We start the tree, as shown in Fig. 3.8, with the root being any one of the variables. We start with x_1. The successors of this vertex are the residues of the function for $x_1 = 0$, and $x_1 = 1$ are 0 and x_2. We then draw two edges, one labeled 0 and the other 1. The edge 0 points to the 0-residue, which is 0. The other edge is labeled 1 and points to the other residue, which is the second variable, x_2. From the latter variable, again, we draw two edges, 0 and 1, pointing to the nodes that represent the residue of the function for x_2, which are 0 and 1. The diagram gives the value of the function for all combinations of the variables. For example, the value of $f(1,1)$ is obtained by starting from the root and following path 1,1 and finding that the value is 1. The BDD for the OR function can be deduced from that of the AND simply by complementing the labels of the edges and the values of the two leaves, 0 and 1. Another way of developing this diagram is to realize that:

If $x_1 = 0$ then $f = 0$,

else $f = x_2$

if $x_2 = 0$ then $f = 0$,

else $f = 1$.

This explains why in some publications, instead of labeling the edges 0 and 1, they may be labeled T (then) and E (else).

Notice that this function is symmetrical in both variables and the same diagram would have been obtained if we started with either variable. However, in general, the diagram is not unique and it is sensitive to the order in which the variables are considered. Next we consider a more complex func-

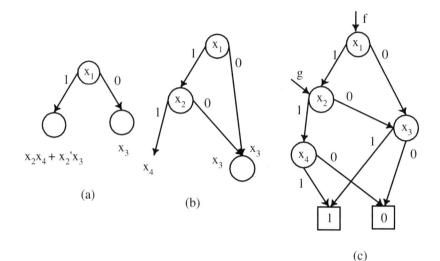

Figure 3.9 Building a BDD for the function $x_1x_2x_4 + x_1'x_3 + x_2'x_3$ using the order $x_1 \leq x_2 \leq x_3 \leq x_4$.

tion: $f(x_1, x_2, x_3, x_4) = x_1x_2x_4 + x_1'x_3 + x_2'x_3$. We first use the following order of variables: $x_1 \leq x_2 \leq x_3 \leq x_4$. The residues of the function for x_1 and x_1' are $f_{x_1} = x_2x_4 + x_2'x_3$ and $f_{x_1'} = x_3 + x_2'x_3 = x_3$, as illustrated in Fig. 3.9a. Next, we form the rsidues of $f_{x_1'}$ with respect to x_2. They are $f_{x_1'x_2'} = x_3$ and $f_{x_1'x_2} = x_4$, respectively. They are added to the graph as shown in Fig. 3.9b. Finally, the residues for x_3 and x_4 are given to complete the diagram. Again, since the function is symmetrical in x_1 and x_2, we would have obtained the same diagram had we started with x_2 instead of x_1. The diagram becomes more complex when we start with one of the other variables. We will follow the order $x_4 \leq x_3 \leq x_2 \leq x_1$. The residues for x_4 are $f_{x_4'} = x_1'x_3 + x_2'x_3$ and $f_{x_4} = x_1x_2 + x_1'x_3 + x_2'x_3 = x_1x_2 + x_3$. For each branch we take the residue with respect to x_3, as illustrated in Fig. 3.10. It is left for the exercises to find the BDD when x_3 is considered before x_4.

For more complex functions, the graph is formed with more than one node for some variables. It then may be reduced by deleting redundant nodes. The development of BDDs by hand is a tedious task. However, there are well-documented algorithms that build such graphs [Hachtel 1996]. The complexity of such an algorithm is of the order $O(2^N/N)$. The graph may include the representation of another function. For example, the function $g(x_2, x_3, x_4) = x_2x_4 + x_3$ is represented by a subgraph in Fig. 3.9b. BDDs can be used to determine various properties of the functions they represent. If we start from the root and trace through a path that leads to 1 (or 0), we find a set of variable values for which the function is 1 (or 0) independent of the other variables. For the diagram in Fig. 3.9, $f = 1$ for $x_1 = 0$ and $x_3 = 1$. This implies that $x_1'x_3$ is a prime implicant of f. Hence finding all paths to 1 would yield the products of the function.

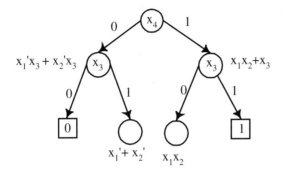

Figure 3.10 Building a BDD for the function in Fig. 3.9 using the order $x_4 \leq x_3 \leq x_2 \leq x_1$.

Test pattern generation is another application where BDDs are useful. We will illustrate this using an example. Consider the function $f(x_1, x_2, x_3) = g + x_3 = x_1 x_2 + x_3$ that we used in Section 3.2.2 to illustrate the application of Boolean difference in test pattern generation. In that method we assume that we have generated a test pattern for a SA0 on the primary input x_1. Of course, we will have $x_1 = 1$. The response to this pattern of the good circuit may be 0 or 1, while the faulty circuit response will then be 1 or 0. Thus to find this test pattern, we can trace on the circuit's BDD the paths to 0 and to 1 starting from the root. Any variable other than x_1, if it appears in both paths, must have the same value, either 1 or 0. For this example, starting from the root we can go through $x_1 = 1$ and $x_2 = 1$ to 1. Also, we trace the path through $x_1 = 0$ to 0 through $x_3 = 0$. The corresponding values of the variables constitute the test pattern, $x_1 x_2 x_3 = 110$. The paths leading to 0 and 1 are shown in Fig. 3.11a in solid lines. For $x_2/0$, we start from node x_2, but we need to recall that this node is reached when $x_1 = 1$. The paths leading to 0 and 1 are shown in Fig. 3.11b. For a stuck-at fault on an internal node, $g = x_1 x_2$, we will rearrange the BDD as shown in Fig. 3.11c. The diagram represents both g and f. We need the paths for $g = 0$ and $g = 1$; however, this dictates that x_3 must be equal to 0.

3.7 NETLISTS

A netlist describes a design as a collection of connected modules. The modules may be logic gates or transistors. An example of a netlist is the SPICE deck listed in Section 3.2.4. A netlist consists of a set of records that represents the modules and their connectivities. A record usually gives the name of the module, its function, and the input and outputs to the module. The function is well defined in a library of components. The library may consist of electrical components such as transistors, resistance, and capacitance. It may be a list of logic primitives such as NAND, NOR, XOR, and so on. However, the components could also be macros that represent more complex components. The component

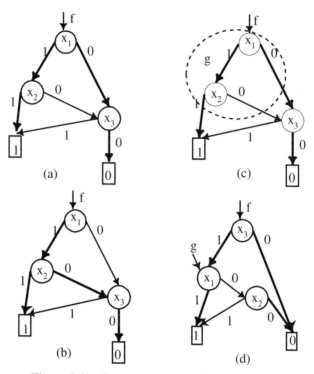

Figure 3.11   Test pattern generation using BDDs.

may be a core that is defined by its I/O and functionality in a core-based system. Similarly, the components on a system level are RAMs, ALU, and so on. For the logic level they are logic gates (NAND, OR) and on the circuit level (NMOS, PMOS, resistance). Almost every CAD tool has its own representation of a netlist.

The logic-level representations shown in Fig. 3.3 for the structural domain can each be represented as a netlist.

$$\text{AND-NOR}(z, b, c, d)$$

$$g_1 \qquad \text{AND}_2(a, b, c)$$

$$g_2 \qquad \text{NOR}_2(z, a, d)$$

$$\text{wire} \quad a$$

"AND-NOR" is the circuit's name. It is associated with a list of primary inputs and outputs b, c, d, and z. Each gate has a name, a logic function, and input and output leads. For example, g_1 is a two-input AND and its inputs are b and c. The output of this gate is an internal node and it is thus declared as "wire" in the last entry of the netlist.

An important netlist representation has been developed for the purpose of facilitating the interchange of designs between different incompatible systems. One representation is the electronic description interchange format (EDIF) that is extremely verbose. Presently, EDIF is a standard language.

3.8   HARDWARE DESCRIPTION LANGUAGES

Hardware description languages (HDLs) are becoming the standard in designing digital systems. In other words, the issue is not whether or not to use to HDLs, but which language is more advantageous to use. Although use of an HDL may be at the gate level, designers now prefer to express their designs on the behavioral or register transfer level (RTL) and defer the detail of the implementation to CAD tools. Logic synthesis tools are available to create circuits that readily become mapped into silicon. HDLs have several advantages over schematics. Two of the main advantages are efficient management of complexity and shortening the design cycle. As a result of the continuous decrease in the minimum feature size of transistors, device density and design complexity have both increased steadily. Densities of several million transistors are now a reality, and to manage such complexity, it is only natural to design on a hierarchical basis.

In addition to their complexity, the life cycle of modern digital systems is becoming shorter than its design cycle. Thus, increasing the efficiency of the designer, reducing costs, and time-to-market are essential attributes that are needed for keeping competitive in the electronics field. The first HDLs were developed at universities in the mid-1970s. APHL [Hill 1981] and ISP [Barbacci 1977] are just two examples of many others. Other languages have been used as proprietary languages in industry. However, because of the open system approach in the 1980s, there was a need to develop a standard for HDLs that could be used on any platform. At present, there are several hardware description languages, the most popular being Hardware C, VHDL, and Verilog HDL.

We concentrate on the two most widely used languages: VHDL and Verilog HDL. Either language mentioned above has the ability of representing (1) a hierarchical description of the design and (2) the design on three main levels: behavioral, RTL, and gate level. These two aspects make the two languages efficient in synthesis. Although designs obtained by synthesis are considered "correct by construction," it is very important to verify them by simulation. In addition, they may be used for simulation at all levels of abstraction. Unlike software programming language, writing correct design with HDL requires correct semantics as well as syntax. A software program may yield the correct results, although it is not very efficient. However, the "yield" of an HDL description is a design that has to be optimal. As with any programming language, an HDL consists of an alphabet to describe variables and other data structures and syntactical constructs. Some of the constructs are global for the entire design and others are for the description of the design flow.

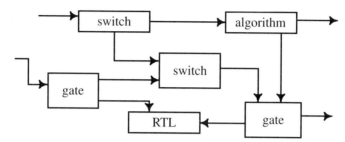

Figure 3.12   Verilog HDL mixed-mode design.

3.8.1   Verilog Language

Verilog HDL was developed by a CAD tools company, Gateway Design Automation, to use with its main product, the Verilog compiler and simulator. The company was later acquired by Cadence, which made the language public domain. A nonprofit organization, Open Verilog International (OVI), was then founded to support the language throughout its standardization. Verilog HDL, which we will refer to in the remainder of this book simply as Verilog, is presently IEEE Standard 1364-1995. This language resembles C programming language. A design can be described in various levels: switch level, gate level, register transfer level, architectural level, and algorithmic level. The various modeling levels may also be used in a mixed style, as illustrated in Fig. 3.12.

With increased complexity of digital circuits, it is more important to concentrate on behavioral level design as described below for a full adder, FA_Bchav [Bhasker 1998]:

```
module FA_Behav (A, B, Cin, Sum, Cout);
   input A, B, Cin;
   output Sum, Cout,
   reg Sum, Cout;
   reg R1, R2, R3;
   always
   @(A or B or Cin) begin
     Sum = (A∧B) ∧ Cin;
     R1 = A & Cin;
     R2 = B & Cin;
     R3 = A & B;
     Cout = (R1|R2)|R3;
   end
endmodule
```

The module defines both the interface (input and output) and the internal structure. In the example, A, B, and Cin are the inputs and Sum and Cout are the outputs. A number of elementary functions are built into the language. They represent basic logic functions such as OR (|), AND (&), and so on. In addition, *user-defined primitive* (UDP) may be defined.

The structure of a circuit is described by making *instances* of modules and primitives within a higher-level module and connecting the instances together using *nets*. A *net* represents an electrical connection, a wire or a bus. A list of *port connections* is used to connect nets to the ports of a module or primitive instance, where a port represents a *pin*. Registers may also be connected to the input ports of an instance. Nets and registers have values formed from the logic values 0, 1, *x* (uninitialized), and Z (high impedance). In addition to logic values, nets also have a strength value. Strengths are used extensively in switch-level models and to resolve situations where a net has more than one driver.

The function of a circuit is described using *initial* and *always* constructs. Statements inside an initial or always are in many ways similar to the statements in a software programming language. They are executed at times dictated by timing controls such as delays, and simulation event controls. Statements execute in sequence in a Begin–End block, or in parallel in a For–Join block. Before simulation, the source code is usually compiled. A compiler or an interpreter builds the data files necessary for simulation or synthesis. The simulation is not necessarily deterministic in the sense that the results may not be the same for different simulators because the order of events on an event queue is not defined by the standard. Usually, a test bench, a Verilog module that invokes the design and applies signals on its inputs and checks the output response versus the expected results, is used to run the simulation. The test bench for the adder module presented above is given in Section 5.2.1.

3.8.2   VHDL Language

VHDL (very high speed integrated circuit hardware description language) was started in Europe in 1981 and later adopted by the U.S. Department of Defense (DOD) as a design language. In 1983, DOD awarded a contract to a team of three companies—IBM, Texas Instruments, and Intermetrics—to develop a language. Version 7.2 was developed and released in 1985. The language was approved in 1987 as IEEE Standard 1076-1987. Nevertheless, VHDL has different dialects. It is an Ada-like language and is highly structured. It describes digital circuit representation in two of the main domains mentioned earlier in the chapter, functional and logic. The language stops short of specifying the physical characteristics of circuit layout.

The language has two principal global constructs: *Entity* and *Architecture*. The entity declaration specifies the name of the design, its interface to other components, and the input and output ports. Hierarchy is defined by means of components, which are analogous to chip sockets. A component is instantiated within architecture to represent a copy of lower-level hierarchical blocks. The structure of a circuit is described by making instances of components within architecture and connecting the instances together using signal. A *signal* represents an electrical connection, a wire, or a bus. A port map is used to connect signals to the ports of a component instantiation, where a port represents a pin.

REFERENCES

Akers, S. B. (1978), Binary decision diagrams, *IEEE Trans. Comput.*, Vol. C-27, No. 6, pp. 509–516.

Antognetti, P., and G. Masobrio (eds.) (1988), *Semiconductor Device Modeling with SPICE*, McGraw-Hill, New York.

Avanti (1997), HSPICE, Avanti Corporation, Milpitas, CA.

Barbacci, M. R., et al. (1977), *The Symbolic Manipulation of Computer Descriptions: ISPS Primer and reference manual*, Research Report, Department of Computer Science, Carnegie–Mellon University, Pittsburgh, PA. Also, *IEEE Trans. Comput.*, Vol. C-30, No. 1, pp. 24–40.

Bhasker, J. (1998), *A Verilog HDL Primer*, Star Galaxy Press, Allentown, PA.

Bryant, R. E. (1986), Graph based algorithms for Boolean function manipulation, *IEEE Trans. Comput.*, Vol. C-35, No. 8, pp. 677–691.

Cheng, Y., et al. (1996), *BSIM3v3 Manual*, Department of Electrical Engineering and Computer Science, University of California, Berkeley, CA.

Fujiwara, H. (1986), *Logic Testing and Design for Testability*, MIT Press, Cambridge, MA.

Gajski, D. D., and R. K. Kuhn (1983), Introduction: new VLSI tools, *IEEE Computer*, Vol. 6, No. 12, pp. 11–14.

Getreu, L. (1976), *Modelling the Bipolar Transistor*, Teknotrix, Inc., Beaverton, OR.

Hachtel, G. D., and F. Somenzi (1996), *Logic Synthesis and Verification Algorithms*, Kluwer Academic, Norwell, MA.

Hennie, F. C. (1974), Fault detection experiments for sequential circuits, *Proc. 5th Symposium on Switching Theory and Logical Design*, pp. 95–110.

Hill, F., and G. Peterson (1981), *Introduction to Switching Theory and Logical Design*, Wiley, New York.

Hodges, D. A., and H. G. Jackson (1983), *Analysis and Design of Digital Integrated Circuits*, McGraw-Hill, New York.

Jeng, M. C., et al. (1983), *Theory Algorithms, and User's Guide for BSIM and SCALP*, Electronic Research Laboratory Memorandum, UCB/ERL M87/35, University of California, Berkeley, CA.

Kohavi, Z. (1978), *Switching and Finite Automata Theory*, McGraw-Hill, New York.

Lee, C. Y. (1959), Binary decision diagrams, *Bell Syst. Tech. J.*, Vol. 38, No. 7, pp. 985–999.

McCluskey, E. J. (1986), *Principles of Logic Design*, Prentice Hall, Upper Saddle River, NJ.

Mealy, G. H. (1955), A method for synthesizing sequential circuits, *Bell Syst. Tech. J.*, Vol. 34, No. 9, pp. 1045–1079.

Michel, P., U. Lauther, and P. Duzy (eds.) (1992), *The Synthesis Approach to Digital System Design*, Kluwer Academic, Norwell, MA.

MicroSim (1994), PSPICE, The Design Center, MicroSim Corporation, Los Angeles, CA.

Miczo, A. (1986), *Digital Logic Testing and Simulation*, Harper & Row, New York.

Moore, E. F. (1956), "Gedanken experiments on sequential machines," in *Automata Studies*, C. E. Shannon and J. McCarthy (eds.), Princeton University Press, Princeton, NJ.

Nagel, L. W. (1975), *SPICE2: A Computer Program to Simulate a Semiconductor Circuit*, Memo ERL-M520, University of California, Berkeley, CA.

Quarles, T., et al. (1994), *SPICE3 Version 3F5 User's Manual*, University of California, Berkeley, CA.

Sheu, B. J., et al. (1988), BSIM: Berkeley short-channel IGFET model, *IEEE J. Solid-State Circuits*, Vol. SC-22, pp. 558–566.

Walker, R. A., and D. E. Thomas (1985), A model of design representation and synthesis, *Proc. 22nd Design Automation Conference*, pp. 453–459.

PROBLEMS

3.1. Use Boolean difference to develop a test set for stuck-at faults on the primary inputs of the circuit in Fig. P3.1. Then, with the help of a fault simulator, determine the fault coverage for all stuck-at faults for this test.

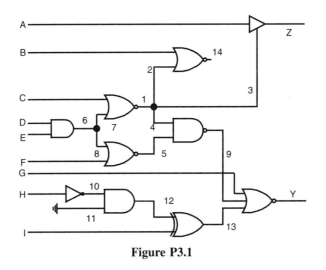

Figure P3.1

3.2. Develop the BDD for the functions **(a)** $f(a, b, c, d, e, g) = ab + cd + eg$ and **(b)** $f(a, b, c, d) = a \oplus b \oplus c \oplus d$.

3.3. Use the BDDs that you developed in Problem 3.2 to determine the test patterns that detect stuck-at faults on inputs a and b.

3.4. Develop the state table and the state diagram for a synchronous sequential circuit with one input line and one output line that recognizes the input string $x = 1111$. The circuit is also required to recognize overlapping sequences, as can be seen in the output string that results from the following input string: $x = 1100111111101$. The output string is 0000000111100.

3.5. Draw the flowchart for the FSM of Problem 3.4.

3.6. For the circuit in Fig. P3.6, develop the BDD (use only one leaf for 1 and one leaf for 0). Then find all test patterns to detect stuck-at faults on the primary outputs.

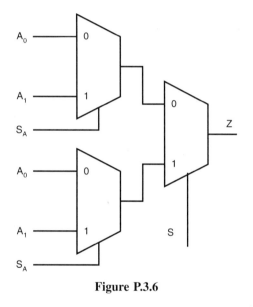

Figure P.3.6

3.7. Follow the example used in Section 3.7 and develop the netlist for the circuit shown in Fig. 2.6 of Chapter 2.

3.8. Write a Verilog HDL structural code for the circuit used in Problem 3.7.

4

VLSI DESIGN FLOW

4.1 INTRODUCTION

Knowledge of the various design processes is vital in understanding testing and design for testability. In this chapter we explore the design process to enhance design testability. For example, appending a DFT structure to the design at its completion may compromise the design goals. Instead, we would like to remove barriers to testing while designing and to make the DFT a natural process in design. Barriers to creating easily testable designs may occur at different levels of the design, from the behavioral specifications to the physical layout. Therefore, we examine the design process at each level. Except for a few circuits or subcircuits, the design process is fully automated in most cases. By *automation*, we mean that software programs are used to perform the tasks. These software programs are known as computer-aided design (CAD) tools. The tools perform operations on the design representations, the models that we discussed in Chapter 3. Any tool is based on an *algorithm* that is developed to perform the task *as accurately as possible within a reasonable amount of time*. In other words, it is not enough to know how to obtain an optimal solution. It is equally important to trade off between the optimality of the outcome and the time to completion.

In this chapter we first discuss the CAD tools and the concept of algorithms. Then we examine the logic design phase and the physical design phase. Both phases of the cycles are shown in Fig. 4.1. Accordingly, we discuss both high-level and logic synthesis as well as the technology mapping, which actually depends on the various styles of design implementations. Often, we refer to these styles as *design methodologies*.

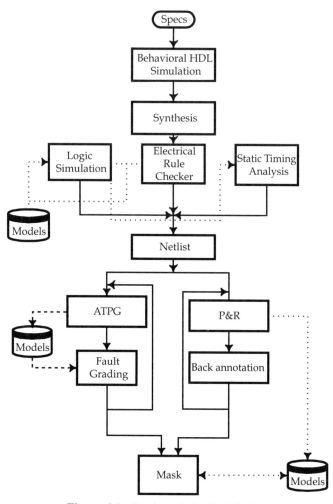

Figure 4.1 Logic and physical design.

4.2 CAD TOOLS

Design automation (DA), also known as the use of CAD tools in design, has played a vital role in circuit design by greatly facilitating the various stages of the design process to the extent that they should be part of a design engineer's skill. These tools are a continually evolving set of software programs that have replaced many of the tedious tasks that a designer used to do. They are the main cause of design cycle reduction, as illustrated in Fig. 1.1. Originally, these tools were relegated to perform repetitive tasks that would take much longer when executed by humans. Hence the physical design and layout of the circuit was one of the first design aspects to be handled with CAD tools. However, these days the tools are used in more elaborate and not, relatively, so straightforward tasks such as layout. Design automation has evolved into a multimillion-dollar business, and many companies have been founded on one or another aspect of the design for test. Commercial synthesis tools then became very common. In the IC design field, the term *synthesis* is synonymous with automated transformation of a design from one level of abstraction to another of lesser degree of abstraction. Any synthesis tool follows an algorithm that includes two elements: (1) mapping from one design description to another that is closer to the physical viewpoint and (2) optimizing the design to suit the technology to which it is mapped.

However, CAD tools perform more than synthesis. They involve all steps in the design cycle from specification entry to parameter extraction from the layout:

- Design entry
- Simulation (at different levels of design abstraction)
- Synthesis behavior (to RTL level), logic (to logic gates)
- Test pattern generation
- Floor planning
- Technology mapping
- Physical design: placement and routing
- Design rule checker (DRC)
- Logic versus schematic (LVS) tool
- Parameter extraction

Design entry tools serve as the interface between users and the other tools. They can be of different styles. *Command modes* were the first means for interaction; however, other modes, such as *graphic user interface* (GUI), are becoming more sophisticated, due to the unprecedented advancement in graphic design. Also, most CAD programs have dual types of entry. Design entry tools enable users to enter a design hierarchically. Other CAD tools are investigated further in subsequent sections of the chapter, but first we discuss the concept of algorithm.

4.3 ALGORITHMS

Any of the CAD tools mentioned above follows an algorithm that often includes a *heuristic* and a *cost function*. Although informal, an algorithm can be viewed as a recipe. In computer science terminology, an algorithm is "any well-defined computational procedure that takes some value, or set of values, as input and produces some value, or set of values, as output" [Cormen 1992]. It is thus a sequence of data processing steps that operates on the model of the circuit to perform one specific task. The task may be a logic simulation, and the model is a netlist that connects some logic gates, attributes of the gates (such as delays), and a set of stimuli. The outcome of the task is the function or timing of outputs of the circuit. Another task is logic synthesis, which transforms an RTL representation of a circuit into logic gates.

In performing this task, the tool may constantly be guided by a *cost function*. The term *cost* here has no relation to the monetary value of the product. It is a local meaning related to the particular algorithm. In performing the task and selecting between alternatives to realize it, certain parameters are evaluated constantly to assess the results. A cost function in synthesis may be to minimize area or power. An example on the physical level is placement of cells on a chip. The attempt is to find a place on the floor of the chip such that connectivities are minimized. Thus the cost function here is the total length of the interconnect wires. In all these examples, the model for the circuit is a graph with appropriate selection of the vertices and the edges (see Chapter 3).

The complexity of algorithms depends on the problem to be solved. It is measured by the space required in the computer memory and the execution time. Usually, both factors characterize the algorithms. However, as computer memory is becoming bigger, time complexity is more important. The execution time of the software realizing the algorithm is a function of the processing steps followed to solve the problem. This is determined mostly by $n = |V|$, where V is the set of vertices of the graph, $G(V, E, W)$, representing the problem. The time complexity is usually expressed as a function, $f(n)$, such that $f(n) \le cg(n)$, where c is a constant. This is described as $f(n)$ is of order $g(n)$, $O(g(n))$. To illustrate the concept, consider the multiplication, M, of two $n \times n$ matrices, A and B, $M = A \times B$. Any element, $m_i \in M$, is given by $m_{i,j} = \sum_{k=1}^{k=n} a_{i,k} b_{k,j}$. Since there are n^2 elements in M, the complexity of matrix multiplication is $O(n^3)$. The complexity is thus an order of a polynomial in n.

Algorithms with such a complexity are called *tractable*. Most problems encountered in testing are *intractable*. They are characterized as being NP-complete problems. NP stands for nondeterministic polynomial. That is, there is a nondeterministic algorithm that solves the problem in polynomial time. Hence the algorithm includes a heuristic that directs the solution in a more efficient manner. A heuristic is a common sense rule (or set of rules) intended to increase the probability of solving some problem without guaranteeing an optimal solution. A heuristic cannot guarantee that the problem will always have a solution. As we will find out in Chapter 6, test pattern generation is an NP-complete problem [Goel 1980, Fujiwara 1982]. This is also true of partitioning a circuit or placement and routing [Ibarra 1975].

Algorithms developed to solve the same problem often differ in their efficiency in solving the problem. This is for one of several reasons. They may be written for different computer platforms, say a PC versus a workstation, although more and more the difference between the two platforms in narrowing. Also, engineers with different capabilities or experiences compose them. For the algorithm to be efficient, three main attributes are needed: the right data structure, the right heuristic, and a capable programmer. In addition, the need for an efficient software toolbox can never be overemphasized.

To appreciate the complexity of algorithms and the need for heuristics, assume, for example, that you are developing an algorithm to simulate a human being playing chess. If you try to consider all possible moves and their consequences in every play, a game would not end within a human lifetime. Instead, a heuristic that utilizes the human experience in playing chess may be devised to avoid unnecessary or trivial moves. In this way, the time complexity is reduced.

4.4 SYNTHESIS

Synthesis is the process that transforms the design from one level of abstraction to another that is closer to the physical circuit level in the physical domain. It is the most challenging task to automate because it is more difficult to capture by an algorithm. There are a number of reasons for this: different design constraints and goals, many design representation levels, or several alternatives for design implementation of a given design.

According to the taxonomy used in Chapter 3 and illustrated in Figs. 3.1 and 4.2*a*, synthesis transforms a design from a high level of representation (point *A*), the algorithmic code in the behavior domain, to the lowest level in the physical domain (point *P*), the mask. The path from *A* to *P* consists of several translations from one level to another in the same domain and from one domain to another at the same level, and optimization at any point along the path [Gajski 1983, Walker 1985]. These translations are few-to-many, meaning that there are usually few structural implementations for each behavioral representation and even more physical configurations for each structural model.

The transitions shown in Fig. 4.2*b* represent different tasks performed on the design [Michel 1992]. Based on this model, *synthesis* is a transition from the behavioral to the structural domain. A self-loop is an *optimization* of the design at the same level. A transition from the structural to the physical level results in generating (placing and routing) the mask for the design, while the reverse transition is known as *parameter extraction*. From the layout, electrical parameters are calculated and can be included in the structural domain for accurate simulation of the design. At present, the extraction is also fed back in the design on the behavioral level.

The different synthesis processes are mapped in Fig. 4.3. Step 1a translates the various operations into concrete subsystems, processors, and so on. Step 1b results in an RTL-level description that relates the component from step 1a. The design is represented by operation on the data path by control steps that have been scheduled according to a certain clock. The *RTL synthesis* consists of transitions 2a, 2b, and 2c. This is not a synthesis task proper; it is more of binding the blocks at the system level to specific blocks and relating the different tasks to the clock. Step 2c refines the gate-level design by optimizing on some attribute, number of gates, or interconnect wires. However, step 3 is the technology mapping. Steps 4a

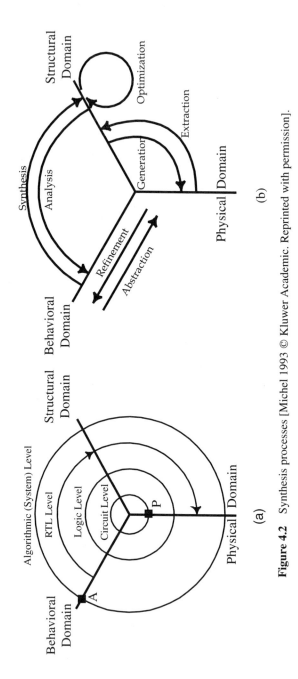

Figure 4.2 Synthesis processes [Michel 1993 © Kluwer Academic. Reprinted with permission].

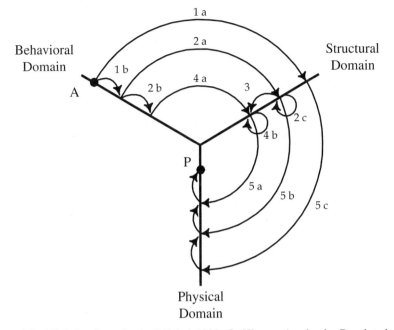

Figure 4.3 High-level synthesis [Michel 1993 © Kluwer Academic. Reprinted with permission].

and 4b are for the logic synthesis and optimization. Any or all of steps 5a, 5b, and 5c generate the masks, a process that is also known as *physical synthesis*. For a design based solely on standard cells, step 5a is sufficient. However, a system on a chip that includes cores would also involve step 5c, depending on the type of core, as we define them in Section 4.6.

Depending on the availability of tools and other factors, the designers may use one or all synthesis types. The more automation that is used, the faster it is to complete the design. A qualitative comparison of the various synthesis types is shown in Fig. 4.4, where the ease of design is given versus performance, cost of development, and completion time. This trend will be clearer as we describe synthesis processes in the remainder of this chapter.

Chronologically speaking, physical synthesis was the first to be automated and is described in detail in subsequent sections of this chapter. Next came logic synthesis, except that it really was more of a transformation of the functional descriptions into optimized, technology-independent logic gates. It is only in the late 1980s that synthesis tools appeared on the market. However, there are companies, such as IBM, that had been using their proprietary tools a decade earlier [Hong 1974]. The first commercial synthesis tools were based on research work done mostly at universities [Brayton 1984, 1987].

Not far behind these logic synthesis tools was the automation process of RTL to gate-level mapping. It has been called *RTL synthesis* to differentiate it from

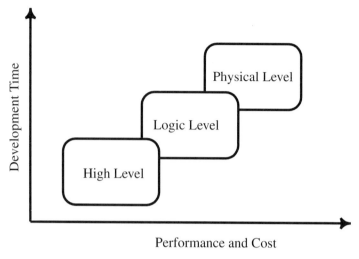

Figure 4.4   Comparing synthesis types.

the previous type. The RTL description consists of cycle-by-cycle behavior, which is still defined in programming-like language in terms of registers and combinational logic blocks such as arithmetic and logical function.

4.4.1   Behavioral Synthesis

Behavioral synthesis has many advantages. First, it shortens the design cycle and minimizes the chance for errors. As a result, it is possible to hit the market at the appropriate market window for that design. The importance of this market window to product profit was discussed in Chapter 1. Another important factor is that it is possible to explore the design space and select the most appropriate implementation for the product. The design space used to be defined solely by area (hardware) and performance (operating frequency). Thus there is a trade-off between the two attributes, but power and testability are becoming very important parts of the metric for design evaluation.

This type of synthesis consists of three main processes: compilation, scheduling, and allocations, which transform the behavioral model to the RT model shown in Fig. 4.5. The behavioral description given in the left side of the figure is translated into the structure to its right through step 2b in Fig. 4.3b. The design seems to require an adder and a decrementer. Since the two operations on the data are mutually exclusive, they must both involve adding C. Thus, according to the value of F, either A or -1 is added to C, as depicted by the rightmost structure in the figure. This optimization step is equivalent to step 2c. The optimization was to reduce the area of the design. Optimization to improve testability is discussed in Chapter 14.

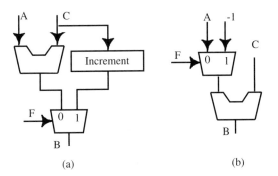

Figure 4.5 RTL optimization: alternative implementations.

4.4.2 Logic Synthesis

The result of behavioral synthesis is an RTL module that needs to be mapped to the logic or even physical level. The RTL model already includes architecture consisting of several basic modules, such as registers, multiplexers, and functional units. While some of these modules are part of the library, the others are to be designed on the logic level. Logic synthesis consists of two stages: technology independent and technology dependent. The latter stage is also known as *technology mapping*. Steps 4a and 4b represent logic synthesis. The second step is an optimization.

In *technology-independent synthesis*, the network is represented by a model that may be any of the possibilities enumerated in Chapter 3: truth tables, algebraic expressions, BDDs, or other forms of graphs [Bryant 1986]. Multilevel synthesis, which is the approach generally used, organizes the Boolean expressions in tree structures and performs several operations on the tree. The operations are decomposition, extraction, substitution, and collapsing, which are analogous to polynomial division [Bartlett 1986, Brayton 1987]. The following expression is in factored form:

$$X = (AB + B'C)[C + D(E + AC')] + (D + E)(FG) = F_1(C + DF_5) + F_3F_4$$

It is represented by the binary tree of Fig. 4.6, where the leaves represent the literals and the other nodes represent the logical operations AND and OR. The main goal is to minimize the number of literals in the Boolean expression. On the tree, these are the edges leading to the leaves and to the five functions F_1 through F_5. In this case the count is 18. These literals represent the interconnections between the gates when the design is implemented. In present technologies, the majority of the IC area is occupied by the wiring connecting the gates.

Technology-dependent optimization transforms a Boolean expression into a gate network using the gates in the target library. This is *library binding*, more

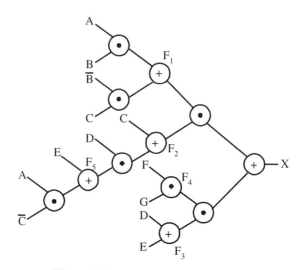

Figure 4.6   Multilevel logic synthesis.

commonly known as *technology mapping*. The library may consist of only NAND gates or of specific generic gates such as multiplexers. At present, some of the cells in a library are designed to facilitate testing. For example, flip-flops are used for scan-path design or LFSRs for BIST. Both scan path and BIST are DFT techniques that were mentioned in Section 1.10 and are described in detail in Chapters 9 and 11.

4.5   DESIGN METHODOLOGIES

Integrated circuits consist of transistors that are placed on the chip and are connected in such a way as to realize the design. Several masks define the locations and connectivity of the transistors. A mask corresponds to one of the silicon compound layers that form the transistors and the interconnect layers. Circuit implementations may be grouped into two main categories: fully custom and semicustom designs, as illustrated in the hierarchy shown in Fig. 4.7. In fully custom design style, all transistors are handcrafted. Since the designer controls all stages of the chip layout, maximum design flexibility and high performance are possible. Consequently, only highly skilled and competent designers are engaged in such design methodology. Also, development time is long and development costs are extremely high. It is thus more suitable for large-volume products. Memory and microprocessor ICs are created using unique masks for all layers during the manufacturing process. The user controls chip density with high utilization. However, the high cost of design and testing can be amortized successfully over the high volume.

   Semicustom design is facilitated by having CAD tools use a library of

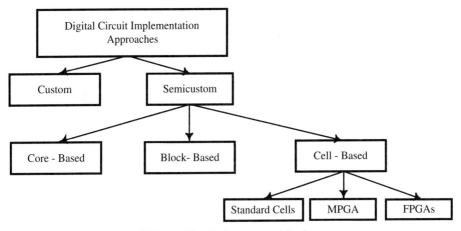

Figure 4.7   Design methodologies.

predesigned components and sometimes of prefabricated chips such as pro-grammable logic arrays (PLAs) and field-programmable gate arrays (FPGAs). It is a restrictive mode of design and results in less compact circuits than that of fully custom design. Its main advantage is shorter time-to-market and rapid prototyping. Therefore, semicustom is more appropriate for moderate and low-volume products. We describe each category in subsequent subsections. A com-parison of the various design methodologies using a metric of ease of design and flexibility is shown in Fig. 4.8. The more automated the steps used in the design, the more restricted the designer is and the design time is shorter. How-ever, the semicustom methodologies are becoming more efficient, as we learn in the next section.

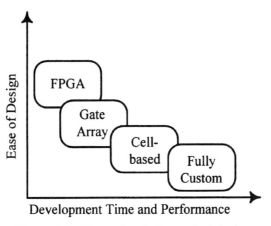

Figure 4.8   Comparing design methodologies.

4.6 SEMICUSTOM DESIGN

Practically speaking, fully custom design is very rare. Even in microprocessors, some components are extracted from a library. Actually, the trend nowadays is toward reuse of already designed and tested cells, usually referred to as *cores* or *intellectual properties* (IPs). This terminology is used because the core that is embedded in the design may be acquired from another vendor. The classification of semicustom design here is based on the size and complexity of the components used in the library. The first category is *cell-based*, for which the library components are small gates such as NAND, NOR, and flip-flops. This category is subdivided in turn into standard cells, FPGAs. The second category is *block-based*; these are larger cells such as an ALU or large multiplier. The last category is *core-based*; the core may be an entire microprocessor or several RAMs.

Cores come in three types: soft, firm, and hard. A *soft core* is usually described in an HDL language and is, accordingly, very flexible; a *firm core* is already mapped to a library; and a *hard core* is given as a layout that cannot be changed. Core-based design is discussed in Chapter 15. Next, we discuss cell-based design, most of which is also applicable to block-based design.

4.6.1 Standard Cell Design

Under this approach, the logic is mapped into a predesigned library. Depending on the cell type, two types of designs are distinguished: standard cells and macros. The design is called *standard cell* if the cells represent logic gates such as multi-input NANDs, NORs, and so on, and the masks of these gates are of the same height. They are usually saved in a database and designers select the appropriate cells to realize their design. The cells are then placed in rows and interconnected as illustrated in Fig. 4.9a. The space between the rows is called the *channels*. The interconnecting wires are laid along tracks in the *channels*. This process of interconnecting them is known as *routing*. If the cells to be connected are not in adjacent rows, the wire passes from one channel to another through cell location. This is known as *feed-through*. The width of the channels varies according to the degree of interaction between the cells and the quality of routing. The routing process is discussed in Section 4.7.2. Placement and routing are done automatically, thereby almost removing designers from the physical design process. Sometimes these cells are larger than simple logic gates; they are then called *macros*. Actually, a macro might be as big as a microprocessor and is referred to as a *core*. However, for macros and cores, the height is not uniform, as illustrated in Fig. 4.9a.

4.6.2 Mask-Programmable Gate Arrays

A gate array consists of a prefabricated two-dimensional array of transistors that are not connected. Thus there are generic masks for all layers except for

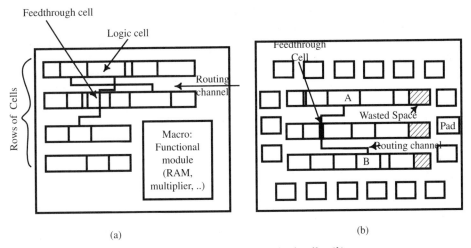

Figure 4.9   Semicustom design: (*a*) standard cells; (*b*) gate arrays.

the metalization mask. The metalization layer is customized to the user's specifications. Adjacent transistors can be connected to form a two-input NAND gate or a flip-flop as shown in Fig. 4.10. User logic is implemented by patterning these transistors into logic functions and connecting the different modules. Connectivities of the modules are along the channels. Once the connectivities

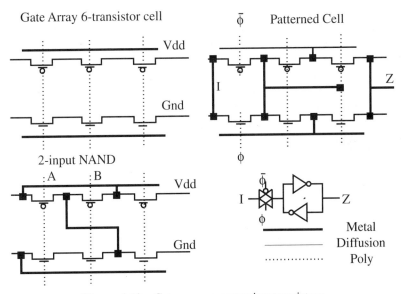

Figure 4.10   Gate arrays: patterning transistors.

of the cells are finalized, the metal mask is used to finalize the fabrication of the gate arrays. This is the reason they are called mask programmable. The image of the gate arrays appears as shown in Fig. 4.9b, which is similar to standard cells except that the channels are of equal width. A cell library, making the designer's expertise less critical than in the case of the fully custom methodology, usually facilitates the design. For the same reasons, MPGAs offer shorter development time and lower development costs than do standard cells and, a fortiori, fully custom ICs. A special class of gate array is channelless; they are known as *sea of gates*.

4.6.3 Programmable Devices

Programmable devices, which are available in a variety of sizes, architectures, and programmability technologies, have played a key role in digital design for awhile. Programmable devices have the advantage of being available as off-the-shelf parts that can be programmed in a few minutes. Another advantage is that some of them are reprogrammable. However, they are not as versatile as standard cell and gate-array styles of design. Because of their regularity and their programmability, they have particular faults in common as well as particular testing problems.

One group of these devices is characterized by the implementation of any design in a two-level AND/OR structure. This category includes the three types listed in Table 4.1. In addition to failures common in conventional ICs, programmable devices are subject to unique failure modes that we identify as *programmability failures*. A faulty chip or the malfunctioning of the device programmer can cause these types of failures. Like any electronic equipment, the device programmer can fail due to noncalibration. Some devices may be added or omitted unintentionally. To account for these extra and missing devices, DFT techniques have been specially developed [Hong 1980; Fujiwara 1981, 1984].

Field-programmable gate arrays constitute the second category of programmable devices. Xilinx introduced them in 1985 [Xilinx 1985]. Like MPGAs, these devices consist of uncommitted logic blocks, which are organized in rows or in a matrix form. Unlike MPGAs, there are vertical and horizontal channels in the matrix architecture that usually intersect in a switchbox. Also, the mapping of the design is on the logic level. The blocks can be con-

TABLE 4.1   Types of AND/OR Arrays

Type	AND Plane	OR Plane	Remark
PROM	Fixed	Program	EPROM and EEPROM
PLA	Program	Program	Are not as used
PLD	Program	Fixed	May include flip-flops

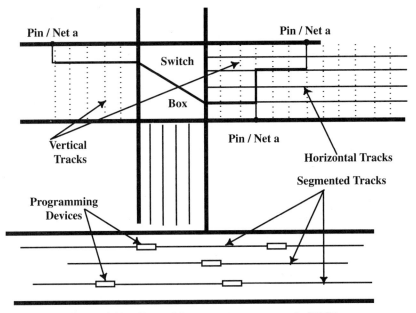

Figure 4.11   General interconnect resources in FPGAs.

nected by a general interconnect resource that comprises segments of wires of variable lengths, *segmented tracks*, as illustrated in Fig. 4.11 [Chan 1994]. These wires are organized in specific geometries in such a way as to accommodate channel or switchbox routing, and they are connected by programmable devices. The three main types of programmable devices used are SRAMs [Xilinx 1985], antifuses [Actel 1991, Quicklogic 1993], and electrically erasable connections [Altera 1992]. FPGAs are described in more detail in Chapter 13.

Similar to PLDs, FPGAs are also subject to programmability failures. However, the magnitude of the problem is increased since the programmable devices are also used in the interconnect resources. This may cause missing and extract connections between the segmented routing resources. In addition, they may produce longer delays than were accounted for at design time. FPGAs are prone to specific physical defects, depending on the underlying architecture and programming technology. For instance, in EPROM-based technology, erasure of programmable connections can occur with exposure to light. The logic cell array's use of volatile static memory makes it prone to problems caused by noise, radiation, and power dropouts. Independent of the technology, failure modes in the interconnection between the segmented metal wires may result in extra and missing connections. These failures result in different possible fault models, as listed in Table 4.2. Testing of FPGAs is one of the main topics in Chapter 13.

TABLE 4.2   Interconnect Failures: Mapping Failure Modes into Fault Model

Interconnect Failure	Fault Model
Extra connection between independent wires	Bridging fault
Missing connection between segments of metal	Stuck-open fault or floating gate
Improper programming of the program elements	Delay fault

Source: Adapted from Chan [1994].

4.7   PHYSICAL DESIGN

Once the design has been synthesized and mapped to a technology, a netlist is obtained and used in the physical design phase, which involves the following main processes: floor planning, placement, routing, and parameter extraction for logic versus schematic (LVS).

4.7.1   Floor Planning

Floor planning involves assigning locations for the various modules on the chip. Figure 4.12 shows a progression of floor plans for the same chip. The square shape is a better plan, as it minimizes area. This helps in decreasing delays and hence improves performance. However, it is really late in the design cycle to consider floor planning for the first time at the physical design phase. Today, floor planning is already considered at the logical phase of the design, as illustrated in Fig. 4.13. At the logic level the size of the modules is estimated, and this results in better chip compaction than when done only on the physical level. Floor planning is an essential design step for hierarchical design methodology. With the emerging paradigm of core-based systems for an SOC, floor planning is considered even at the specification level of the product. Briefly speaking, core-based design relies on incorporating already designed modules, a DSP

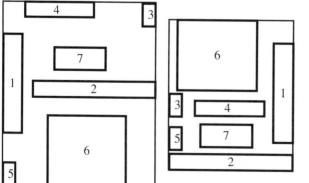
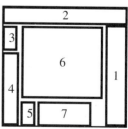

Figure 4.12   Floor planning.

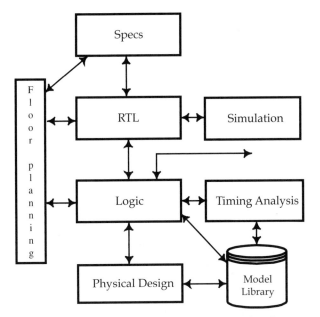

Figure 4.13   Another perspective of IC design flow.

core, or an entire microprocessor. A floor plan is thus needed to decide about the location on the chip of the reusable cores and their relation to other components yet to be designed. In addition, floor planning is very important to the testing strategy, as outlined in Chapter 1 and illustrated in Fig. 1.15. Core-based design and testing is the topic of Chapter 15.

Placement and routing (P&R) are very closely related processes. They can be defined as: Given a netlist describing the interconnections of the logic blocks and design rules, it is required to find (1) locations for the logic blocks (placement) and (2) paths for all interconnections between blocks (routing). Once the design is placed and routed, it is important to resimulate it with the actual information about the circuit. For this a parameter extraction is done as indicated symbolically on the Y-chart in Fig. 4.2b. The information is used for optimization of P&R or even for resynthesis.

4.7.2 Placement

The placement task is concerned with finding a location on the chip floor for each cell obtained from the technology mapping. It is equivalent to floor planning under very specific circumstances: The exact size of the cells is known and the general organization of the cell is predetermined, say, in rows. The goal of placement is to decide about these locations such that the wiring length of all interconnects is as short as possible. This constraint allows smaller chip and propagation delays. To facilitate placement, the floor is organized in a grid

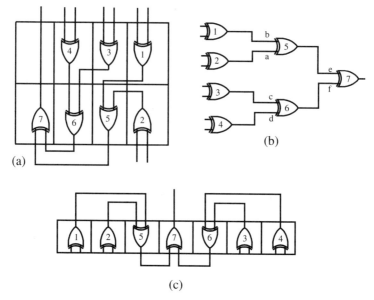

(a)

(b)

(c)

Figure 4.14 Placement configurations.

consisting of placeholders, $P = \{p_1, p_2, \ldots, p_N\}$ and objects, $M = \{m_1, m_2, \ldots, m_r\}$ to be placed in the holders. Each object, $m_i \in M$, is associated with a set of signals, S_i. For example, in Fig. 4.14, we have eight gates (objects) and six nets to be placed. Each gate is connected to a number of these nets varying between one and three nets. Two different placements of these cells are also shown. How can one judge which of the two configurations is better? This is to be assessed by the cost function used by the placement algorithm. To optimize on the cost function, we need to consider all possible alternatives that are $N!(N - r)!$ The time to accomplish the task will be of order $O(N!)$, where $N \geq r$. Placement is thus an NP-complete problem, and therefore a heuristic must be used. As we mentioned in Section 4.3, a heuristic will yield a nearly optimal solution in a reasonable amount of time.

Algorithms for placement are of two main types, *constructive* and *iterative improvement*. A commonsense way of doing constructive placement is first to place the most connected module in the middle of the space. Then proceed with one of the lower connectivities. Following this approach rgarding the example in Fig. 4.14 resulted in arrangement shown in part (a). The other approach starts from an initial placement of the modules, which can be done in a random fashion, proceeds in an iterative improvement, and is performed until a final state is reached such that the wiring length is minimized.

There are several possible heuristics for iterative improvement. We describe one of them, the min-cut [Lee 1961]. The algorithm operates on a graph model in which the vertices are the cells to be placed and the edges are the interconnect

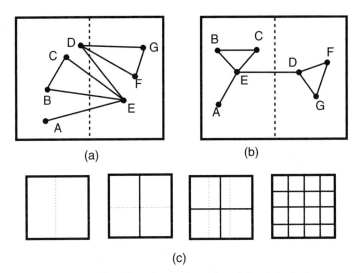

Figure 4.15 Placement using the mincut algorithm: (*a*) original placement; (*b*) final placement; (*c*) plan for iterative improvement.

wires between the cell pins. With the initial placement shown in Fig. 4.15*a*, a min-cut approach to improving the placement will separate the floor area in two halves and use a cost function, the number of wires crossing the separation line. The effort of the algorithm is to move some of the vertices (cells) in such a way as to minimize the cost function. If a move of a component, j, results in a cost function $C_j < C_{j-1}$, the move is accepted; otherwise, the acceptance or rejection will depend on the type of strategy followed, as we discussed in Section 4.3. The two halves of the design are each divided into two parts and the vertices are moved to locations that minimize the cost function. The process is repeated as illustrated in Fig. 4.15*c* until every object has its place on the floor and no further subdivision of the space is possible. For this example, moving nodes D and E reduces the cost function from 6 to 1 as illustrated in Fig. 4.15*b*.

4.7.3 Routing

Routing is the process of formally defining the precise conductor paths necessary to interconnect the cells so as to minimize the routing area or the delays. As for placement, routing starts with initial routing, which used to be called *loose routing* and later, *global routing*. This is then followed by a detailed routing which depends on the configuration: *channel routing* or *maze routing*. Before describing some of the routing algorithms, let us list some premises. A *net* is a collection of pins to be connected electrically. Figure 4.16 shows a channel with six pins: A, B, C, D, E, and F. The conductors connecting the pins are under some electrical constraints, such as a minimum conductor width w, and minimum spacing between the conductors, s. Both parameters are a function

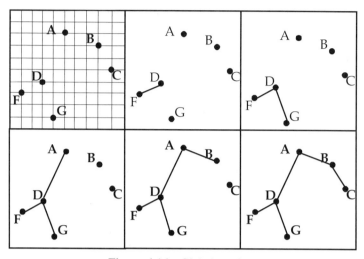

Figure 4.16 Global routing.

of the technology used. For a generic technology, 2λ, assume that $w = s = 3\lambda$. It is usually convenient to combine the two parameters into a center-to-center spacing of $\delta = w + s$. This will be considered the grid of spacing. All wires will be running along equidistant vertical and horizontal lines of the grid as shown in the upper left corner of Fig. 4.16. All horizontal tracks will be in one plane which is different from that spanned by the vertical tracks. An immediate consequence of this grid is that the length of the wire is measured in δ along the vertical and horizontal tracks. Thus two adjacent points along the horizontal and vertical track have a distance of δ. Points A and B are 6 δ apart, and so are the pair D, E. This is the actual shortest length of the wire to connect the two corresponding points. This distance is called the *Manhattan distance*, owing to the well-known street configuration of most of Manhattan, one of the New York City boroughs: avenues along the north–south direction and streets perpendicular along the east–west direction. In addition, it is an easier way of measuring the distance than using the Euclidean distance, along the diagonal.

Refer to the example in Fig. 4.16, in which the six pins A, B, C, D, E, and F of the same net are to be connected. The problem will be modeled with a graph, $G(V, E,$ and $w)$: V is the set of pins, E is the set of the nets connecting the pins, and w is the set of the distances between the pins. Since every pin is connected electrically to every other pin, the graph would be a complete graph. However, we know very well that if we connect pin A to pin B to C, then A is connected to C electrically without having an additional wire connecting these two pins. Thus what is needed is to figure out a tree, $T(V_t E_t w_t)$, that spans all pins. In such a case, $V_t = V$, $E_t \subseteq E$, and w comprises the lengths on the edges in E. Since we need to minimize the interconnection wire length, it is desirable to obtain a *minimal spanning tree* (MST), that is, a tree for which the sum of

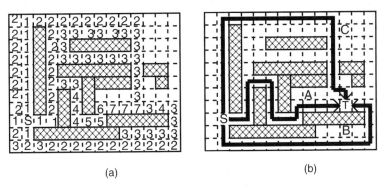

(a)   (b)

Figure 4.17   Maze routing.

the edges is minimal. There are several algorithms to find a minimal spanning tree [Kruskal 1956, Prim 1957]. The simplest algorithm will follow a greedy strategy. First, the edges in the set E are sorted in increasing order according to their distances (Manhattan). Start from a vertex that is connected to the shortest edge and place this edge in E_t. In our example, one such vertex is F; delete it from V and place it in V_t. Select from E the shortest edge that connects any vertex in V_t to a vertex in V. It is D in this case. Place it in V_t and delete it from V. Repeat the last step until V is empty. The total length of the edges in E_t is the minimal distance. The algorithm is illustrated step by step in Fig. 4.16. The last frame in the figure shows the MST.

Notice also that the global routing is not actually a layout of the wires connecting the pins but is just an order in which the pins (vertices) will be connected. Once the global routing is completed for all nets, the detailed one follows. Consider connecting two pins that we label as S and T in Fig. 4.17. The shaded parts of the routing space are obstacles that cannot be used for connecting the two points. The detailed routing finds the specific path of connection. One of the main algorithms finds its roots in applications in operation research [Lee 1961]. Several variations of this algorithm have improved the detailed routing between the two points in such way as to minimize (1) the distance between the two points, (2) the number of turns in the route since a via is needed at each turn, and (3) the number of obstacles removed, as this means more space taken for routing another wire with ease.

As shown in Fig. 4.17, there are three routes, and a greedy approach would select route C, the shortest. However, there are more corners. Each turn is associated with a via that connects the two segments of wires that are usually on separate layers. Each via contributes to an increase in the resistance of the path. A large number of vias also increases the chance of defects such as opens, as discussed in Chapter 2. Similarly, net A can be connected to minimize the corners or to minimize resources, as illustrated in Fig. 4.18a and b, respectively. All these criteria are important, but they do not include the effects of routing on testing. For example, in the case where we conserved resources, net A is routed

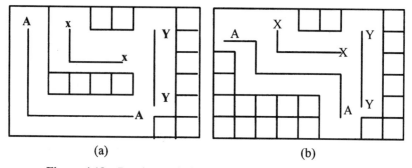

Figure 4.18   Routing optimization for (*a*) noise and (*b*) space.

very closely to net *X*. If the two nets are long enough, their signals might interact and result in crosstalk noise, as described at the end of Chapter 2. This can affect operation of the circuit for deep-submicron technology circuits.

A concrete example of the importance of placement and routing in testing is the placement of flip-flops in design that includes the scan-path method. Scan-path design is one of the DFT constructs explained in Chapter 9. In this design, some or all of the flip-flops will be connected, at testing time, in one shift register. If placement is done without regard to these design goals, adjacent flip-flops in the same shift register might be placed far away from each other and in any order, causing very long interconnects between them. This affects adversely the performance of the design and deters designers from using this technique. Usually, the placement is completed without concern to this structure, and it is important to route them for minimal path. Thus instead of finding the MST as discussed above, it is important to find the minimum spanning path. This problem is addressed in Chapter 9. The importance and complexity of the problem is well explained in [Beenker 1995].

It is preferable to include scan-path configuration during synthesis (one-path design) instead of after synthesis (two-stage design approach), and in this fashion, the P&R process will take testability into consideration in addition to area and performance. This second approach is discussed in Chapter 14.

4.7.4   Back-Annotation

Back-annotation is the process of extracting information about the circuit from the layout. The layout of ICs, the circuit artwork, consists of a collection of polygons that are connected, overlapped on the same layer and different layers. A circuit extractor estimates various electrical parameters, such as resistance and inductance of lines, capacitances of nodes, and dimensions of the devices. This is accomplished in two phases. The first phase, called *netlist extraction*, consists of the identification of the devices and the determination of electrically connected regions. The second phase consists of the estimation of the electrical parameters of the devices and the nets.

Although the shapes extracted are regular, they are of varied dimensions and make the extraction process very elaborate, and there are efforts to expedite their operation using parallel algorithms [Banerjee 1994]. In addition to planar capacitance, the extractor should have wall-to-wall capacitances, which are becoming predominant in present submicron feature sizes. The annotated netlist represents the real circuit and, upon simulation, supplies the actual timing and delays. The information may be fed back to any of the synthesis stages, as illustrated in Fig. 4.19. As a result, the circuit may be resynthesized or optimized to meet the specifications. CAD tools used for parameter extraction are usually called *technology CAD* (TCAD), and interfacing them with CAD tools for design and test is pursued vigorously at present.

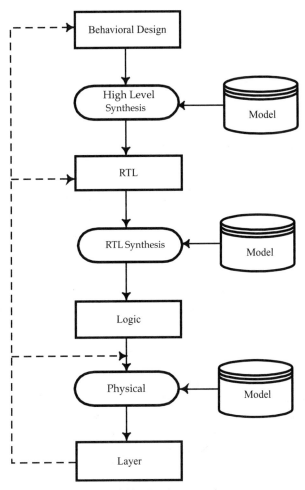

Figure 4.19 Back-annotation.

REFERENCES

Actel (1991), *ACT Family Field-Programmable Gate Array Databook*, Actel Corporation, Santa Clara, CA.

Altera (1992), *Applications Handbook*, Altera Corporation, San Jose, CA.

Banerjee, P. (1994), *Parallel Algorithms for VLSI Computer-Aided Design*, Prentice Hall, Upper Saddle River, NJ.

Bartlett, K., et al. (1985), Synthesis and optimization of multi-level logic minimization under time constraints, *IEEE Trans. Comput.-Aided Des.*, Vol. CAD-4, No. 4, pp. 582–596.

Beenker, F. P. M., R. G. Bennetts, and A. P. Thijssen (1995), *Testability Concepts for Digital ICs*, Kluwer Academic, Norwell, MA.

Brayton, R. K., et al. (1984), *Logic Minimization Algorithms for VLSI Synthesis*, Kluwer Academic, Norwell, MA.

Brayton, R. K., et al. (1987), MIS: a multiple-level logic optimization system, *IEEE Trans. Comput.-Aided Des.*, Vol. CAD-6, No. 11, pp. 1062–1081.

Bryant, R. E. (1986), Graph based algorithms for Boolean function manipulation, *IEEE Trans. Comput.*, Vol. C-35, No. 8, pp. 677–691.

Chan, P., and S. Mourad (1994), *Logic Design with FPGAs*, Prentice Hall, Upper Saddle River, NJ.

Cormen, C., C. E. Leiserson, and R. L. Rivest (1992), *Introduction to Algorithms*, MIT Press, Cambridge, MA.

Fujiwara, H., and K. Kinoshita (1981), A design of programmable logic arrays with universal tests, *IEEE Trans. Comput.*, Vol. C-30, No. 11, pp. 823–828.

Fujiwara, H. (1982), The complexity of fault detection for combinational logic circuits, *IEEE Trans. Comput.*, Vol. C-31, No. 6, pp. 555–560.

Fujiwara, H. (1984), A new PLA design for universal testability, *IEEE Trans. Comput.*, Vol. C-33, No. 8, pp. 745–750.

Fujiwara, H. (1986), *Logic Testing and Design for Testability*, MIT Press, Cambridge, MA.

Gajski, D. D., and R. H. Kuhn (1983), Introduction: new VLSI tools, *IEEE Computer*, Vol. 6, No. 12, pp. 11–14.

Goel, P. (1980), Test generation cost analysis and projections, *Proc. Seventeenth Design Automation Conference*, pp. 77–84.

Hong, S. J., and D. L. Ostapko (1980), FITPLA: a programmable logic array for functional independent testing, *Digest Tenth International Symposium on Fault-Tolerant Computing*, pp. 131–136.

Hong, S. J., R. G. Cain, and D. L. Ostapko (1974), MINI: a heuristic approach for logic minimization, *IBM J. Res. Dev.*, Vol. 18, No. , pp. 443–458.

Ibarra, G. H., and S. K. Sahni (1975), Polynomially complete fault detection problems, *IEEE Trans. Comput.*, Vol. C-24, No. 3, pp. 242–249.

Kruskal, J. B. (1956), On the shortest spanning subtree of a graph and the travelling salesman problem, *Proc. AMS*, Vol. 1, No. 1, pp. 48–50.

Lee, C. Y. (1961), An algorithm for path connections and its applications, *IRE Trans. Electron. Comput.*, Vol. EC-10, No. 3, pp. 346–365.

Michel, P., U. Lauther, and P. Duzy (eds.) (1992), *The Synthesis Approach to Digital System Design*, Kluwer Academic, Norwell, MA, 1992.

Prim, R. C. (1957), Shortest connection networks and some generalizations, *Bell Syst. Tech. J.*, Vol. 36, No. 6, pp. 1389–1401.

Quicklogic (1993), *Very High-Speed Programmable ASIC*, Quicklogic, Santa Clara, CA.

Walker, R. A., and D. E. Thomas (1985), A model of design representation and synthesis, *Proc. 22nd Design Automation Conference*, pp. 453–459.

Xilinx (1985), *The XC4000 Data Book*, Xilinx, Inc., San Jose, CA.

PROBLEMS

4.1. What is the order of complexity of the multiplication of three matrices of order $N \times M$, $M \times K$, and $K \times L$?

4.2. Consider four algorithms with time complexities n, n^3, 2^n, and $n!$, respectively. The time taken to solve a problem of size 10 by the first algorithm is 10 ms. What is the time needed for the other algorithms? What are these times for a problem of size 100?

4.3. Suppose that you are to design a circuit from a behavioral description given in the form of Boolean equations. Outline on the Y-chart in Fig. 3.1 various processes to accomplish your design.

4.4. Change the factored Boolean function given in Section 4.4.2 and illustrated in Fig. 4.6 to a sum of products and determine the number of literals of the obtained expression.

4.5. Put the steps used to develop the MST in Section 4.7.3 in algorithmic form.

4.6. You are to determine a global route for the different components shown in Fig. P4.6. Develop a minimal spanning tree. Is this tree unique? Justify your answer.

4.7. Modify the algorithm developed in Problem 4.6 to obtain the shortest path instead of the spanning tree.

4.8. In which process of the DFT cycle would you recommend that care needs to be taken to avoid crosstalk noise?

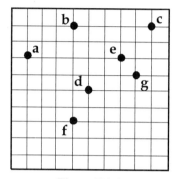

Figure P4.6.

PART II

TEST FLOW

5

ROLE OF SIMULATION IN TESTING

5.1 INTRODUCTION

Design verification is necessary to check its conformity to specifications. Simulation is the most used means of verification, although a trend is starting toward formal verification. Simulators are essential programs in the CAD toolbox since they are used at different stages of the DFT cycle and at different levels of abstraction ranging from behavioral, RTL, structural, and switch level to circuit and device level. Each level is a finer process of simulation, requiring more information from the designer and more computer time. A *mixed-mode simulator* allows the simulation of different parts of the design at different levels of abstraction. For example, whereas most of the design is simulated on the functional level, a critical part is simulated on the circuit level and switch level [Bryant 1984]. Usually, we distinguish between functional and timing simulation. The first checks the correct operation of the circuit, and the second helps to determine timing violation and critical paths. For present high-complexity circuits, a unified design and test platform that facilitates the interaction between high-level representation and the circuit level is becoming necessary [Moundanos 1998].

To perform its task, the simulation program requires the following components, illustrated in Fig. 5.1: (1) design models, (2) a component library, (3) stimuli to be applied to the design, and (4) expected responses. The various types of design descriptions have been presented in Chapter 3. The input stimuli can be applied using one of the following formats: logic values, waveforms, pseudorandom pattern generator, and test benches. Different libraries correspond to the level of design abstraction. At the functional level, each entry in

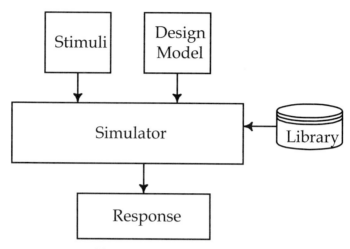

Figure 5.1 Simulation process.

the library is represented by its logical function without details of the internals of the block. The library for the structural level consists of a set of standard cells described in terms of basic logic functions and their propagation delays. However, no information is available regarding the transistor level. Detailed simulations are then required before fabrication to determine critical paths. For a large circuit, this type of simulation is not practical. A special class of timing simulator is used, the *static timing analysis*. Here the simulator takes as input a netlist connecting transistors and metal and poly paths, and constructs a model of the circuit in terms of resistance, capacitances, and whenever applicable, inductance. Path delays are then calculated, and the largest path, the critical path, is determined without the need of generating stimuli for the circuit. Although the emphasis is on digital testing in this book, it is important to mention that many circuits used at present include analog subcircuits. In such cases an analog simulation is also expected. A mixed signal or analog–digital simulator is more appropriate.

For testing, as opposed to verification, the simulator is used at different stages of the design [Fujiwara 1985, Miczo 1986, Abramovici 1990]. After incorporation of a DFT structure, the circuit is simulated to evaluate the effectiveness of this structure. Also, after generating a test set, say, for stuck-at faults, its fault coverage is calculated using a *fault simulator*. Since the final test to be applied to the chip is usually done on the gate level, fault simulation is also performed on the gate model of the circuit. However, fault injection in circuits described at higher level of abstraction is also possible.

The first topic we address is the effect of design size on the simulation process. Then we discuss logic and timing simulation. In the remainder of the chapter we concentrate on fault simulation.

5.2 SIMULATION OF LARGE DESIGNS

When designing circuits with millions of gates, one can no longer rely on traditional gate-level simulation to complete the design on time. The only way to contain the verification time is to adopt a new verification that is more compatible with large designs.

- It is important to shift the verification effort to the RTL level instead of the gate level. The main advantage is that the simulation time is a fewfold faster.

- It is more practical to use the appropriate verification technique for different parts of the verification process. Early in the design cycle, it is more suitable to use formal verification to check the result of synthesis transformation from RTL to gate level. However, formal verification is still not as effective as simulation in verifying RTL design functionality against design specifications.

- *Static timing analysis* has replaced simulation for timing verification. This methodology is a more exhaustive verification that does not require the generation of a large set of data (stimuli), but which is much faster.

5.2.1 Test Benches

To simulate a design described in HDL, irrespective of its level of abstraction, a test bench is used to (1) apply the stimuli to the design to be stimulated and (2) collect the responses and compare them to expected values. The following test bench invokes a design called *adder* and performs the operations on this design described above. It is written in Verilog HDL.

```
`timescale 1ns/1ns               // Time unit is 1ns
module adder;                    // Design Test bench
reg PA, PB, PCI;
wire PCO,PSUM;
FA_Behav F1(PA, PB, PCI, PSUM, PCO); // Instantiate module under test
initial
  begin: ONLY_ONCE
  reg[3:0] Count;                // Count hold the count of stimuli
                                 //4 bits are needed to accommodate
up to 8
  for (Count=0; Count < 8; PAL =Count +1)
    begin
    {PA, PB, PCI} =Count;        //The stimuli are the values of Count
    #5 $display ("PA, PB, PCI=%b%b%b", PA, PB, PCI,":::PCO,
              PSUM=%b%b", PCO, PSUM);
                                 //Creation of the response display
```

```
       end
    end
endmodule
```

The output of a simulation is requested in binary form (%b) but may be expressed in a hexadecimal number, or, of course, be displayed as waveforms.

5.2.2   Cycle-Based Simulation

As the design passes through various phases in the design cycle, it is important to check that it is still performing the function intended from the onset. Thus it is important to compare the simulation results at these different phases. Early in the cycle different components are designed independently, then they are integrated, and ultimately they form a large design with a large test bench. For simulation on the structural level, it is not possible to consider the entire design; therefore, only parts of the design may be simulated at one time. To compare the results of simulation, the function of the circuit is synchronized to a master clock. That is, the design is evaluated when the clock edge occurs, and all other timing internal to the circuit is ignored. In this fashion it is the functionality at the different design phases that can easily be compared.

5.3   LOGIC SIMULATION

Logic simulation is divided primarily into three types: software-based, hardware accelerators, and hardware emulators. The first type is the most widely used. We concentrate on it in this chapter. Since circuits are growing continually larger, simulation takes more time and there are attempts to use parallel algorithms to speed up the operations of software simulators [Banerjee 1994]. In the second type of simulation, the simulation algorithm is *hardwired* instead of implementing it in a software program. This improves its performance and is particularly useful for a large design with a large volume of stimuli.

Emulation is the third means of verification. It is similar to prototyping. However, while in the past it was possible to prototype by wire-wrapping SSI and MSI components, the present technology builds emulators with powerful field-programmable gate arrays (FPGAs). For example, the Pentium chip was emulated using 3500 Xilinx 3000 FPGA chips. There are two main advantages of hardware emulators. First, it is not possible to verify an entire processor using software simulator because the simulator is not fast enough to handle the enormous amount of stimuli. Second, the emulator allows the verification of the design within the system environment. An emulator speeds up the simulation by two to three orders of magnitude over software simulation.

5.4   APPROACHES TO SIMULATION

Software simulators may be of the compiled or interpretive types [Breuer 1972, Miczo 1986]. The latter is known mostly as event-driven.

5.4.1   Compiled Simulation

In the compiled type, the circuit netlist is converted into a sequence of machine language instructions, which executes the various logic and arithmetic functions of the circuit. The following procedure for a compiled code simulator is a variation of that given in [Banerjee 1994]:

```
Procedure Compile-code simulator
   Read the circuit description
   Break feedback loops, if any
   Order the gates into levels
   Generate compiled code
   Read in the initial value of each line
   Read in next input vector and update values
   FOR each new input data DO
      FOR each level of logic DO
         For each gate in the level DO
            Execute the compile code for the gate
         END FOR
      END FOR
      IF new value of outputs of feedback lines same as old values
         THEN output results
      ELSE set new input value of inputs of feedback lines same as
outputs
   END FOR
End procedure
```

The algorithm implies operating on all gates on one level before proceeding on the next level. The levelization of the circuit is illustrated in Fig. 5.2. All primary inputs and feedback lines are at level 0. All gates at a given level have inputs only from lower levels. In other words, no gate can get inputs from a gate at the same or higher level than itself.

The primary advantage of compiled code is its fast execution. However, it has a few shortcomings. The circuit needs to be compiled each time a design change is needed. In addition, for each input pattern, activities of all parts of the circuit are evaluated, although many signals remain unchanged. For example, consider a multi-input AND gate with one of the inputs being 0. No matter what the changes are on the other inputs, the output will remain 0, and hence it is unnecessary to execute any instructions that evaluate the output of this gate. Another problem with this type of simulator is related to the delays, which is discussed in detail in the next section.

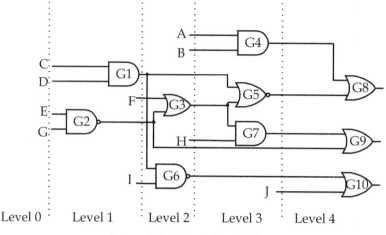

Figure 5.2   Circuit levelization.

Compiled simulators are usually used for zero delay and are useful for combinational circuit and synchronous sequential circuits. However, they cannot handle delay-associated problems such as hazard and races. To alleviate the problems of compiled simulation, event-driven simulators are more effective, particularly for large circuits.

5.4.2   Event-Driven Simulation

In event-driven simulation, gates are evaluated only when there are events. An *event* is a change in the value of the signal. A gate is evaluated only if at least one of its inputs changes value. Otherwise, no evaluation is performed. We illustrate the concept using the circuit of Fig. 5.2, annotated as shown in Fig. 5.3. A change of E $(0 \rightarrow 1)$ has the potential of affecting all gates indicated by an asterisk (*) in the following order: G2, G3, (G5 and G7), and finally, (G8 and G9). Since the output of G2 changes $(0 \rightarrow 1)$, the output of the gate on the next level, G3, has to be evaluated. Given that this output has not changed, there is no need to evaluate G5 or G7 and, consequently, gate G8. The output of G9, though, has to be evaluated since it may be affected by the change on G2. However, being an OR gate with one input already equal to 1, G9 will not be affected by the change on G2. As we proceed from one level to another, a delay δ is assumed before evaluation. Events on the same level are evaluated at the same time interval. To allow appropriate scheduling, a time wheel is used. After the change in G2, G3 is scheduled for processing in time 2δ and G9 in time 4δ. After evaluation of G3, then G5 and G7 are scheduled for time slot 3δ, and so on. Scheduling ahead may exceed the number of the slots in the wheel, and accommodation must be made to keep the schedule correct. The process can be summarized as follows [Banerjee 1994]:

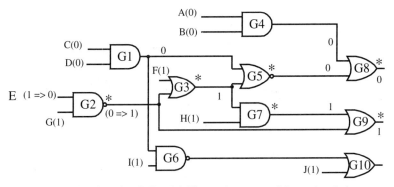

Figure 5.3 Circuit of Fig. 5.2 illustrating event-driven simulation.

```
Procedure Event-driven simulator
  Read circuit description
  Read input vectors
  FOR each input vector to be simulated DO
    Process new inputs
    Update input nodes
    Schedule connected elements on timing wheel
    WHILE elements left for evaluation
      Evaluate element
      IF change on the output
        THEN update all fan-outs and schedule
        Connect element on timing wheel
    END WHILE
  END FOR
End procedure
```

Given that only a fraction of the signals in a circuit change value in response to the application of a stimulus, event-driven simulation causes a significant reduction in computation time.

5.5 TIMING MODELS

In addition to the levels of simulation, there are several operational modes that reflect different accuracy of the results. Higher accuracy of the results require longer simulation time. The delay in the circuit can be represented as shown in Fig. 5.4a, where every square represents the delay of the gates it follows. In the least accurate simulation, all components are having the same *unit delay*. No actual time is associated with this delay. This is the simplest mode of simulation (Fig. 5.4b), because most factors, such as the loading effects of fan-in and fan-out, are ignored.

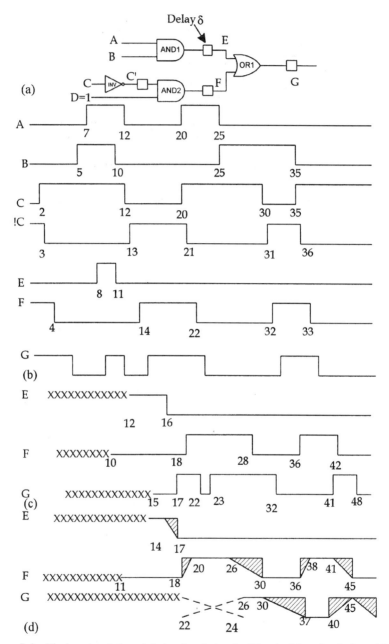

Figure 5.4 Timing simulation: (*a*) simulated circuit; (*b*) unit-delay simulation; (*c*) varied delay simulation; (*d*) multivalued simulation.

If the components of the circuit are assigned different delay values, the simulation results are more realistic, as illustrated in Fig. 5.4c. Notice that before the propagation of the input values through the different gates in the circuit, the outputs of these gates were undetermined. This is indicated in the figure by x, which means *undetermined* or *unknown* value. For this simulation we extend the logical values of the nodes from $\{0, 1\}$ to $\{0, 1, x\}$ [Eichelberger 1965]. The accuracy can be improved further by assigning to each component multiple delays: rise and fall times. The effect of these times is shown in Fig. 5.4d. Actual timing simulation is performed taking into account all parasitic capacitances, in addition to those of the transistors. The parasitic data can either be estimated or extracted after layout, as mentioned in Chapter 4. In this fashion one observes not only the logic level of the signal but also its shape. Because the rise and fall times are not unique, they are in a range of values from t_{min} to t_{max}. Therefore, the slew rate of the signals can vary and cause these signals to remain for an appreciable amount of time in a logic range that is rising (u) or falling (d). It is possible that the end result will be an erroneous signal that is undetermined (e). Thus for accurate simulation we extend the logic values to $\{0, 1, x, u, d, e\}$.

The high capacitive load of MOS technology and its bidirectional signal flow have imposed special requirements on simulators. This has led to the concept of *strength of the signal*, which may be forcing, nonforcing, or of high impedance. Forcing strength is the status of a node that is pulled up or down at a fast rate. A nonforcing strength is the status of a node that is connected to power or ground through a resistor that is allowing any charge to be added or removed at a finite rate. When signals of different strengths collide, the strongest signal always overcomes the weaker. Signals A and B may flow in either direction through the transmission gate (Fig. 5.5) whenever S is high. Thus either inverter 1 or 2 drives the output. The strength of the signal driving any of these inverters depends on the strengths of the output signals of 1 and 2. If the strength of G1 is greater than G2, the signal flows from A to B, and vice versa. Using logic levels and strength, a nine-value logic system has been adopted by the IEEE as Standard 1164-1993. This standard may be invoked in HDL simulators.

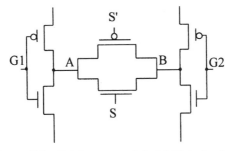

Figure 5.5 Driving strength in MOS technology.

5.5.1 Static Timing Analysis

In static timing analysis (STA), the concern is with determining the critical path delay. The delays are calculated using the data sheets of the various circuit components under worst-case analysis conditions. The critical path delays reported by the STA are not usually the same as those provided by functional simulation. Models used in the function simulator do not generally include the loading capacitances and hence they do not yield results as accurate as those obtained by STA. Most STA programs do not consider the logic function as the circuit is analyzed, and the delay calculation may yield false *paths*. For example, consider a circuit that implements a function $f(a, b, s)$ that under some values of a and b is represented by $f = ss'$. The path from s to f is impossible to activate; nevertheless, an STA will produce a value for it.

5.5.2 Mixed-Level Simulation

Mixed-level simulation mode is a useful feature since it is possible to exercise real timing simulation on the most critical parts and multidelay on other parts.

5.6 FAULT SIMULATION

Fault simulation is performed during the design cycle to achieve the following goals:

- Testing specific faulty conditions
- Guiding the test pattern generator program
- Measuring the effectiveness of the test patterns
- Generating fault dictionaries

To perform its task, the fault simulation program requires, in addition to the circuit model, the stimuli, and the responses of a good circuit to the stimuli, a fault model and a fault list. This is illustrated in Fig. 5.6. As we learned in Chapter 2, there are different fault models, and the most widely used is the stuck-at model. Test patterns generated for this model have proven to be useful for other types of models, such as multiple stuck-at, bridging, and delay faults. The responses deduced by the fault simulator are used to determine the fault coverage.

The fault simulation process is illustrated in Fig. 5.6. A fault is considered from the list and a pattern is applied to the circuit. If the fault is detected, it is dropped from the fault list and the next fault is considered. Otherwise, another pattern is applied, and if the fault is not detected when all patterns are applied, the fault is then considered undetectable by the test and is removed from the fault list. The process is continued until the fault list is empty.

As for logic simulators, fault simulators are of the compiled type or the event-

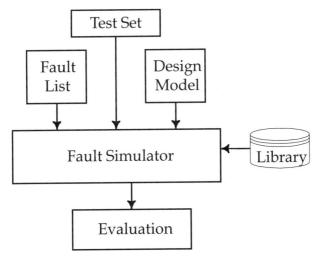

Figure 5.6 Elements of fault simulation.

driven type. The compiled code is executed for each fault for as many times as there are test patterns or only as many times as it takes to detect the fault in the case where *fault dropping* is allowed. The execution time has an upper bound of $N_f N_p$, where N_f is the number of faults in the fault list and N_p is the number of test patterns. It is possible to reduce this execution time. To represent the faulty value, the simulation program uses one computer word. However, the logic value on any of the nets is either 0 or 1; it is thus sufficient to store this value in only one bit of the computer word. In this fashion, the different bits can represent several faults. For example, a 16-bit word can be used to represent 15 faults, leaving one bit for the fault-free value of the circuit. Simulators using this scheme are called *parallel fault simulators*.

There are three main approaches to event-driven simulators: (1) parallel single fault, (2) deductive fault simulation, and (3) concurrent fault simulation. We illustrate each one with an example. Any of these approaches starts with a fault list and generates the good circuit response to the stimuli. We call this the *fault-free copy* or *image* of the circuit. It is the manner in which they handle faults from the fault list that differentiates among these approaches.

5.6.1 Parallel Fault Simulation

In parallel fault simulation, only one copy of a good circuit is calculated for a given test pattern [Thomas 1975, Iyenger 1988]. By one copy we mean the collection of values of all signals in the circuit resulting from application of a test pattern. If the computer word size is N, then $N - 1$ copies of faulty circuit are also generated. For a total M faults in the circuit $\lceil M/(N - 1) \rceil$, simulation runs are then necessary. The symbol $\lceil K \rceil$ is equal to K if K is an integer, or it

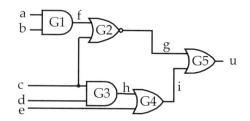

Figure 5.7   Circuit used to illustrate fault simulation.

is equal to the smallest integer greater than K. If the circuit has P primary outputs, the number of runs has a lower limit given by $\lceil M/[(n-1)P] \rceil$.

Next we illustrate the detection of faults taken $(N-1)$ at a time for the example shown in Fig. 5.7. First, a fault list is generated for the circuit. The circuit has 10 lines, excluding fan-outs. This results in 20 stuck-at faults. Notice, however, that this number can be reduced by collapsing the fault list. Second, we decide about the test pattern to be simulated. In this example we used the pattern $abcde = 10010$. We assume arbitrarily that $N = 16$ and we calculate the fault-free copy and up to 15 faulty copies of the circuits. All this information is summarized in Table 5.1. The faults are listed in the first row of the table. The first column of this table gives the signal on each line in the circuit, including the primary inputs, a, b, c, d, and e, and the primary output, u. For the other lines, the signals are expressed in terms of the logical operation performed by the corresponding gate. The second column lists the fault-free (ff) circuit response at each of the lines represented in the first column. Each of the other 15 columns represents the responses under the fault indicated in the first row. Notice that for the primary inputs, the faults selected are limited by the test pattern. That is, we could have, for example, listed $a/0$ and $a/1$; however, the pattern requires that $a = 1$. Thus it is not necessary to check whether or not the fault $a/1$ is detected. If it is to detect a fault on a, this pattern would be $a/0$. The third column represents the copy of the circuit under the fault $a/0$. The next column is the copy for $b/1$, and so on for the rest of the columns. To facilitate examination of the table, the faulty response is shown in boldface type whenever it is different from the corresponding ff response at any line.

We recognize whether or not the fault is detected by comparing the fault-free value of the primary output, u, to the corresponding values in the column representing the copies of the faulty circuit. We then recognize that only the following faults are detected by the test pattern applied: $b/1, f/1, g/0, u/0$. A quick investigation of this set of faults indicates that $f/1$ and $g/0$ are equivalent; $f/1$ dominates $b/1$, and $u/0$ dominates $g/0$. Thus we could have represented these faults by one of them, $b/1$. Detection of this fault implies detection of $f/1$ and its equivalent $g/0$, as well as $u/0$, which dominates $g/0$. Equivalence of dominance were discussed in Sections 2.6.1 and 2.6.2. We could actually have reduced the fault list by applying equivalence and dominance relationships before simulating.

TABLE 5.1 Parallel Fault Simulation

	ff	a/0	b/1	c/1	d/0	e/1	f/0	f/1	g/0	g/1	h/0	h/1	i/0	i/1	u/0	u/1
$a = 1$	1	0	1	1	1	1	1	1	1	1	1	1	1	1	1	1
$b = 0$	0	0	1	0	0	0	0	0	0	0	0	0	0	0	0	0
$c = 0$	0	0	0	1	0	0	0	0	0	0	0	0	0	0	0	0
$d = 1$	1	1	1	1	0	1	1	1	1	1	1	1	1	1	1	1
$f = ab$	0	0	1	0	0	0	0	1	0	0	0	0	0	0	0	0
$g = f + c$	0	0	1	1	0	0	0	1	0	1	0	0	0	0	0	0
$h = cd$	0	0	0	1	0	0	0	0	0	0	0	1	0	0	0	0
$e = 0$	0	0	0	0	0	1	0	0	0	0	0	0	0	0	0	0
$i = e + h$	0	0	0	1	0	1	0	0	0	0	0	1	0	1	0	0
$u = g + h$	0	0	1	1	0	0	0	1	0	1	0	1	0	0	0	1

5.6.2 Deductive Simulation

Deductive fault simulation requires one image of the circuit, the fault-free response to the test pattern [Armstrong 1972]. Fault detection is deduced from the current image. All detectable faults are deduced in only one pass. Such a pass takes longer than a pass in parallel simulation. However, there are fewer passes than parallel simulation. It has been proven that for small circuits, fewer than 500 gates, parallel simulation is more efficient than deductive simulation [Chang 1974]. These experimental results were dependent, of course, on using much slower computers having 4 bits per word. Although most present circuits are much larger than 500 gates, deductive simulation is more realistic.

This type of fault simulator creates the fault-free image for the test pattern. Then a list of faults is associated with every line in the circuit. These faults are sensitizable to the output through this line. The faults are propagated from the primary inputs of the circuit to its primary outputs, one level at a time. We illustrate the process using the example in Fig. 5.7 and the same pattern as that used in the preceding section. Propagation of the faults is illustrated on the circuit shown in Fig. 5.8a. The results are also displayed, level by level, in Table 5.2. The list of a and b is passed to f, then from f and c to g, and so on.

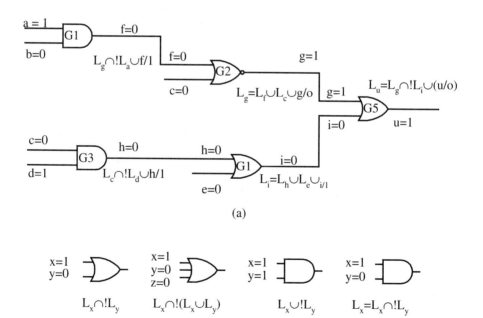

(a)

(b)

Figure 5.8 Deductive simulation: (*a*) operation on elementary logic gates; (*b*) application to the circuit in Fig. 5.6.

TABLE 5.2 Deductive Simulation Level by Level[a]

Node	ff	Level 0, Primary Input	Level 1	Level 2	Level 3	Level 4	Level 5, Primary Output
a	**1**	a/0					
b	**0**	b/1					
c	**0**	c/1					
d	**1**	d/0					
e	**0**	e/1					
f	0		$Lb \cap (!La \cup f/1$	b/1, f/1			
g	1			$L_f \cup L_c \cup g/0$	b/1, f/1, c/1, g/0		
h	0			$L_c \cap L_d \cup h/1$	c/1, h/1		
i	0				$Lh \cup Le \cup i/1$	c/1, e/1, i/1	
u	1					$Lg \cap !L_i \cup u/0$	b/1, f/1, c/1, g/0, u/0

$La \cap !Lb = La - La \cap Lb$. This can be written as all elements of La minus the elements in the intersection of La and Lb. So, $Lu = Lg - Lg \cap Li \cup u/0$
$= (b/1, f/1, c/1, g/0) - (c/1) + u/0 = b/1, f/1, g/0, u/0$ as indicated in the last column.

The fault list at the output of a gate is deduced from the values of the fault-free image and the type of gate. Let us refer to the lists associated with lines a and b as L_a and L_b. Each list includes only one fault, $a/0$ and $b/1$, respectively. Since $a = 1$ and $b = 0$, only L_b is sensitizable to f. The faults on a are masked by the 0 value on b. Thus $L_f = [L_b \cap !L_a] \cup \{f/1\}$, where ! indicates the complement, $\cap$ the intersection operation on sets, and $\cup$ the union operation. Also recall that $[L_b \cap !L_a] = L_b - L_b \cap L_a$. In other words, they are the faults in L_b that are not in L_a. Therefore, $L_f = \{b/1, f/1\}$. Had $b = 1$, L_a could have been sensitized to f, that is, $L_f = L_a \cup L_b \cup \{f/1\}$. The operation on the sets will thus depend on the logic gate and the values on its inputs. They are all presented in Fig. 5.8b.

We know from the fault-free image that since $f = 0 = c$, the fault list on g will be $L_g = L_f \cup L_c \cup \{g/1\} = \{b/1, f/1, c/1, g/0\}$. In a similar fashion, we can find the fault list at input i of G5. It is $L_i = \{c/1, h/1, e/1, i/1\}$. Since the output gate, G5, is an OR gate and its inputs g and i are 1 and 0, faults detected at the output, u, are given by $L_g \cap !L_i \cup \{u/1\} = [L_g - L_g \cap L_i] \cup \{u/1\} = \{b/1, f/1, g/0, u/1\}$.

5.6.3 Concurrent Fault Simulation

Concurrent fault simulation combines features of parallel and deductive simulations [Ulrich 1974, Abramovici 1977, Rogers 1987]. Faulty gates are simulated when their output differs from that of the fault-free circuit. It is faster than other algorithms and uses dynamic storage allocations. We illustrate this simulation using the circuit in Fig. 5.7 and the same pattern as that used to illustrate the other types of simulation. The results of the various operations are shown in Table 5.3. The first row of the table includes all gates and their respective inputs. The second row shows the values of the test applied on the primary inputs, a, b, c, d, and e, and the fault-free values on the outputs of the five gates. In subsequent rows, the values of the faulty signals are given. From the values on the inputs and outputs of the gate, the possible detectable faults are deduced. For example, for gate G1, the values on a, b, and f are 1, 0, and 0. The faults that are potentially detectable are $a/0$, $b/1$, and $f/1$. The behavior of this gate for these faults is shown in the fourth through sixth rows. If any of these faults is to be detected, it has to be observable on the output of the gate, f. That is, the faulty values of f have to be different from the fault-free values shown in the second row. Thus only two faults, $b/1$ and $f/1$, have the potential to be detectable and are considered in the next steps of the simulation. Fault $a/0$ is dropped immediately.

Gate G2 is a NOR gate with inputs c and f (the output of G1) and output g. In addition to the two faults carried out from the previous gate, $b/1$ and $f/1$, the possible detectable faults are $c/1$ and $g/0$. Since for all faults the output of the faulty gate is different from the fault-free value (row 2, column 8), all faults on gate G2 are propagated to the next gate, G5. But before exploring these faults any further, we need to examine the other gates, G3 and G4, since signals from these gates are required to propagate the fault to the primary output, u.

TABLE 5.3 Concurrent Fault Simulation for Circuit in Fig. 5.7

G1	a	b	f
	1	0	0
a/0	0	0	0
b/1	1	1	1
f/1	1	0	1

G2	c	f	g
	0	0	1
b/1	0	1	0
c/1	1	0	0
f/1	0	1	0
g/0	0	0	0

G3	c	d	h
	0	1	0
c/1	1	1	1
d/0	0	0	0
h/1	0	1	1

G4	e	h	i
	0	0	0
c/1	0	1	1
e/1	1	0	1
h/1	0	1	1
i/1	0	0	1

G5	g	l	u
	1	0	1
b/1	0	0	0
c/1	1	1	1
e/1	1	1	1
f/1	0	0	0
g/0	0	0	0
h/1	1	1	1
i/1	1	1	1
u/1	1	0	0

Gate G3 is an AND gate with inputs $c = 0$ and $d = 1$ and thus $h = 0$. The potentially detectable faults are deduced as $c/1$, $d/0$, and $h/1$ (column 12). Only the first and third faults have a chance to be detectable eventually, and they are carried on in investigating the next gate, G4. The fault $d/0$ is dropped for the same reason that we dropped $a/0$ in the earlier pass. The results of the similar operations on gate G4 propagate the faults $c/1$, $e/1$, $h/1$, and $i/1$ to gate G5. Next, we combine the faults sensitized at g, $b/1$, $c/1$, $f/1$, and $g/0$, and those sensitized on line i, $c/1$, $e/1$, $h/1$, and $i/1$. The two sets of faults are detected if the faults on the inputs of G5 are detected. The faulty responses at u for all of these faults are given at the rightmost column of the table. Comparison of these values with the fault-free value of 1 implies that the only faults detected by the pattern are $b/1$, $f/1$, $g/0$, and $u/0$. As expected, these are the same faults detected using parallel fault simulation with the same test pattern (10010). The results can be summarized as follows.

1. The test pattern used is *abcde* = 10010.
2. Output of G1: Fault $a/0$ is dropped.
3. Output of G2: All faults are possibly detected.
4. Output of G3: Fault $d/0$ is dropped.
5. Output of G4: All faults are possibly detected.
6. Output of G5: Faults $i/1$, $e/1$, $h/1$, and $c/1$ are dropped and the remaining faults are detected: $b/1$, $f/1$, $g/0$, and $u/0$.

5.7 FAULT SIMULATION RESULTS

After fault simulation, the faults are either *detected* or *undetected*. The fault is detected if it has been controllable and observable by one of the patterns in the test set. In such a case, at least one of the primary outputs of the faulty circuit is different from the good circuit. Otherwise, it is not detected by any of the patterns of the test set. However, the output of a fault simulator separates faults into more than these two categories. Stuck-at faults on redundant logic are not testable and are referred to as *untestable*. That is, there are no patterns that can detect these faults. Other unstable faults occur on lines in the circuit that are *tied*, pulled up to V_{dd}, or pulled down to GND. In the circuit of Fig. 5.9, node 11 is held low, and thus SA0 on this line cannot be detected. Such tied lines may also mask or block other faults. In this case, faults on input H and node 10 are masked as well as the SA0 on node 12, the output of this AND gate. A fault on a line that is unused is not testable. This is the case of node 14 in Fig. 5.9. As a result, faults on line *B* are also undetectable because they are not observable. The other categories vary from one vendor to another. If the PO of the good circuit is either 1 or 0 but the corresponding PO of the fault circuit is x, we have a *possibly detected fault*, which is sometimes called a *potentially detected fault*. In some situations the injected fault causes the circuit

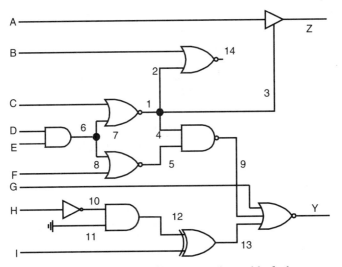

Figure 5.9   Circuit to illustrate undetectable faults.

to oscillate, and we have an *oscillatory fault*. The oscillation probably occurs because of the feedback in a combinational circuit with zero-delay models. An *oscillatory fault* that affects a large part of a circuit is called a *hyperactive fault*. Such faults cause a serious problem to the fault simulator since they may make simulation time unnecessarily long.

Usually, faults are dropped from the fault list as soon as they are definitely detected. However, sometimes a fault simulator includes the option of dropping it after it has been detected by n test patterns, where n is usually selected by the user of the fault simulator. This terminology is not standard and every simulator vendor utilizes its own terminology and attributes its own interpretation to this terminology [Mourad 1993].

5.7.1   Fault Coverage

The effectiveness of a test set is quantifiable. It is the percentage of the faults detected by a test and is known as the *fault coverage*, defined as

$$\text{fault coverage} = \frac{\text{faults detected}}{\text{total number of faults}}$$

A more realistic expression is

$$\text{fault coverage} = \frac{\text{faults detected}}{\text{detectable faults}}$$

where detectable faults = all faults − untestable faults. It is also possible to report the results in terms of fault classes. That is, after collapsing faults, the entries in the fault list represent fault classes rather one single fault. The calculations of fault coverage based on individual faults, T_f, versus fault classes, T_c, may yield disparate results. Consider a circuit with 36 faults that are collapsible into 20 classes where a class may include 1 to 4 faults. Assuming that a fault class is not detected, then $T_c = 19/20 = 95\%$. The corresponding values of T_f are $35/36 = 97\%$ and $32/36 = 89\%$ [Mourad 1993].

As we mentioned in Section 2.2 a single fault may represent many physical detects, but SAFs do not model all types of defects. Therefore, even if a fault coverage of 100% is reachable, this does not guarantee that the circuit is *defect-free*. The relationship between fault coverage and defect level is shown in Fig. 1.14. The defect level is usually measured in defects per million (DPM). Thus a defect level of 0.1% is equivalent to 1000 DPM. Inspection of the figure indicates that high fault coverage is required to ensure low defect levels, 1 to 100 DPM. To reach such a level, a SAF coverage of 99.9% is required. To improve defect coverage, current testing is used to complement SAF voltage, as discussed in Chapter 7.

5.7.2 Fault Dictionary

In addition to the fault coverage, a fault simulator may provide a *fault dictionary*. The fault dictionary associates with each fault a set of test patterns that uniquely identifies the fault. For example, consider the information in Table 5.4, where an entry of $a_{jk} = 1$ indicates that fault k is detected by pattern j. Now, if we know that a circuit passes all test patterns except the fourth pattern, we can deduce that the failure was due to fault $f8$. If, however, the circuit fails both test patterns 3 and 4, the cause is either $f5$ or $f8$. To distinguish the two faults, more information is needed, such as the output signal associated with each fault and pattern. This information is used at test application time and is helpful in identifying the defect that caused the fault. This is known as *fault diagnosis* and is very important in enhancing the process yield. The effectiveness of the diagnosis depends on that of the fault model in representing defects. Defects

TABLE 5.4 Fault Dictionary

Pattern	Fault							
	1	2	3	4	5	6	7	8
1	1		1			1		
2		1	1	1				
3			1		1		1	
4			1				1	1

that are not modeled by the stuck-at fault cannot then be identified. Current testing help improve the diagnosis.

REFERENCES

Abramovici, M., et al., M. A. Breuer, and K. Kumar (1977), Concurrent fault simulation and functional level modeling, *Proc. 14th Design Automation Conference*, pp. 128–137.

Abramovici, M., M. Breuer, and A. D. Friedman (1990), *Digital Systems Testing and Testable Design*, IEEE Press, Piscataway, NJ.

Armstrong, D. B. (1972), A deductive method for simulating logic faults in logic circuits, *IEEE Trans. Comput.*, Vol. C-21, No. 5, pp. 464–471.

Banerjee, P. (1994), *Parallel Algorithms for VLSI Computer-Aided Design*, Prentice Hall, Upper Saddle River, NJ.

Breuer, M. A. (ed.) (1972), *Design Automation of Digital Systems*, Prentice Hall, Upper Saddle River, NJ.

Bryant, R. E. (1984), A switch-level model and simulator for MOS digital systems, *IEEE Trans. Comput.*, Vol. C-33, No. 2, pp. 160–177.

Chang, H. Y., et al. (1974), Comparison of parallel and deductive fault simulation methods, *IEEE Trans. Comput.*, Vol. C-23, No. 11, pp. 193–200.

Eichelberger, E. (1965), Hazard detection in combinational and sequential circuits, *IBM J. Res. Dev.*, Vol. 9, No. 1, pp. 90–99.

Fujiwara, H. (1986), *Logic Testing and Design for Testability*, MIT Press, Cambridge, MA.

Iyenger, V. S., and D. T. Tang (1988), On simulating faults in parallel, *Digest of Papers 18th International Symposium on Fault-Tolerant Computing*, pp. 110–115.

Michel, P., U. Lauther, and P. Duzy (eds.) (1992), *The Synthesis Approach to Digital System Design*, Kluwer Academic, Norwell, MA.

Miczo, A. (1986), *Digital Logic Testing and Simulation*, Wiley, New York.

Moundanos, D., J. A. Abraham, and Y. V. Hoskote (1998), Abstraction techniques for validation coverage analysis and test generation, *IEEE Trans. Comput.*, Vol. C-47, No. 1, pp. 2–14.

Mourad, S. (1993), Computer-aided testing systems: evaluation and benchmark circuits, *VLSI Design*, Vol. 1, No. 1, pp. 87–97.

Rogers, W. A., J. F. Guzolek, and J. Abraham (1987), Concurrent hierarchical fault simulation, *IEEE Trans. Comput.-Aided Des.*, Vol. CAD-6, No. 9, pp. 848–862.

Sahni, S., and A. Bhatt (1980) The complexity of design automation problems, *Proc. Design Automation Conference*, pp. 402–410.

Thomas, E. W., and S. A. Szygenda (1975), Digital logic simulation in a time-based, table-driven environment, 2: Parallel fault simulation, *Computer*, Vol. 8, No. 3, pp. 38–49.

Ulrich, E. G., and T. Baker (1974), The concurrent simulation f activity in digital networks, *Computer*, Vol. 7, No. 2, pp. 39–44.

PROBLEMS

5.1. Design a transistor level for the logic gate in Fig. 5.4a and estimate the delays at the output. Check your results using a SPICE-like simulator.

5.2. Determine the stuck-at faults in the circuit of Fig. 5.2 that can be detected by the pattern 1110011101. Use the parallel fault simulation algorithm.

5.3. Repeat Problem 5.1 using deductive fault simulation.

5.4. Which, if any, of the faults detected in Problem 5.1 are equivalent?

5.5. For the circuit of Fig. P5.5, use concurrent fault simulation to determine the faults detected by an exhaustive test set. Plot the detectability profile of the circuit. (See Problem 2.2 for a definition of the detectability profile.)

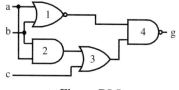

Figure P5.5

5.6. With reference to Table 5.4, determine, if possible, the most likely fault causing the failure if **(a)** the circuit passes the entire test except for pattern 1; **(b)** the circuit does not pass any of the patterns in the test.

6

AUTOMATIC TEST PATTERN GENERATION

6.1 INTRODUCTION

In Chapter 1 we explained why fault models are important in making testing more manageable and effective. We also gave a brief description of the three types of testing that are fault-independent: exhaustive, pseudoexhaustive, and pseudorandom. Exhaustive testing is not a realistic approach for large circuits due to its test application time. Pseudoexhaustive testing is more of a design for testability approach than a test pattern generation technique. Pseudorandom test pattern generation is deferred to Chapter 11.

In this chapter we present test pattern generation that targets specific faults. Test sets of this type are called *fault-oriented* or *deterministic test sets*. The fault model used here is the *stuck-at* type. We concentrate on the faults detected by observing the logic level (voltage) of the primary outputs at the gate level. Current testing is the topic of Chapter 7.

A deterministic test set may be *algebraic* or *algorithmic*. An example of an algebraic method is the use of the Boolean difference that we introduced in Chapter 3. We applied it to combinational circuits, but it has been used also for sequential circuits [Hsias 1971]. However, such a method is not efficient for test pattern generation of large circuits [Larrabee 1989], but it gives a good understanding of path sensitization [Sellers 1968]. It has been shown that test pattern generation for a combinational circuit is an NP-complete problem [Ibarra 1975, Fujiwara 1982]. This suggests that there are no test pattern generation algorithms with polynomial time complexity. Goel argues that the execution time of an algorithm grows proportionately to the square of the number of gates in the circuit [Goel 1980]. Because of such complexity, several heuristics have been proposed. The best known are the D-algorithm [Roth 1967], critical path algorithm, PODEM [Goel 1981], FAN [Fujiwara 1983], and SOCRATES [Schultz 1988]. The D-algorithm is based on D-calculus and guarantees a solution (a test pattern) if one exists. The other algorithms cannot guarantee a solution, but they do work in most situations.

All algorithms are based on the four main operations processes of excitation, sensitization, justification, and implication, all described in the next section. Each algorithm starts from a different part of the circuit. The D-algorithm starts at the faulty line, and its main difficulty is in reconverging fanout through XOR gate as we clarify later in the chapter. The critical path algorithm, which was devised as an alternative to simulation, starts from the primary outputs of the circuit and generates a test pattern for several faults, while PODEM starts from the primary inputs. FAN is an improvement over PODEM because of the use of new heuristics that minimize on backtracking. Also, SOCRATES is an improvement over FAN and utilizes testability measures to facilitate backtracking.

In the rest of the chapter we discuss the D-algorithm, the critical path, and PODEM. However, we first define path sensitization and justification, which we have used informally in generating patterns in Chapter 2. In describing automatic test pattern generation algorithms, we also make use of graphical representation of the circuit on the gate level.

6.2   Terminology and Notation

In this section we revisit some concepts and define some terms and operations that will be helpful in understanding the deterministic test pattern generation.

6.2.1   Basic Operations

To generate a pattern for a stuck-at fault on a line, we need to *provoke* or *excite* the fault, *sensitize* the results to a primary output, and *justify* the logic values required on the other lines in the circuit. In performing these operations, values are assigned to the lines in the circuits. We need to find the *implications* of these values on other gates.

To provoke or *excite* a line is to *control* it to a logic value that is the complement of the value at which it is stuck; this is equivalent to placing the faulty signal on the line. This signal is a discrepancy from the fault-free circuit. For example, to provoke the stuck-at 1 fault on line W, $W/1$, of the circuit in Fig. 6.1a, we must put 0 on this line, $W = 0$.

It is necessary *to sensitize* or *propagate* the fault to a primary output in order to observe it. The path from the faulty location to the primary output is a *sensitizing* or *propagation path*. A fault may have more than one sensitizing path to the same output or to different outputs. The fault $W/1$ has one sensitizing path: through G3, G4, and G6. To sensitize the fault to the output of G3, we must have $E = 1$. Finally, to propagate the fault to the primary output, Z, we

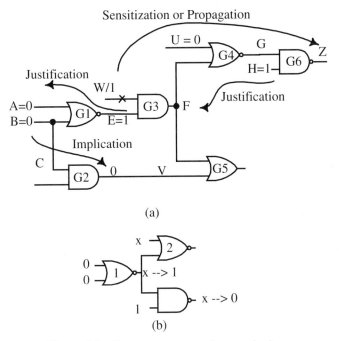

Figure 6.1 Test pattern generation terminology.

need to have $H = 1$. The values on E and H need to be *justified* to the primary inputs. We justify 1 on E by having $A = B = 0$. Next we find the *implication* of B on gate G2. Sometimes in propagating and justifying we encounter a conflict because some of the lines we need to control have values already assigned. In such cases it is said that we encountered an *inconsistency* [Roth 1966].

6.2.2 Logic and Set Operations

Provoking the fault on a line is equivalent to placing the complement of the faulty signal on the line. Since this faulty signal may be 0 or 1, we distinguish it by denoting the SA0 (SA1) fault by D (D'). This *discrepancy* notation was developed for use in the D-algorithm, but we will also adopt it throughout the chapter independent of the algorithm. The sensitization of the fault from one of the inputs of a gate to its output is determined by the signals on the other inputs of the gate and the type of gate. Before applying any patterns, we assume that all the nodes of the circuit are at an unknown logic value, x. In Fig. 6.1b, the 0 values on the inputs of the first NOR gate change its output from x to 1. The implication of this value on the second NOR is to change its output from x to 0, while the output of the AND gate remains at the unknown value x. Thus the logic values are now extended from $\{0, 1\}$ to the set $\{0, 1, D, D', x\}$. Table 6.1 defines the three main logic operations on this 5-tuplet of logic values.

TABLE 6.1   Logic Operations Using the Set of Logic Values 0, 1, D, D', and x

AND	0	1	x	D	D'	OR	0	1	x	D	D'	A	A'
0	0	0	0	0	0	0	0	1	x	D	D'	0	1
1	0	1	x	D	D'	1	1	1	1	1	1	1	0
x	0	x	x	x	x	x	x	1	x	x	x	x	x
D	0	D	x	D	0	D	D	1	x	D	1	D	D'
D'	0	D'	x	0	D'	D'	D'	1	x	1	D'	D'	D

During sensitization (propagation toward the output) and justification (tracing backward to the inputs), the same line may be assigned different logic values. Sometimes these values may be compatible; other times, they are not. To decide if we can conciliate the two assignments, we make use of the *intersection* of the two assignment sets that are defined in Table 6.2, where ϕ is the empty set and ψ is an undetermined operation. Consider an assignment that requires, for example, logic 0 on a line and another assignment requiring x on the same line. Since x may be interpreted as a 0 or a 1, the result of intersecting 0 with x is 0. On the other hand, if the second assignment is 1, there are no values to satisfy both assignments, and we have an inconsistency [Roth 1966].

6.2.3   Fault List

To generate a test set for a circuit, we need first to develop a fault list as defined in Chapter 2. Instead of listing all the possible faults in the circuit, we may use fault equivalence and dominance (see Chapter 2) to collapse the list to the checkpoints of the circuit. For a fanout-free circuit, the checkpoints are the primary inputs. Otherwise, the checkpoints are comprised of the primary inputs and the branches of all fan-outs in the circuit. Circuit S1 in Fig. 6.2 has 12 lines, including the branches of the stem, F. The total number of SSA faults is then 24. There are seven checkpoints: A, B, U, W, Y, $F1$, and $F2$. Therefore, the total number of faults is reduced to 14. However, using equivalence and dominance, we can reduce this list to contain fewer faults. For a large circuit, collapsing faults is not as easily done, but, of course, any reduction in the number of faults is highly desirable.

TABLE 6.2   Intersection Operation

$\cap$	0	1	x	D	D'
0	0	ϕ	0	ψ	ψ
1	ϕ	1	1	ψ	ψ
x	0	1	x	D	D'
D	ψ	ψ	D	D	ϕ
D'	ψ	ψ	D'	ψ	D'

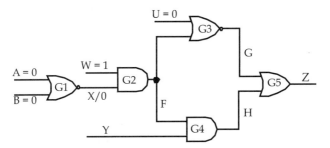

Figure 6.2 Circuit S1 to illustrate the D-algorithm.

In the remainder of the chapter we start with the D-algorithm and then present the main features of other heuristics which are an improvement over this algorithm. We assume that a fault list has been compiled for the circuit under test and that this fault list has been collapsed to the shortest possible number.

6.3 THE D-ALGORITHM

The D-algorithm is based on set theory and is explained formally in [Roth 1966]. Instead of using mathematical derivation, we outline it here with the flowchart shown in Fig. 6.3. We illustrate the operations defined in Section 6.2.1 using the D-algorithm terminology. To provoke a fault, we construct its *primitive D-cube for failure* (PDCF). This is simply the set of logic values on the inputs and outputs of this gate that *provoke* the fault on its output. Thus if the gate in question is a three-input AND gate and the fault is an SA0, the PDCF is the set $\{1, 1, 1, D\}$. This is illustrated in Fig. 6.4. Similarly, for a two-input NOR gate and an SA1 fault, it may be one of the three cubes $\{0, 1, D'\}$, $\{1, 0, D'\}$, or $\{1, 1, D'\}$ since it is sufficient to have a 1 on any of the inputs of the NOR gate to drive the output to 0. The PDCF for $X/0$ is $\{A, B, X\} = \{0, 0, D\}$.

The next step is to enumerate the possible sensitizing paths to primary outputs. The PDCF is then propagated along any of these paths, one gate at a time, utilizing the logical operations defined in Table 6.1. Propagating this cube to the output is called the *D-drive*. To propagate the fault on W in Fig. 6.4*b*, we form the *D*-cube $\{W, E, F\} = \{D, 1, D\}$. For this we use the logical operation in Table 6.1. As we continue propagating the *D*-cube, we will reach a primary output on which a D or D' will appear; then the signals assigned on the nodes during propagation are justified to the primary inputs. One possibility of the justification of F is illustrated in Fig. 6.4*c*. Both propagation and justification operations require performing the intersection on the cubes according to the operations defined in Table 6.2. Next we illustrate the algorithm by applying it to faults on an internal node, on a primary input, and on a primary output of

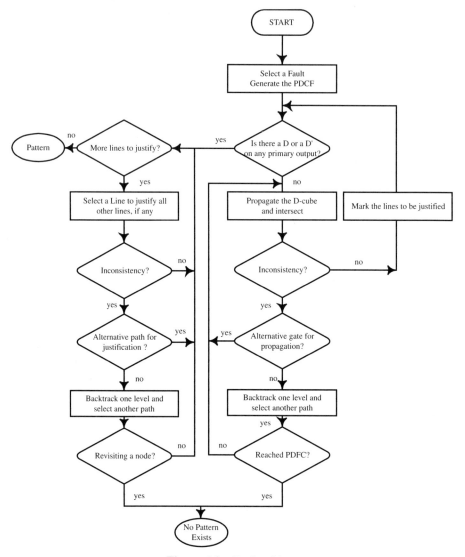

Figure 6.3   D-algorithm.

circuit S1 in Fig. 6.2, which is the same circuit we use to illustrate other ATPG algorithms.

6.3.1   Case of an Internal Node

We selected line X as an illustration for an internal node and we assume that it is stuck at 0. We make a list of all possible paths, which can be sensitized, from

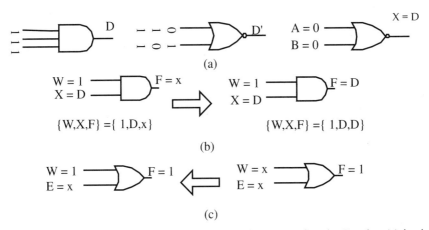

Figure 6.4 D-algorithm operations: (*a*) PDCF; (*b*) propagating the *D*-cube; (*c*) justification.

the fault to a primary output. There are two sensitization paths: one through G2, G3, and G5 and the other through G2, G4, and G5.

As we develop the cubes, the PDCF, and the propagation and justification cubes, we list the results, as shown in Table 6.3. The first row of the table lists all the nodes, and the second column lists the various gates used in propagation and the lines used in justification.

1. Initialize the circuit by placing *x* on each node (row 2).
2. For a SA0, $X = D$ and hence $A = B = 0$. The PDCF is then $\{0,0,D\}$, as shown in the table in row 3.

TABLE 6.3 D-Algorithm for $X/0$ in Circuit S1 of Fig. 6.2

1	Operation	Gate	A	B	X	Y	W	U	F	G	H	Z
2	Initialization		x	x	x	x	x	x	x	x	x	x
3	PDFC	G1	0	0	D							
4	∩		0	0	D	x	x	x	x	x	x	x
5	D-drive	G2			D		1		D			
6	∩		0	0	D	x	1	x	D	x	x	x
7	D-drive	G3					0	D	D′			
8	∩		0	0	D	x	1	0	D	D′	x	x
9	D-drive	G5								D′	0	D′
10	∩		0	0	D	x	1	0	D	D′	0	D′
11	Justification	H = 0					x			0		0
12	∩		0	0	D	0	1	0	↯	D′	0	D′
13	Justification	H = 0					0		x		0	
14	∩		0	0	D	0	1	0	D	D′	0	D′

3. Now we take the intersection of these two cubes in the two rows according to the operations of Table 6.2 and obtain the cube in row 4.

4. Propagating D through G2, we form the D-cube $\{W, X, F\} = \{1, D, D\}$ shown in row 5 and intersect it with row 4. The primary input, W, which used to be indeterminate, is now forced to logic 1. This is consistent with the operations in Table 6.2.

5. Select a sensitizing path from the list of possible sensitization paths. We selected the path through G3. Thus our aim is to propagate the D through this path to the primary output, Z.

   5.1. To propagate through G3, we form the D-cube $\{U, F, G\} = \{0, D, D'\}$.

   5.2. Again we intersect with the previous cube and obtain the results in row 8.

   5.3. The value of Z is still x, so we need to propagate the frontier.

   5.4. To propagate through G5, we construct the cube $\{G, H, Z\} = \{D', 0, D'\}$ given in row 9.

   5.5. The intersections with row 8 yields row 10.

6. Since we reached the primary output, it is time to justify the signals on H, Y, U, and W.

7. To justify H, we place the cube $\{F, Y, H\} = \{0, x, 0\}$ and intersect to obtain row 10. We run into a conflict since we cannot force F, which carries the fault, to 0. Here we reach an inconsistency and we need to select another cube, if there is one, to justify H. We then select another cube $\{x, 0, 0, \}$ and intersect with row 10.

Since we justified successfully all values that we encountered as we were propagating, the set of signals on the primary inputs forms the test pattern $ABUWY = 00010$. Notice also that every time we construct a cube, we need to intersect with the preceding one. We can do the two steps and represent them on the same row. Table 6.3 can thus be reduced to the form shown in Table 6.4. For subsequent examples we use this more compact representation.

TABLE 6.4   Compact Version of Table 6.3

1	Operation	Gate	A	B	X	Y	W	U	F	G	H	Z
2	Initialization		x	x	x	x	x	x	x	x	x	x
3	PDFC	G1	0	0	D	x	x	x	x	x	x	x
4	D-frontier	G2	0	0	D	x	1	x	D	x	x	x
5	D-frontier	G3	0	0	D	x	1	0	D	D'	x	x
6	D-frontier	G5	0	0	D	x	1	0	D	D'	0	D'
7	Justification	$H = 0$	0	0	D	x	1	0	ψ	D'	0	D'
8	Justification	$H = 0$	0	0	D	0	1	0	D	D'	0	D'

TABLE 6.5   D-Algorithm for $Y/1$ of the Circuit S1 in Fig. 6.1

1	Operation	Gate	A	B	X	Y	W	U	F	G	H	Z
2	Initialization		x	x	x	x	x	x	x	x	x	x
3	PDCF	Y	x	x	x	D'	x	x	x	x	x	x
4	D frontier	G4	x	x	x	D'	x	x	1	x	D'	x
5	Implication	$F = 1$	x	x	x	D'	x	x	1	0	D'	x
6	D frontier	G5	x	x	x	D'	x	x	1	0	D'	D'
7	Justification	$F = 1$	x	x	1	D'	1	x	1	0	D	D'
8	Justification	$X = 1$	**0**	**0**	**1**	D'	**1**	x	**1**	**0**	D	D'

6.3.2   Case of a Primary Input

The stuck-at-1 fault on Y is selected to illustrate application of the algorithm on a primary input. The results of the process are shown in Table 6.5. For the fault $Y/1$, D' is placed on this line as shown in row 3 of the table. This value is propagated through gate G4 and the corresponding D = cube would then be $\{Y, F, H\} = \{D', I, D'\}$. The implication of $F = 1$ is indicated in the fifth row of the table and results in $G = 0$. The propagation is then through G5: $\{H, G, Z\} = \{D', 0, D'\}$. Since $G = 0$, we need only justify $F = 1$, which requires that $W = X = 1$ and consequently, that $A = B = 0$. The test pattern, $ABUWY = 00x10$, is indicated in the table in boldface. There are therefore two patterns to detect this fault.

6.3.3   Case of a Primary Output

Consider the fault $Z/0$ in circuit S1 of Fig. 6.2. It is directly observable and no sensitization is needed, only justification. First, we place a D on this line, but we have several choices for PDCF. It is sufficient to have one of the gate inputs be equal to 1. If we select $G = 1$, H can remain indeterminate and it does not have to be justified. The PDCF is $\{U, H, Z\} = \{1, x, D\}$ and shown in row 2 in Table 6.6. To justify $G = 1$, we need both inputs of G3 to be 0. Next, we justify $F = 0$. Here again we have choices: $W = 0$, $X = 0$, or both. Since W is a primary input, it is better to select W, which results in a test pattern: $ABUWY$

TABLE 6.6   D-Algorithm for $Z/0$ of the Circuit S1 in Fig. 6.1

1	Operation	Gate	A	B	X	Y	W	U	F	G	H	Z
2	PDCF	G5	x	x	x	x	x	x	x	1	x	D
3	Justification	G	x	x	x	x	x	0	0	1	x	D
4	Justification	F	x	x	x	x	0	0	0	1	x	D
5	Justification	F	x	x	0	x	x	0	0	1	x	D
6	Justification	X	x	**1**	0	x	x	**0**	0	1	x	D

= *xx*0*x*. If, instead, *W* = 0, we have *X* = 0; this value then has to be justified also as shown in rows 5 and 6 or 7. The corresponding patterns would then be *ABUWY* = *x*10*xx* and 1*x*0*xx*, respectively. The two patterns are shown in rows 4 and 6 of Table 6.6. Thus few test patterns are available to detect this fault. If you consider the three patterns and you substitute 0 and 1 for *x*, you find that there are 14 patterns. These patterns indicate that it is sufficient to specify two of the five inputs to detect the fault. The other three inputs are unspecified, don't care values. It is advantageous to have some of the inputs uncommitted because when patterns have been generated for all faults, they can be merged in a shorter test. Compacting the test set length by taking advantage of the don't care values is not a trivial matter, and there are static and dynamic methods to perform the compaction.

6.3.4   Alternative Strategies

In the examples of Section 6.3.3, we consistently propagated to a primary output, stepping through several gate levels in the circuit. This was then followed by justification by stepping backward over the same levels. Whenever we encountered an inconsistency, we backtracked and selected the other *D*-frontier, or another alternative for justification. To accelerate the process, it is possible to follow another strategy. For example, it is possible to propagate the frontier over only one level of logic (one gate), then immediately justify the lines backward for one level (one or several gates, depending on the fanin). In this fashion, when inconsistency is encountered, it is handled without having to go through many levels of propagation and justification.

6.4   CRITICAL PATH

This algorithm traces a path from a primary output to primary inputs [Abramovici 1984]. In this process, segments of the path are critical. A *critical line* is such that if its logic value is changed, it will change the logic value of the corresponding output. For example, if a line of the path has a value 1 and it changes to 0 because of a stuck at 0 on the line, this fault will be observed at the output. Stuck-at faults of all such critical segments are detected by the same pattern.

The process is similar to justification; it consists of stepping backward on the circuit, one gate at a time. It would help to know how to determine the critical path for elementary gates. Figure 6.5*a* shows the critical values for simple logic gates: AND, OR, and XOR. In all cases the critical values are indicated in bold. For an AND gate, a 1 critical at the output requires that all inputs be 1. If any one input changes to 0, this will be observed on the output of the gate. Thus all inputs are 1 critical. For a 0 critical at the output of the gate, only one of the inputs at a time is set to 0, with all other inputs set to 1. If this input is changed to 1, it will be observable at the output. But if any of the other inputs is changed to 0, the output is not affected.

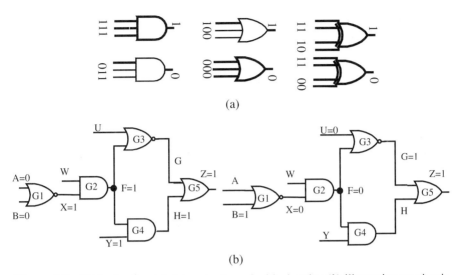

Figure 6.5   Critical paths: (*a*) determination of critical paths; (*b*) illustration on circuit S1 in Fig. 6.2; (*c*) another critical path for S1.

For an OR gate, the results are, of course, the dual of the AND gate. A 1 critical on the output of an XOR gate implies an odd number of inputs with 1, and the other inputs are 0. A change on *any of the inputs* will change the output to 0. Similarly, a 0 critical on the output requires an even number of inputs with 1. Here zero is also considered as an even number. To determine a critical path, we follow four steps:

1. Select a primary output and define 1 (or 0) as the critical value.
2. Justify the critical value to the input of the gate and determine the critical inputs.
3. Repeat step 2 until all primary inputs are determined.
4. If the justification is inconsistent, backtrack and repeat steps 2 and 3.

We apply this algorithm to circuit S1 in Fig. 6.2 and summarize the results in Table 6.7. The headings of the table are the same as those used for the D-algorithm. Let us start with $Z = 1$. For this, it is necessary but also sufficient that either G or H be 1; let's specify that $H = 1$ and $G = 0$. In this fashion, H is critical since, if it is changed to 0, Z also changes and a fault on H is observed on Z. Had we set both H and G equal to 1, neither would be critical since changing one of the them to 0 would not change the output. Since G4 is an AND gate, $H = 1$ implies that $Y = F = 1$, and they are all critical. The justification of G is straightforward since it is not critical, but since $F = 1$, there is need to assign 1 to U. Justifying $F = 1$ results in $W = X = 1$ (row 5 in Table 6.7). If $X = 1$, then $A = B = 0$. All critical values in the table are indicated in

TABLE 6.7   Critical Path for $Z = 1$ in Circuit S1

1	Gate	Z	B	X	Y	W	U	F	G	H	Z
2	G5	x	x	x	x	x	x	x	0	1	1
3	G4				1			1		1	
4	∩	x	x	x	1	x	x	1	0	1	1
5	G3								0		
6	∩	x	x	x	1	x	0	1	0	1	1
7	G2				1		1	1			
8	∩	x	x	1	1	1	0	1	0	1	1
9	G1	0	0	1							
10	∩	0	0	1	1	1	0	1	0	1	1

bold. The pattern obtained by the steps just described is $ABUWY = 00x11$. The critical values are indicated on the corresponding lines of the schematic in Fig. 6.5b. The faults detected by this pattern are $A/1$, $B/1$, $Y/0$, $X/0$, $F/0$, $W/0$, $H/0$, and $Z/0$.

In addition, we could have selected G instead of H to be critical ($G = 1$ and $H = 0$); then both U and F are critical 0. Justification of F results in a critical 0 for either W or X. Selecting X is more advantageous since it will, in turn, imply either A or B being critical 1. The more critical segments we have, the more faults are detected by the pattern. The critical path in this case is indicated on the circuit in Fig. 6.5c by the values assigned to the corresponding lines. The test pattern is $ABUWY = 1x0xx$. The faults detected would then be $A/0$, $X/1$, $F/1$, $U/1$, $G/0$, and $Z/0$. In justifying F, we selected $X = 0$. If, instead, we selected $W = 0$ and $X = 1$, we would have included $W/1$ instead of $X/1$ in the list of faults detected and the pattern would have been $ABUWY = 0000x$.

6.5 BACKTRACKING AND RECONVERGING FANOUT

Both algorithms discussed so far are inefficient when backtracking in a class of circuits that uses reconvergent fanout with XOR gates. We illustrate this with the example shown in Fig. 6.6. Consider the fault $H/0$. The PDCF is $\{A, B, H\} = \{1, 1, D\}$. To propagate this frontier to the output, R, we need, for example, $N = Q = 1$. However, this is not possible since N and Q realize complementary functions. Assume that we are not aware of this fact and we proceed with the justification. We first justify N arbitrarily with $J = 0$ and $K = 1$. Next, we proceed with Q. We can have either $L = M = 1$ or $L = M = 0$, as illustrated in the search tree shown in Fig. 6.7. The implication of the first assignment is that $F = G$ and, consequently, $K = 0$. This is inconsistent with the assumption made earlier in the justification of N. Now we turn to the second case, which requires that F and G are complementary to each other ($K = 1$) as well as to C and E. This implies that $J = 1$, which is inconsistent with the original value of J obtained

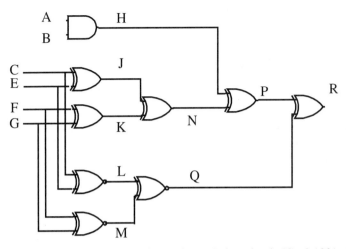

Figure 6.6 Circuit S2: error correction and translation circuit [Goel 1981 © IEEE. Reprinted with permission].

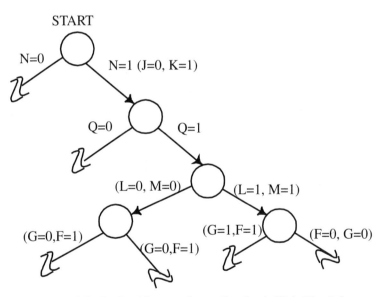

Figure 6.7 D-algorithm search tree for circuit S2 in Fig. 6.6.

when N is justified. Tracing back will take us again to node Q, and now we try to propagate the D-frontier using $Q = 0$ instead and obtaining D' at the output. This excessive backtracking motivated [Goel 1981] to develop a new test pattern generation algorithm that is known to run an order of magnitude faster than the D-algorithm [Agrawal 1988].

6.6 PODEM

The path-oriented decision-making (PODEM) algorithm starts the search for the test pattern at the primary inputs of the circuit. The algorithm is outlined by the flowchart in Fig. 6.8. Starting with an objective, a specific fault to be detected, a search tree is created in which two choices are available, 0 and 1, for the primary inputs. The choice is random. Evaluate the implications of this choice on the subsequent gates to the output. If it furthers the objective—controlling the fault site to the intended value—accept it and select another PI. If an inconsistency occurs, the algorithm backtracks and selects another input combination. The search stops whenever a pattern is generated or no patterns are possible (undetectable fault).

The advantage of PODEM is that it cuts down on the backtracking, as will be demonstrated using the circuit S2 in Fig. 6.6. This is the same circuit that was used to illustrate the difficulty encountered by the D-algorithm with reconvergent fanout. The target fault is $H/0$ and the search tree is shown in Fig. 6.9.

1. Assign x to all inputs.

2. Assigning 0 to the primary input, A, causes $H = 0$; hence this choice does not further our goal and it is discarded. Then $A = 1$.

3. Similarly, $B = 0$ is rejected and $B = 1$ is selected.

4. We proceed with $C = 0$. The implication of this value is not furthering the objective, nor is it blocking it.

5. Thus we select $E = 0$; this then results in $J = 0$ and $L = 1$.

6. Similarly, we select $F = G = 0$, which implies that $K = 0$, $M = 1$, $N = 0$, and $Q = 1$.

7. We can propagate D on P, then D' on the primary output, R.

From the search tree in Fig. 6.8, the primary inputs are then $\{ABCEFG\} = \{110000\}$. Compared to the D-algorithm, the backtracking is definitely reduced. The algorithm was also applied to two other circuits, as illustrated in Figs. 6.10 and 6.11.

6.7 OTHER ALGORITHMS

PODEM has been effective in reducing the occurrences of backtracking, but two new strategies have helped in reducing them even further: FAN [Fujiwara 1983] and SOCRATES [Schultz 1988]. These algorithms do extensive analysis of the circuit connectivities before backtracking. Also, the search is aided by testability analysis.

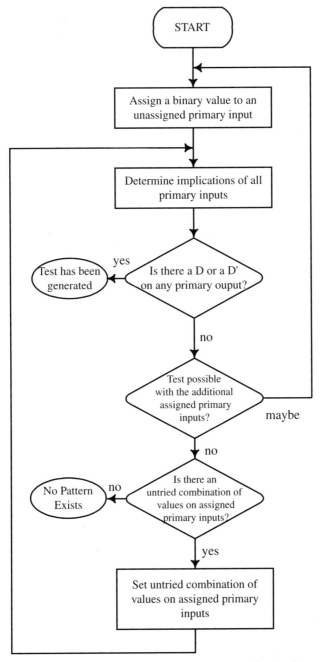

Figure 6.8 PODEM algorithm [Goel 1981 © IEEE. Reprinted with permission].

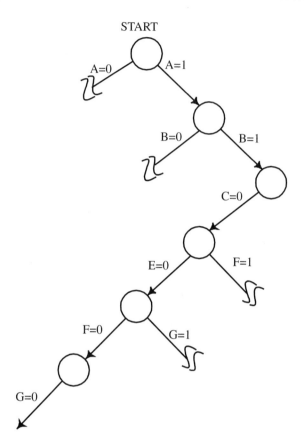

Test has been generated

Figure 6.9   PODEM search tree for circuit S2 in Fig. 6.6.

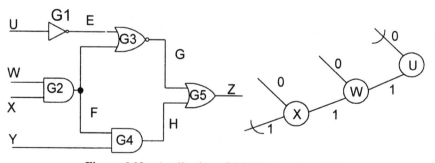

Figure 6.10   Application of PODEM to circuit S3.

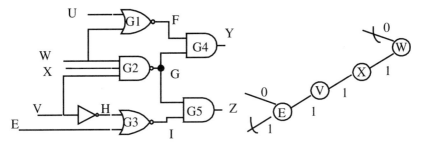

Figure 6.11   Illustrating the use of PODEM on circuit S4.

6.7.1 FAN Algorithm

The fan-out-oriented test generation (FAN) algorithm remedies the exhaustive approach of PODEM by pruning from the search tree any branching that would not yield a solution. FAN uses the following strategies [Fujiwara 1986]:

1. In each step of the algorithm, determine as many signal values as possible that can be implied uniquely.
2. Assign a faulty signal D or D' that is uniquely determined or implied by the fault in question.
3. When the D-frontier consists of a single gate, apply a unique sensitization.
4. Stop the backtrack at a headline, and postpone the line justification for the headline to later.
5. Multiple backtracking is more efficient than backtracking along a single path.
6. In the multiple backtrack, if an objective at a fanout point, p, has a contradictory requirement, stop the backtrack so as to assign a binary value to the fan-out point.

Using these strategies, FAN minimized the backtracks and reduced the test generation time as illustrated in a study that compared FAN to other algorithms [Fujiwara 1983]. Only data relevant to FAN and PODEM are presented in Table 6.8 for five benchmark circuits. The last two columns are of particular interest because they indicate the effectiveness of the algorithm. These columns give the number of faults that were not detected after 1000 backtracks; they are referred to as the *aborted faults*.

6.7.2 SOCRATES

The SOCRATES (structure-oriented cost-reducing automatic test pattern generation) algorithm is based on FAN, but uses an improved implication procedure and a unique sensitization procedure. In addition, this algorithm reduces

TABLE 6.8   Comparison of PODEM and FAN Algorithms

Circuit	Computing Time		Average Backtracks		Faults Aborted (%)	
	PODEM	FAN	PODEM	FAN	PODEM	FAN
1	1.3	1	4.9	1.2	0.32	0.37
2	3.6	1	42.3	15.2	2.26	3.13
3	5.6	1	61.9	0.6	2.42	4.00
4	1.9	1	5.0	0.2	0.99	1.10
5	4.8	1	53.0	23.2	0.82	1.02

Source: Data from [Fujiwara 1983].

backtracking by early recognition of conflicts, and it utilizes several heuristics, which are applied at different states of the test generation process.

6.8   TESTING SEQUENTIAL CIRCUITS

There are several reasons why test pattern generation for sequential circuits is more difficult than for combinational circuits. [Hennie 1974], [Breuer 1976], and [Miczo 1983], among others, have addressed these reasons, which we state briefly here.

- Most real circuits are sequential in nature.
- The major difficulty in testing sequential circuits is that the output response of the circuit depends not only on the input patterns, but also on the internal states of the circuit. Notice that this sequential circuit may be synchronous or asynchronous and the states are not always directly observable.
- There is a need to drive the circuit first in a known state before applying the test pattern that will sensitize the fault to a primary output. Initializing the circuit might itself require more than one pattern. Therefore, the order in which the test patterns are applied is critical to fault detection. Several techniques have been used to initialize the circuit to a known state before testing. The most forward approach is the use of master set/reset. Another alternative is the use of a synchronizing sequence that forces the circuit in a final state regardless of its initial state [Hennie 1968]. However, not all circuits have synchronizing sequences. Also, under faulty conditions, there is no guarantee that either initialization technique will work.
- Timing is another factor that complicates test pattern generation. For proper operation of sequential circuits, adequate consideration needs to be given to setup and hold times. If timing restrictions are not followed, the circuit may behave in an unpredictable manner. In addition, even when the test is applied according to the timing specification, it is possible that the delays of the different components of the circuit produce *hazard* and race conditions.

Hazards can exist in both combinational and sequential circuits because of component delays (logic hazard) or as an inherent function of the circuit logic itself (functional hazard). Whereas for a combinational circuit, the signal will eventually end in its correct value before strobing, this may not be the case for sequential circuits. A hazard condition may put the circuit in a faulty state. For example, in a master-slave flip-flop, the slave may capture the static hazard on the master. This phenomenon, known as *essential hazard* [McCluskey 1986], can prevent detection of the fault for which the test was intended.

In the remainder of this section, we examine sequential test pattern generation using a functional approach and a fault-oriented deterministic method that has been adapted to sequential circuits.

6.8.1 Functional Testing

It is possible to test a sequential circuit by applying a specific input sequence that exercises its function. However, such a test does not guarantee a complete stuck-at fault testing of the circuit unless simulation is used. Another approach is to verify the operations of the circuit in accordance with its state table [Moore 1956, Hennie 1968]. This method requires the enumeration of all possible fault machines and yields an experiment that not only detects faults, but also identifies which fault has occurred. The methodology is to sequential circuits what exhaustive testing is to combinational circuits. Although it is not practical for large FSMs, it gives an insight into the validation of FSMs, as we find out next.

6.8.1.1 Checking Experiment. A fault will transform an FSM, M, into some other state, M_f. Assuming that the number of states has not increased, the checking sequence will distinguish M from M_f. In general, only the primary inputs can be controlled and only primary outputs can be observed. We have seen that for a completely specified FSM, the mathematical formulation gives a unique next state and an output for each transition. The states are not observable directly, but they can be recognized from the output sequences.

To verify the correctness of an FSM, we apply a specific input sequence and we observe the output sequence. For the fault-free machine, the sequence initializes M into a known state, then forces it to pass through all possible transitions. The input sequence consists of three main sequences: a *synchronizing* or *homing sequence*, a *distinguishing sequence*, and a *transition sequence*.

- The synchronizing sequence (SS) places the M in a known state.
- The homing sequence (HS) places M in a known state that is identifiable from the output response sequence.
- The distinguishing sequence (DS) produces an output sequence that defines uniquely the initial state of M at which the sequence was applied.

TABLE 6.9   State Table for FSM *M*

Present State	Next State	
	$I = 0$	$I = 1$
A	*C*, 1	*B*, 0
B	*C*, 0	*B*, 1
C	*D*, 1	*C*, 3
D	*A*, 1	*C*, 0

- The transition sequence (TS) indicates the transition from one state to any of the other states.

Let us illustrate the development of these sequences using machine *M*, as defined algebraically in Section 3.2.3 and the state table form shown in Table 3.3 and reproduced in Table 6.9. Finding the transition sequence is very easily developed from either form. In passing from state *A* to any of the other states, *B*, *C*, and *D*, we need to apply the sequences 1, 0, and 01 as indicated on the binary tree in Fig. 6.12. In this representation, the line segments labeled 0 and 1 represent the transitions from one state to another for input 0 and 1. We identify the transition from *A* to any of the other states by the unique output response shown in the table.

To develop the other sequences, we use a *successor tree*. Initially, *M* may be in any of the four states {*ABCD*}. We are not certain of the exact state and we call this collection of states the initial *uncertainty* of the machine [Kohavi 1978]. If we apply to *M* input 0, the next states will be either (*C*) with output 0 or any of the states (*ACD*) with output 1. (*C*) and (*ACD*) are the 0-*successors* of the initial uncertainty {*ABCD*} and they form an *uncertainty vector*. Similarly, we can generate the 1-successors by applying 1 to {*ABCD*}. We then obtain the 1-successors, which are (*BC*) with 0 and (*BC*) with 1. To each uncertainty, we

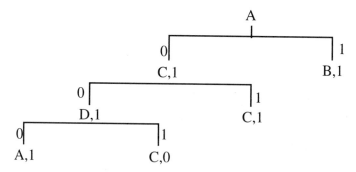

Figure 6.12   Transition sequence for FSM *M*.

Level

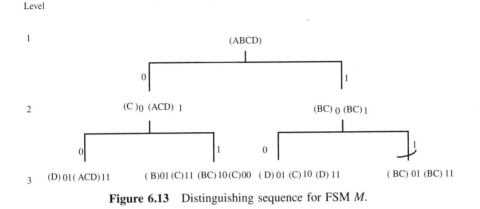

Figure 6.13 Distinguishing sequence for FSM *M*.

again apply inputs 0 and 1 and generate higher-level uncertainties. The formation of these uncertainties is shown in Fig. 6.13. The subscripts indicate the output sequence for the machine transitions. The process of building the tree is terminated unsuccessfully, as we obtain an uncertainty that has already been obtained. In this example the rightmost branch with 11 terminates in $(BC)_{11}$ and need not be explored any further. We elaborate later on successful termination conditions, which depend on the sequence sought, homing or distinguishing. The vectors that have components with single states are called *trivial uncertainties*; those that consist of single states and identical states are called *homogeneous uncertainties*. For example, $(BB)CD$ is a homogeneous uncertainty, whereas $(A)(B)(D)$ is a trivial uncertainty.

For the *homing sequence*, the process is terminated successfully when a trivial or homogeneous vector is obtained because at least one single state is identified. The tree in Fig. 6.13 shows four different homing sequences, which are listed in Table 6.10*a*. The first sequence is of length 1 and the others are of length 2. Any of these sequences will guarantee the final state with a unique output sequence, but it does not identify the initial state. Notice that the homing sequence does not distinguish the initial states.

Once we initialize the FSM, we can apply the *distinguishing sequence* to verify the correctness of the machine. For the distinguishing sequence, we follow the same procedure as for the homing sequence except that we terminate the process when we obtain a trivial uncertainty vector. If we reach a homogeneous uncertainty, we abandon the search along this path. For machine *M*, there is one distinguishing sequence, which is 10, one of the homing sequences. A distinguishing sequence is a homing sequence, but not vice versa. The response of the machine to this *DS* sequence is summarized in Table 6.10*b*. The output strings distinguish the initial states uniquely.

The initialization of the FSM may also be accomplished using the synchronizing sequence. This sequence is developed using a successor tree; however, this time, there is no need to separate the uncertainties by their output. A path

TABLE 6.10 (*a*) **Homing Sequences and** (*b*) **Application Distinguishing Sequences 10**

HS	Final State	Output	Initial State	Final State	Output Sequence
0	C	0	A	C	00
00	D	01	B	C	10
01	B	01	C	D	11
	C	11	D	D	01
10	C	00			
	D	01			

| (*a*) | (*b*) |

in the tree is terminated when a repeated uncertainty is obtained and terminates successfully when a trivial uncertainty with a singleton state is obtained. Usually, this SS is used to initialize the FSM. However, the existence of such a sequence cannot be guaranteed. The successor tree, which yields the synchronizing sequence 101, is shown in Fig. 6.14; it will put *M* in state *C*. This sequence is 3 bits long, whereas the homing sequences are at most 2 bits long. In general, it is more beneficial to use shorter sequences.

Next, we use the various sequences generated for *M* to verify the correctness of the machine:

1. Initialize the *M* using an SS or HS to state S_0.
2. Apply a DS to verify this state.
3. Let the final state obtained in the preceding step be S_i.
4. Apply a DS to verify S_I.
5. Repeat steps 3 and 4 until all states have been identified.
6. If in this process, a state S_j is not reachable, apply a TS to get you to this state and apply a DS to verify it.

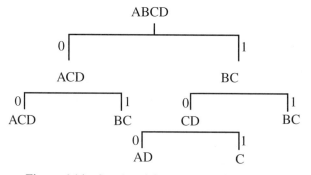

Figure 6.14 Synchronizing sequence for FSM *M*.

7. Use the TS to verify all transitions except for those already verified in step 5.

Applying this process on FSM *M*, we obtain the checking sequence given in Table 6.11. This table is organized in time steps. For each step, an input is applied on the state in the same column. The next state and the corresponding output are then entered in the next time-step column. Initially, at time 0, the FSM is represented by its total uncertainty {*ABCD*} and the output is not defined. Then 0 is applied on the input to home the machine to state *C*. Thus *C* and its corresponding output, 0, are entered for the state and the output in time step 1. Now, to verify state *C*, we apply the DS (10) in time steps 1 and 2 and observe the output sequence {11}, which according to Table 6.10*b*, is the signature of state *C* for the distinguishing sequence used.

The interest in the use of the checking experiment has continued through the 1980s when it has been recommended in conjunction with built-in self-test (BIST). Since the checking sequence is usually very long, schemes have been proposed to compact the circuit's response to the sequence. We describe such schemes in Chapter 11.

6.8.1.2 *Exhaustive Testing.*
As for the combinational parts, it is possible to use exhaustive and pseudoexhaustive testing for sequential circuits. The advantages of these types of test pattern generation are simplicity and low cost. However, for an *S*-state FSM with *N* primary inputs, an exhaustive test is 2^{S+N} long. A pseudoexhaustive approach in which exhaustive patterns are created for partitions of the circuits was recommended to reduce the test length and proved to be successful for combinational circuits [McCluskey 1981]. This approach was also applied to sequential circuits that include DFT constructs [Wunderlich 1989]. Because exhaustive test patterns do not use the functionality of the circuits, they are extremely long.

6.8.2 Deterministic Test Pattern Generation

Very few algorithmic techniques exist for testing sequential circuits. Some of the methods have been devised for asynchronous circuits but can be adjusted for use with synchronous circuits. For the latter type, we use the Huffman model for FSM that has been described in Chapter 3 [Huffman 1954]. The circuit consists of a combinational part and a set of flip-flips as illustrated in Fig. 6.15*a*. For simplicity, only D flip-flops are used since it is not difficult in any case to transform other flip-flops to this type.

[Seshu 1965] developed the first algorithm for testing sequential circuits. This technique, which predates the D-algorithm, does not require state tables as it is simulation-based. It starts with a vector and simulates both the good and faulty machines. Then the input sequence is changed one bit at a time, and the effectiveness of the changes is evaluated to select the change that detects

TABLE 6.11 Checking Sequence for FSM *M*

Time	0	1	2	3	4	5	6	7	8	9	10	11	12	13	14	15	16	17	18	19	20	21	22	23	24
Input	0	1	0	1	0	0	1	0	0	0	1	1	0	1	1	0	0	0	1	1	1	0			
State	ABCD	C	C	D	C	D	A	B	C	D	A	B	B	C	C	D	C	D	A	B	A	B	B	C	
Output	—	0	1	1	0	1	1	0	0	1	1	0	1	0	1	1	0	1	1						

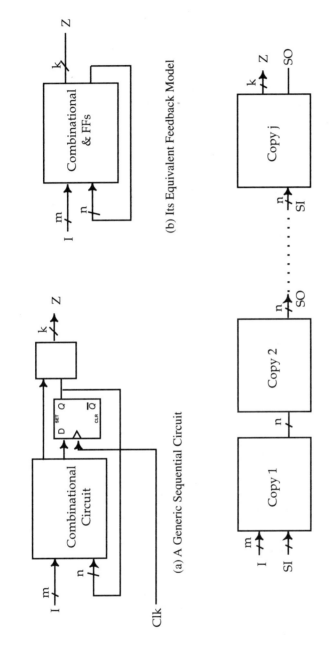

(a) A Generic Sequential Circuit

(b) Its Equivalent Feedback Model

(c) Time Frame Based Model

Figure 6.15 Sequential test pattern generation: iterative logic array.

155

most of the faults. The effectiveness of the patterns was determined according to four heuristics that are described concisely by [Miczo 1986].

A heuristic bearing resemblance to Seshu's consists of cutting all feedback loops of a sequential circuit and thus transforming it into a combinational circuit [Kubo 1968]. It is assumed that all flip-flops in the circuit are fault-free. Test patterns are generated for the transformed circuit using the D-algorithm. The technique is applied only to synchronous machines and, in this manner, race and hazard problems are avoided. Once a test pattern is generated, it is important to check that the feedback lines have the same value at the inputs and the outputs. If this is the case, a test pattern has been generated; otherwise, another pattern needs to be generated.

[Putzolu 1971] proposed the iterative test generation (ITG) heuristic. This technique extends and improves Kubo's work. It was intended for asynchronous circuits, but it is also applicable to synchronous circuits. In addition, it does not require that the storage devices be fault-free. However, as a heuristic, it does not guarantee finding a test for the circuit. The ITG algorithm consists of the following steps:

1. Cut the feedback loops according to a heuristic that assigns weights to the different feedback lines and then determines those lines that would be cut.
2. Use the D-algorithm to generate patterns.
3. Run a simulation to check the effectiveness of the test sequence to detect the fault.

The cutting strategy is such that a minimal number of feedback lines are selected to make the circuit acyclic. Each feedback line is then represented by input and output to the circuit. They are labeled the pseudo-inputs (SIs) and pseudo-outputs (SOs). The resulting circuit shows the behavior of the original circuit at a given moment in time. To examine the behavior in subsequent moments, different copies of the original circuit are produced as illustrated in Fig. 6.15c. The copies are snapshots at different time intervals. If the SOs of one copy are connected to the corresponding SIs of the next copy by a delay, the modified circuit should behave as the original circuit. The D-algorithm is applied to the modified circuit with the following restrictions:

- The fault is detected when it is propagated to a primary output.
- The test patterns cannot be allowed to depend on any particular values of the SI. That is, the intersection of x and any other value is always ϕ on the SI.

We illustrate the method by generating the pattern to detect the SA1 fault on the output of gate 7 (G7) of the circuit $S5$ shown in Fig. 6.16. This circuit is the example used in [Putzolu 1971] and in which all NOR gates with their inputs

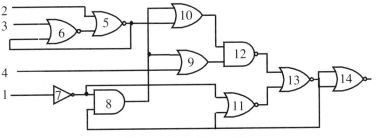

Figure 6.16   Circuit S5.

tied together are replaced by inverters. Two time frames are used as illustrated in Fig. 6.17. The second frame is used to place the PDCF and propagate the fault to a primary output. A justification that requires nodes with feedback to have other values than those assigned by the propagation are then placed on the previous frame, frame 1. In explaining the test generation, we refer to a line that is the output of a gate by the number of this gate.

The PDCF is propagated through G11, then G13, and finally to the primary output, G14, as indicated on frame 2 of Fig. 6.17. This propagation requires that 0 be placed on nodes 12 and 13. This terminates the D-drive. We then move to justification on frame 1. A 0 on 13 is obtained by placing 1 on 11 or on 12. A 1 on 11 implies a zero on PI13. This is an inconsistency per the second restriction stated above. We attempt next to place 1 on 12, which implies a 0 on 9 or on 10. Any of the two possibilities causes the output of G8 to be 0. However, a 0 on 8 requires that a 0 be on either PI13 or on 7. Both assignments cause inconsistencies since PI13 is pseudoinput and 7 is stuck at 1 and should be held high. Therefore, we need to backtrack and find another D-drive.

Another sensitization path for the fault to the primary output 14 is through G8, G9 (or G10), G12, and G13, as indicated on frame 2 of Fig. 6.17. This propagation requires 1 on the other input of G8, 0 on 4, 1 on 10, and 0 on 11. We move next to frame 1 and justify these values. To get a 1 on 13, we need 0 on 11 and on 12. To realize this value on 11, a 1 is placed on 7, which implies a 0 on 1; for 12, on the other hand, both 9 and 10 should be 1. This requirement implies that a 1 be put on 4 and 5. Consequently, we need a 0 on 2 and a 1 on 3. Thus the justification of 13, 11, and 4 is complete and the pattern is {0011}. The implication of this pattern and, in particular, having 1 on 5, is placing a 0 on 6 in frame 2. Finally, we justify 1 on 10 by placing 1 on 5. But since the value of 6 is 1, we need to place 0 on 2. The second pattern becomes {10x0}. The test sequence is then extracted from the values assigned to the primary inputs in the two frames: 0011 and 10x0.

The approach is also applicable to synchronous circuits, as we illustrate on the FSM shown in Fig. 6.18. The path from the fault, $c/1$, to the primary output, Z, is through *gates* G5, G3, F1, and G4. This requires clocking once while keeping e and f equal to 1. However, we first need to provoke the fault. There-

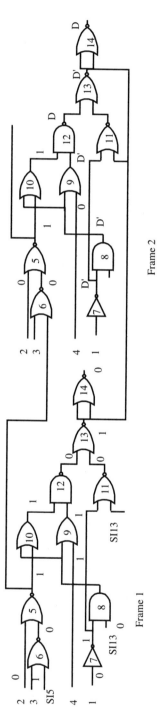

Figure 6.17 Time frames for circuit S5.

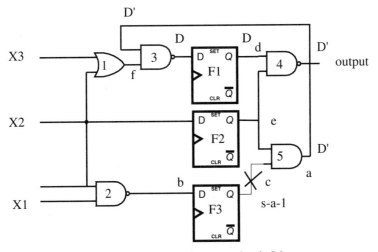

Figure 6.18   Synchronous circuit S6.

fore, we place a 1 on *b* and *c*, and clock once. We clock a second time to propagate the fault. This is equivalent to two time frames, as illustrated in Fig. 6.19. The test sequence is then $\{x11, x1x\}$. In this case, it is sufficient to use the pattern and clock twice.

With the advent of scan-path design in the early 1970s, the interest in sequential test-patterns generation has slowed down since this design technique has reduced the problem of testing combinational circuits. This is discussed in Chapter 9. Only few algorithms were proposed [Mallela 1985, Marlett 1986]. As the use of scan design became widespread and synthesis for testability (Chapter 14) started to become popular, the emphasis was on modifying sequential circuits to make them easily testable.

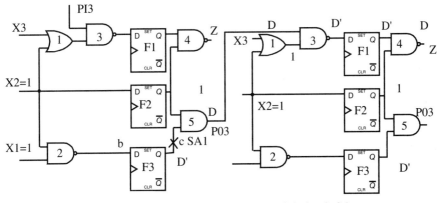

Figure 6.19   Time frames of sequential circuit S6.

REFERENCES

Abramovici, M., et al. (1984), Critical path tracing: an alternative to fault simulation, *IEEE Des. Test Comput.*, Vol. 1, No. 1, pp. 83–93.

Agrawal, V. D., and S. C. Seth (1988), *Test Generation for VLSI Chips*, IEEE Computer Society Press, Los Alamitos, CA.

Breuer, M. A., and A. D. Friedman (1976), *Diagnosis and Reliable Design of Digital Systems*, Computer Science Press, New York.

Fujiwara, H., and S. Toida (1982), The complexity of fault detection for combinational logic circuits, *IEEE Trans. Comput.*, Vol. C-31, No. 6, pp. 555–560.

Fujiwara, H., and T. Shimono (1983), On the acceleration of test generation algorithms, *IEEE Trans. Comput.*, Vol. C-32, No. 12, pp. 1137–1144.

Fujiwara, H. (1986), *Logic Testing and Design for Testability*, MIT Press, Cambridge, MA.

Goel, P. (1980), Test generation cost analysis and projections, *Proc. 17th Design Automation Conference*, pp. 77–84.

Goel, P. (1981), An implicit enumeration algorithm to generate tests for combinational logic circuits, *IEEE Trans. Comput.*, Vol. C-30, No. 3, pp. 215–222.

Hennie, F. C. (1968), *Finite-State Models for Logic Machines*, Wiley, New York.

Hennie, F. C. (1974), Fault detection experiments for sequential circuits, *Proc. 5th Symposium on Switching Theory and Logical Design*, pp. 95–110.

Hsiao, M. Y., and D. K. Chia (1971), Boolean difference for fault detection in asynchronous sequential machines, *IEEE Trans. Comput.*, Vol. C-20, No. 11, pp. 1356–1361.

Huffman, D. A. (1954), The synthesis of sequential circuits, *J. Franklin Inst.*, Vol. 257, pp. 161–190, and 273–303.

Ibarra, G. H., and S. K. Sahni (1975), Polynomially complete fault detection problems, *IEEE Trans. Comput.*, Vol. C-24, No. 3, pp. 242–249.

Kohavi, Z. (1978), *Switching and Finite Automata Theory*, 2nd ed., McGraw-Hill, New York.

Kubo, H. (1968), A procedure for generating test sequences to detect sequential circuit failures, *NEC Res. Dev.* Vol. 12, No. 1, pp. 69–78.

Larrabee, T. (1989), Efficient generation of test patterns using Boolean difference, *Proc. IEEE International Test Conference*, pp. 795–801.

Mallela, S., and S. Wu (1985), A sequential circuit test generation system, *Proc. IEEE International Test Conference*, pp. 57–61.

Marlett, R. (1986), An effective test generation system for sequential circuits, *Proc. 23rd Design Automation Conf.*, pp. 250–256.

McCluskey, E. J., and S. Bozorgui-Nesbat (1981), Design for autonomous test, *IEEE Trans. Comput.*, Vol. C-30, No. 11, pp. 860–875.

McCluskey, E. J. (1986), *Principles of Logic Design*, Prentice Hall, Upper Saddle River, NJ.

Miczo, A. (1983), The sequential ATPG: a theoretical limit, *Proc. IEEE International Test Conference*, pp. 143–147.

Miczo, A. (1986), *Digital Logic Testing and Simulation*, Wiley, New York.

Moore, E. F. (1956), Gedanken experiments on sequential machines, in *Automata Studies*, C. E. Shannon and J. McCarthy (eds.), Princeton University Press, Princeton, NJ.

Putzolu, G., and J. P. Roth (1971), A heuristic algorithm for the testing of asynchronous circuits, *IEEE Trans. Comput.*, Vol. C-20, No. 6, pp. 639–647.

Roth, J. P. (1966), Diagnosis of automata failures: a calculus and a method, *IBM J. Res. Dev.*, Vol. 10, No. , pp. 278–281.

Roth, J. P., W. G. Bouricius, and P. R. Schneider (1967), Programmed algorithms to compute tests to detect and distinguish between failures in logic circuits, *IEEE Trans. Electron. Comput.*, Vol. EC-16, No. 10, pp. 567–580.

Schultz, M. H., et al. (1988), SOCRATES: a highly efficient automatic test pattern generation system, *IEEE Trans. Comput.-Aided Des.*, Vol. CAD-8, No. 1, pp. 126–137.

Sellers, F. F., et al. (1968), Analyzing errors with Boolean difference, *IEEE Trans. Comput.*, Vol. C-17, No. 7, pp. 676–683.

Seshu, S. (1965), On an improved diagnosis program, *IEEE Trans. Electron. Comput.*, Vol. EC-14, No. 2, pp. 76–79.

Wunderlich, H.-J., and S. Hellebrand (1989), The pseudo-exhaustive test of sequential circuits, *Proc. IEEE International Test Conference*, pp. 19–27.

PROBLEMS

6.1. Collapse the faults on the circuit in Fig. P6.1. Then use the D-algorithm to generate test patterns for the circuit. If possible, use a fault simulator to assess the fault coverage of the test set.

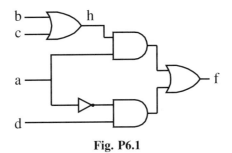

Fig. P6.1

6.2. Use Boolean difference to detect faults on the primary inputs of the circuit in Fig. P6.2 and compare the results with test patterns generated for the same faults in Problem 6.1.

6.3. Use PODEM algorithm to generate test patterns for the circuit in Fig. 5.2.

6.4. Determine test patterns that detect the SA0 faults on lines *f* and *b* of circuit S6 in Fig. 6.18.

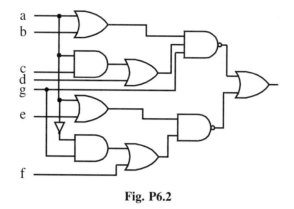

Fig. P6.2

6.5. For the state machine listed in Table P6.5, determine **(a)** a synchronizing sequence, if any; **(b)** a homing sequence; and **(c)** a distinguishing sequence. **(d)** Develop a checking sequence.

TABLE P6.5

Present State	Next State	
	$I = 0$	$I = 1$
A	C, 0	A, 1
B	A, 1	B, 0
C	D, 0	B, 1
D	B, 1	D, 0

6.6. Use the state table that you developed for Problem 3.4 and determine its synchronizing sequence.

7

CURRENT TESTING

7.1 INTRODUCTION

Functional testing detects faults on the logic level. The logic values are determined by measuring the voltage on the primary outputs. Voltage testing is effective in detecting stuck-at faults, particularly in bipolar technology. However, this type of testing does not guarantee detection of other faults beyond stuck-at faults in CMOS technology, which is the most popular technology at present. Another issue with voltage testing is the complexity of test pattern generation for large circuits. As we will see in this chapter, test sets for current testing are typically shorter than stuck-at fault tests. In addition, current testing is also effective in detecting stuck-at faults.

The concepts of fully complementary MOS (CMOS) have been known since the early 1960s [Wanlass 1961]. The interest in this technology was mainly because of its low power characteristics. The circuit draws current only during switching. Under static conditions, the quiescent current is due to leakage and it is too small to cause thermal problems. The first device produced consumed just a few nanowatts of standby power [Riezeman 1991]. RCA announced successful fabrication of a CMOS IC in 1968, and the first CMOS microprocessor was announced in 1974 [Novce 1991]. From the onset, testing of CMOS circuits included monitoring of the quiescent current as part of parametric testing. Strong correlation has been found between this parametric testing and functional testing [Daniels 1990]. Delays in the circuit [Nelson 1975] have resulted in promoting current testing called I_{DDQ} testing. This is the current drawn from the power supply, V_{dd}, and Q stands for quiescent. This current was shown to be elevated for various CMOS defects, including bridging faults and SAFs [Levi 1981, Malaiya

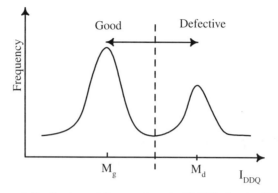

Figure 7.1 General distribution of I_{DDQ} values in CMOS circuits [Williams 1996 ©
IEEE. Reprinted with permission].

1982]. However, only in the late 1980s did current testing became very popular
[Hawkins 1985] and become integrated in CAD tools [Fritzemeier 1991]. At
present it is also used in conjunction with DFT constructs such as scan-path
testing, built-in self-test (BIST), and memory testing [Soden 1992].

I_{DDQ} testing relies on the fact that CMOS devices have very low leakage
current when the network is in the quiescent state. By monitoring this current,
any manufacturing process that causes excessive leakage current is detected.
To detect the higher currents due to a defect with an I_{DDQ} test, it is important
that the defective network's quiescent current be significantly higher than that
of the good network. This should take into account the variation in quiescent
current for the good network due to process variation and the inaccuracies in
measuring the quiescent current. In Fig. 7.1 the first normal distribution depicts
the quiescent current of the good product with a mean value denoted by M_g. The
defective parts, which may draw excessive current, have their own distribution,
with higher values for the quiescent current (mean M_d). If the two distributions
are far apart, a quiescent current limit can be selected that demarcates the good
product from the defective product.

Current testing is not, however, a panacea—it has its problems. To measure the current adequately is possible only by using appropriate instrumentation. In addition, for deep-submicron technology, the quiescent current becomes
too high to be differentiated due to an increase in the subthreshold component
[Righter 1996, Williams 1996].

7.2 BASIC CONCEPT

An IC consists of several million transistors forming different logic gates. Any
of these gates can be, without loss of generality, modeled by an inverter. Let
us consider a CMOS inverter and observe the switching currents for fault-free
and faulty conditions. As the input voltage switches from 0 to V_{dd}, the PMOS,

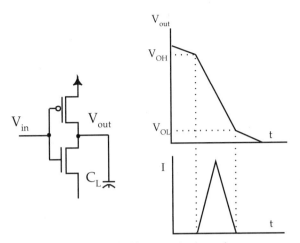

Figure 7.2   Switching current in an inverter.

which is initially on, turns off and the NMOS transistor, which is initially off, turns on. During the switching time t_f, the gate voltage is at values that cause both transistors to be on simultaneously. It is during this and only this period that a current flows between the power and ground rails as approximated in Fig. 7.2. Here we assume that when the transistors are not switching, the quiescent current of the circuit is negligible. In subsequent sections, we take a closer look at the actual value of this current.

A more realistic waveform for the switching voltage and current in a CMOS inverter with minimal dimension in 1.2-μm technology is shown in Fig. 7.3.

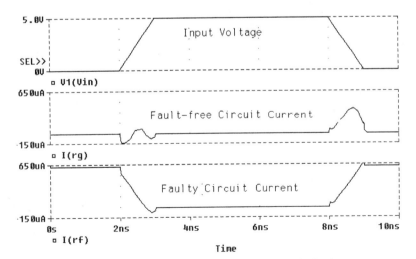

Figure 7.3   SPICE simulation for fault-free and faulty inverter.

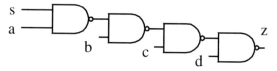

Figure 7.4   NAND tree.

This is the result of PSPICE simulation. The upper curve is the input waveform. It is followed by the corresponding current in a fault-free inverter. If the NMOS transistor is stuck on and the input voltage is high, the output is pulled down to ground through the *n*-channel. This condition masks the fault on the pull-down transistor. However, when $V_{gs} = 0$ ($V_1(V_{in})$ in Fig. 7.3), the PMOS transistor is on and the current will flow steadily between V_{dd} and GND since the NMOS is stuck-on. As shown in Fig. 7.3, this current is noticeably larger than the switching current of the healthy inverter. Notice that it was sufficient to provoke the fault to make it observable. The test required to detect the fault is therefore independent of the position of the gate in the circuit. Current testing does not require propagating the fault to a primary output.

Normally, a circuit will consist of more than one gate switching in response to the pattern applied. The switching may be simultaneous or in succession. For example, in the NAND tree shown in Fig. 7.4, all inputs are kept high except for *S*, which is changed from low to high. Consequently, the output of all gates in the circuit will switch. The switching currents for all gates of the tree are shown in Fig. 7.5. For the first NAND gate, the switching started at T1 = 5 ns

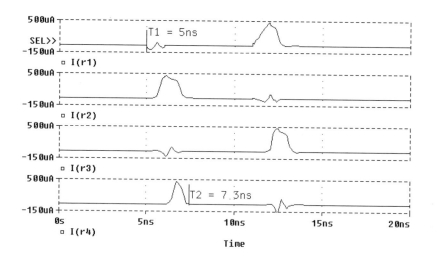

Figure 7.5   Switching current in the NAND tree.

and the last gate in the tree ended switching at T2 = 7.3 ns. Therefore, it is not accurate to measure the current for faulty behavior before time T2. That is, it is important to wait until the slowest gate switches before measuring the current. The duty cycle of the signal is to be long enough to ensure the switching of all gates. Thus it is critical to assess the fault-free current and the time taken for the circuit to settle down.

7.3   FAULT-FREE CURRENT

7.3.1   Switching and Quiescent Currents

The switching current in a CMOS inverter is governed by

$$I = 0.5\ \beta_k[2(V_{gs} - V_t)V_{ds} - V_{ds}^2] \tag{7.1}$$

where

$$\beta_k = \frac{\mu_k \epsilon}{t_{ox}} \left(\frac{W_k}{L_k} \right) \tag{7.2}$$

As V_{gs} switches from low to high, or vice versa, the current passes through a peak that is the maximum of the expression in Eq. (7.1) as $V_{ds} = V_{gs} - V_t$:

$$I_{max} = 0.5\ \beta_k(V_{gs} - V_t)^2 \tag{7.3}$$

where β_k is the device conductance, ϵ the permittivity, t_{ox} the gate oxide thickness, μ_k the electron mobility, and W_k and L_k are the channel width and length of the transistor. The index k stands for the n-channel and p-channel, respectively. The current is at its maximum as $V_{ds} = V_{gs} - V_t$. At such an instant, there is a direct path between the power and ground rails. As we realized in Section 7.2, this current is much higher than the quiescent current.

When the transistors are not switching, one of them will be in the on state, $V_{gs} > V_t$, and the others are in the off state, $V_{gs} = 0$. In the off state, the transistor is still conducting. This is known as *subthreshold* or *weak inversion conduction*. The change of subthreshold current with V_{gs} is shown in Fig. 7.6, which indicates the exponential characteristic of this region. The actual quiescent current is the sum of all off-state transistor currents. Measurements of this current for representative ICs are listed as a function of the number of transistors in Table 7.1. A mean value of the current is given since different chips of the same size will yield a distribution around the mean.

As a MOSFET's dimensions enter the submicron region, the subthreshold current increases due to scaled-down threshold voltages and short-chan-

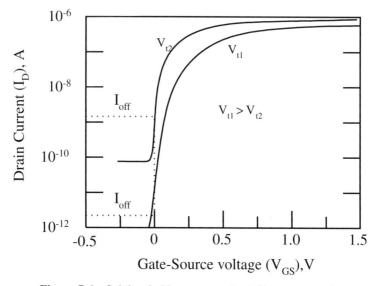

Figure 7.6   Subthreshold current or the (off-state current).

nel effects, such as *drain-induced barrier lowering* (DIBL) and *punch-through* [Soden 1996]. These phenomena, as well as the increase in the number of transistors integrated in a single chip, cause the defect-free I_{DDQ} current to be elevated, as we describe in Section 7.7.

7.3.2   Switching Delays

The duration of current flow depends on the parameters of the transistors and the load on the output. The relation between the current, its duration, and voltage across the load capacitance is governed by

$$\int_{V_1}^{V_2} C \, dv = \int_0^\tau i \, dt$$

TABLE 7.1   Typical Values for I_{DDQ}

Transistor Count (thousands)	Mean I_{DDQ} (nA)
70	0.4
1500	320
6000	980

Source: Data from [Soden 1996].

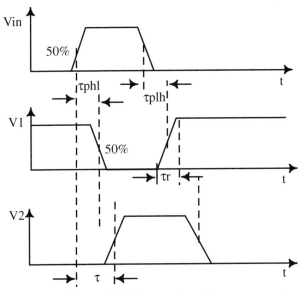

Figure 7.7   Propagation delays.

where τ is the time taken for the voltage to switch from V_1 to V_2. The values of V_1 and V_2 depend on whether the gate is switching from low to high, or vice versa. In the first case the load capacitance is charging through the p-channel, and in the second case it is discharging through the n-channel. Also, depending on the values of V_1 and V_2, we define various delays that are shown in Fig. 7.7:

τ_{phl} Propagation delay high to low is the time taken by the output voltage to decrease from V_{OH} to 50% of the logic swing.

τ_{plh} Propagation delay low to high is the time taken by output voltage to increase from V_{OH} to 50% of the logic swing.

τ The propagation delay within one gate is measured from 50% of the logic swing of the input to 50% of the logic swing for the output. The propagation delay within a pair of successive inverters is the sum of τ_{phl} (τ_{plh}) for the driver and τ_{phl} (τ_{phl}) of the load inverter.

τ_f The fall time is the time from the output to fall from 90% to 10 of its high value.

τ_r The rise time is the time for the output to rise from 10% to 90% of its high value.

7.4   CURRENT-SENSING TECHNIQUES

Conceptually, current testing is simple, but it requires the following steps after test pattern generation:

1. Apply the test pattern.
2. Wait for the transient to settle.
3. Check the static I_{DDQ} against a certain threshold value.

Current measurement is done externally (off chip) or internally (on chip) to the circuit. The second alternative is preferable with BIST structure. However, measurement of the current is not that simple since the measurement structure may interfere with the measured quantity. A successful current probe should have the following characteristics [Keating 1987]. It must be:

- Easily placed between the CUT and the bypass capacitor at the power supply pin
- Capable of measuring small static currents
- Nonintrusive; it does not cause a drop in V_{dd} of more than a few tenths of millivolts
- Capable of fast measurement, no more than 500 ns per pattern

Switching currents in CMOS circuits are very large and, because of the package pin, inductance may cause a large surge in the voltage at the power and ground rails. For this, a bypass capacitor is needed at the power supply. These large currents tend to mask the small I_{DDQ} and require a long time to settle down the transience.

7.4.1 Off-Chip Measurement

A generic representation of current measurement is shown in Fig. 7.8. In this arrangement, a resistance is placed between the V_{dd} pin of the IC and the bypass capacitor of the power supply. The value of this resistance depends on the resolution of the voltmeter and the magnitude of the I_{DDQ}. The voltage drop across the resistance may cause a significant reduction in the voltage at the IC power pin (V_{dd}). Therefore, it is important to build a sensing structure that (1) bypasses the transience and (2) compensates for the voltage drop. Usually, a sense ampli-

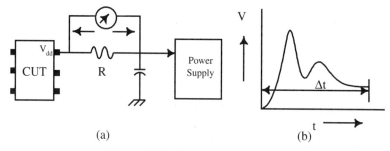

Figure 7.8 Current measurement.

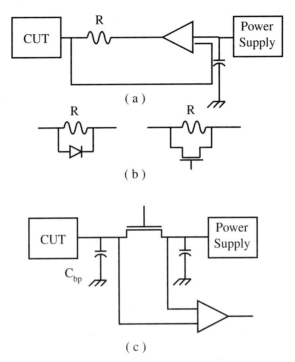

Figure 7.9   Sensing structures: (*a*) use of op amp; (*b*) bypassing techniques; (*c*) eliminating the resistance [Keating 1987 © IEEE. Reprinted with permission].

fier with sufficient gain is used to measure the quiescent current, as shown in Fig. 7.9*a*. The op amp has to be designed such that it compensates for the drop across the probe resistor. It is possible to overcome this voltage drop by shunting the resistor with a diode, but this will cause a drop of 0.6 to 0.8 V. Instead the diode, a pass transistor can be used. Both shunting methods are shown in Fig. 7.9*b*. This transistor is on until the transient effect subsides. At such time, the quiescent current will pass through the resistance. To filter the noise at high frequencies, a bypass capacitance, C_{bp}, is used at the output pin of the CUT. This filtering structure will also cause a delay in the stabilization of the circuit for current measurement. However, the testing is not done at speed, and the limiting frequency will depend on the RC_{bp} value. A more optimal structure is shown in Fig. 7.9*c*; it is possible to eliminate the resistor and keep the transistor on long enough to bypass the transient current. Afterward, the quiescent current is strobed. By eliminating the resistance, the circuit becomes faster and more useful over a larger range of currents. While switching transient currents, the transistor offers a short circuit between the two capacitors and maintains full V_{dd} to the CUT. After the transient currents have settled down, the transistor is turned off and the static current is supplied by C_{bp}, as governed by the relation $I = C_{bp}V/\Delta t$, where Δt is the time necessary for the transience to subside, as illustrated in Fig. 7.8*b*. For a recommended resolution of 25 μA per 10 mV, in 500 ns, the bypass capacitance is then $C_{bp} = I\,\Delta t/V = 1250$ pF. To limit the voltage drop across the CUT to 1 V (for $V_{dd} = 5$ V), the current measurement can be done at any time within 50 ms from turning the transistor off. This measurement structure can also be used to measure transient current, I_{DDT}, and has been improved for higher-speed probing, as described in [Isawa 1997].

7.4.2 On-Chip Measurement

To overcome the limitations of off-chip current probing, on-chip sensors have been proposed [Maly 1988]. They are more commonly known as built-in current (BIC) sensors and they have the advantage of increasing the speed and resolution of the measurements. Figure 7.10 shows a basic BIC structure that consists of a voltage-drop device and a voltage comparator. The voltage-drop device is inserted between the ground terminal, V_{GND}, of the device under test and the ground pin of the chip. The voltage V_{GND} is compared to a reference voltage, V_{ref}, which is adjusted such that $V_{GND} < V_{ref}$ for the fault-free circuit. When the circuit is defective, $V_{GND} > V_{ref}$ and this is flagged at the output of the comparator. Several other BIC designs have been proposed [Miura 1992, Rius 1992, Shen 1992, Hurst 1997].

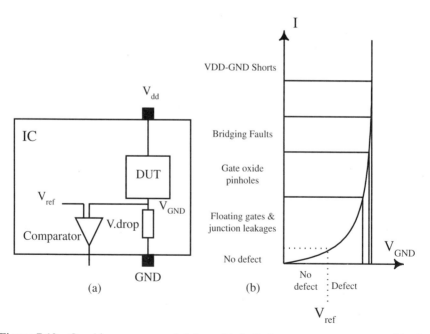

Figure 7.10   On-chip measurement: (*a*) on-chip built-in current measurement; (*b*) relation of the measured current to the voltage drop [Maly 1992 © IEEE. Reprinted with permission].

So far, the voltage-drop device related the current to the corresponding volt-age drop as a linear function. However, the magnitude of the faulty I_{DDQ} varies widely with the type of defect, as illustrated by Fig. 7.10b. If the voltage-drop device measures accurately the small current due to, say, floating gates, the corresponding voltage drop should be large enough to be measurable by the comparator. Therefore, if a linear voltage-drop device is used, it is accurate either with small currents or with large currents, but not with both. A nonlinear element such as a bipolar device is more appropriate for accuracy along a larger range of current [Maly 1992].

7.5 FAULT DETECTION

Consider the NAND implementation of the function $AB + C'$ that is shown in Fig. 7.11a. Assume that the p-channel transistor with input B is stuck on. This is equivalent to having the output of the corresponding NAND gate SA1. To provoke the fault from a voltage testing point of view, we must have $AB = 11$.

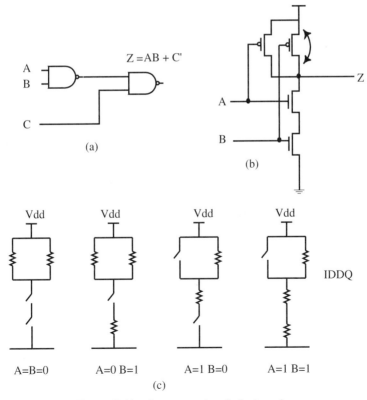

Figure 7.11 Current testing fault detection.

However, to propagate the fault to the output of the circuit (Z), we must, in addition, have $C = 1$. The pattern is, therefore, $ABC = 111$. Next, we examine the effect of the same fault on current flow. The behavior of this NAND gate for all possible values of A and B is shown in Fig. 7.11c. For all cases where $A = 0$ or $B = 0$, the circuit is isolated from ground and the output is pulled up to V_{dd}. Thus the gate appears to behave correctly. For the last case, $AB = 11$, there is a direct path between the V_{dd} rail and ground and the path is active as far as $A = 1$. Therefore, the fault is directly observable [Malaiya 1982] and there is no need to sensitize it through the second NAND gate to the primary output of the circuit. Test vectors generated to activate these faults are equivalent to stuck-at fault patterns that guarantee the propagation of the fault to the output of the gate, and they are called *pseudo-stuck-at patterns*. In addition to stuck-at and stuck-on faults, I_{DDQ} testing is effective in detecting bridging faults [Acken 1983] and some types of stuck-open faults. Most important, I_{DDQ} uncovers defects that cause leakage current regardless of the fault that models these defects. The concept of a leakage model has been developed and we discuss this first since it encapsulates the underlying principles for other fault detection.

7.5.1 Leakage Faults

Defects, which may not affect the logical operation of the circuit, cannot be represented by any of the traditional faults. Some of these defects, such as resistive shorts, may be represented and detected by delay faults. However, it is more likely that testing will combine voltage and current testing rather that delay testing. Instead of using delay fault testing, it is more efficient and effective to use current testing and the *leakage fault model* [Mao 1990, Nigh 1990].

In a MOS transistor, current may leak to any other terminal from any of its four terminals: gate (g), source (s), drain (d), and bulk (b). Therefore, there are six leakage faults per MOS transistor in the circuit:

$$\begin{array}{ll} F_{gs} & \text{leakage between gate and source} \\ F_{gd} & \text{leakage between gate and drain} \\ F_{ds} & \text{leakage between drain and source} \\ F_{bs} & \text{leakage between bulk and source} \\ F_{bd} & \text{leakage between bulk and drain} \\ F_{bg} & \text{leakage between bulk and gate} \end{array}$$

Leakage faults between source and drain are SON faults that were used earlier as examples to illustrate the basic concepts of current testing. We consider next other leakage faults, such as F_{gs} and F_{gd}. In the NOR gate of Fig. 7.12, a resistive short between B and the source of the NMOS transistor cannot be detected within the cell itself since there is no path from V_{dd} through B (the gate) and ground (the source). The path has to be traced through the driver of B and, more specifically, its PMOS transistor, as shown in Fig. 7.12a. For this we have $A = 0$, which drives B to 1 and the path is formed independently of

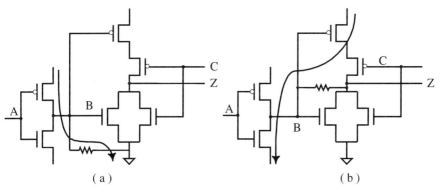

Figure 7.12   Detecting leakage faults.

the other input, C. For an F_{gd} of the same transistor, the path is from V_{dd} to ground through the PMOS transistors, C and B, and the NMOS transistor of the driver. This model will be used to develop I_{DDQ} test patterns in a subsequent section.

7.5.2   Bridging Faults

Bridging faults, discussed first in Chapter 2, are unintentional shorts between nodes in the circuit. The nodes may be (1) the terminals of a transistor (intratransistor faults), or (2) the terminals of different transistors (intertransistor faults). It is possible to view the latter as nodes on the logic gate level. On the transistor level, they may be due to gate oxide shorts or punchthrough [Hawkins 1985]. In such cases they can be represented by stuck-at (SAF), and stuck-on (SON) faults and some as stuck-open (SOP) faults. For example, a gate-to-source short, where the source is connected to ground or to V_{dd}, is a SAF fault, whereas a short between source and drain is a SON fault. A short between gate and source when the source is not connected to ground or V_{dd} acts as a SOP fault. The effectiveness of current testing SAF and SON faults was discussed earlier.

In Chapter 2 we distinguished between those bridging faults that cause an unintended feedback (FB) and those that do not (NFB). A wired-AND or a wired-OR gate generally represents the latter type of faults. Arguably, SSF test pattern can then be used to detect these faults [Millman 1988]. The advantage of using these SSF test patterns is that they are more readily available. Current testing, however, resulted consistently in higher fault coverage in an extensive experiment [Storey 1990]. For a certain circuit, the fault coverage increased from 98.88% to 99.97%. The coverage for SSF patterns is high given that SSFs do not adequately represent many CMOS faults. However, the higher quality of current testing, which is very likely to cost more than SSF testing, would eventually result in a higher defects per million (DPM) that pass the production test. DPM was defined in Chapter 1 [McCluskey 1989]. Using the results plotted in Fig. 1.13, if we use a yield of 90%, the increase in fault coverage by 1% results in a reduction of the DPM from 1000 to 100. Because bridging faults are a predominant problem with CMOS circuits and current testing is very effective in unveiling many defects that are not detectable by voltage testing, efforts has been expended to generate current testing for bridging faults [Storey 1990, Ferguson 1991, Isern 1993, Bollinger 1994]. FB bridging faults, on the other hand, may cause the faulty circuit to exhibit stable or oscillatory sequential behavior. These faults can be detected by voltage testing using a sequence of two patterns [Xu 1985] and by current testing [Rodriguez 1993, Roca 1995].

7.5.3 Stuck-Open Faults

Open defects in circuits are the results of errors in manufacturing, as described in Chapter 2. These discontinuities in the interconnections between the terminals of the transistors may cause different effects, depending on the defect location and size. An open defect at the gate of a transistor, as distinguished from the other locations, is usually called a *floating gate*. If the defect is small so that tunneling current can still flow through, the rise and fall times of the signal increase. For a larger open, the input is decoupled from the gate it is supposed to drive. The resulting floating gate acquires a bias voltage that depends on the electrical structure. This voltage can range in any value between 0 and V_{dd} and may turn on the transistor. Thus, depending on the size of the defect and the layout, open faults may or may not be detected.

7.5.4 Delay Faults

In has been concluded from several studies that (1) most CMOS defects cause a timing delay effect rather than catastrophic failures, (2) most of the defects cause elevation of the quiescent current, and (3) any defect that alters the circuit signal path that cause an increase in current will alter the rise and fall times of the signal. This strong correlation between the increase of delays and of quiescent current makes I_{DDQ} testing a feasible means to detect delay faults. When the signal is delayed, it takes longer for the switching current to decay to its quiescent value. It is this differential in timing that helps detect delay faults. It is possible, however, to detect these faults with I_{DDT} testing. With voltage testing, we realized in Chapter 1 that delay faults require a sequence of two vectors for detection. Current testing, however, takes fewer patterns for this detection. A list of some faults and defects uncovered by current testing is shown in Table 7.2 [Soden 1993].

7.6 TEST PATTERN GENERATION

Test patterns for current testing are signals to create one or more low-resistance paths from V_{dd} to ground in the presence of defects in CMOS circuits.

TABLE 7.2   Relation of Defects to SAF and I_{DDQ} Testing

Defect	SAF	I_{DDQ}
Gate oxide short	Rarely	Yes
Punchthrough (D-S short)	Rarely	Yes
Leaky junctions	Rarely	Yes
Parasitic leaks	Rarely	Yes
Shorts to rails	Yes	Yes
Open drain or source	No	Sometimes
Open gate with $V_{in} > V_{th}$	Yes	No
Transmission gate	No	Yes
Resistive bridge	No	Yes

We describe two approaches. The first uses a switch-level model of the circuit [Rajsuman 1992], and the second uses the leakage fault model given in Section 7.5 [Mao 1990].

7.6.1   Switch Level Model–Based

The switch level model uses a connection graph. In this undirected graph, the vertices are the signals V_{dd}, GND, and any intermediate node (including the output), and the edges represent the transistors. Each edge is named for the signal controlling the corresponding transistor. The model bears a one-to-one relation to the circuit, and it consists of two subgraphs, the p-graph and the n-graph. This is the graph that was used in Chapter 4 for the layout of standard cells. An example of such modeling is shown in Fig. 7.13. A two-input NAND gate consists of three vertices and two edges for the n-graph and two vertices and two edges for the p-graph. The combined graphs are shown in Fig. 7.13c. The dual of this graph represents the two-input NOR gate.

To generate patterns for stuck-on and stuck-open transistors, it is important to trace the path from V_{dd} to GND. For this, all paths from V_{dd} to the output (p-graph) or the output to GND (n-graph) are identified. Generating test patterns for shorts in the circuit can be summarized as follows: For a short in an n-network (p-network), a 0 (1) is applied to the faulty edge while 1's (0's) are applied to all other edges. Application of this algorithm to an inverter, a NAND gate, and a NOR gate is shown in Table 7.3.

We examine next the circuit shown in Fig. 7.14. For each logic gate we can create the patterns in the fashion outlined above. However, it is important next to justify the signal to the primary inputs of the circuit [Rajsuman 1992]. There is, of course, no need to sensitize the output of the gate to a primary output. It will be left as an exercise to develop the graphs for the network and generate the patterns. This testing approach is useful for SON and SAF models, but it is not adequate for all possible shorts in the transistors.

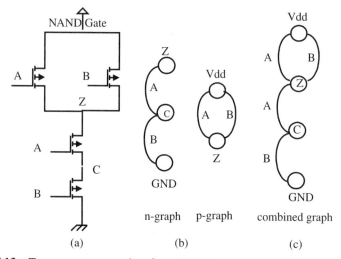

Figure 7.13 Test pattern generation for a NAND gate [Rajsuman 1994 © Artech. Reprinted with permission].

TABLE 7.3   Patterns for Detection of Shorts on the Transistors

	Inverter		NAND		NOR	
	Transistor	*A*	Transistor	*AB*	Transistor	*AB*
p-graph	*A*	1	*A*	11	*A*	10
			B	11	*B*	01
n-graph	*A*	0	*A*	01	*A*	00
			B	10	*B*	00

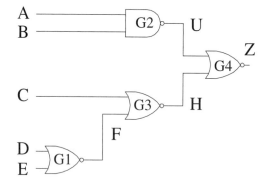

Figure 7.14   Test pattern generation, another example.

7.6.2   Leakage Fault Model–Based

This is a hierarchical test pattern generation approach that utilizes gate-level simulation instead of switch-level simulation. To begin, each cell type is characterized for the detection of all leakage faults within it. This characterization is done only once. The NAND gate shown in Fig. 7.15 will be used to illustrate the characterization [Mao 1990]. The results are stored in a table. For each cell with k I/O pins, an exhaustive 2^k test set is used. Each pattern represents a state of the cell. For each pattern, the faults detected on each transistor are entered in the table. Notice that some of these patterns represent illegal cell states. For example, $K = 4$ means that although only one n-channel (A) is on, the output, O, is pulled down. This is not physically possible for a fault-free NAND gate. For similar reasons, patterns 0, 2, and 7 are also illegal. Such patterns are eventually removed from the table. Figure 7.15 shows a NAND gate and its fault table. The first row of the table lists the transistors in the circuit, and K is the (A, B, O). The entries in the table are in octal and represent each the 6-tuple $\{(F_{bg}, F_{bd}, F_{bs}), (F_{ds}, F_{gd}, F_{gs})\}$. For example, in the sixth row of the table, the pattern applied is $K = 5$; that is, $(A, B, O) = (1,0,1)$. For transistor N1, the entry is $(70)_8 = (111000)$, indicating that the pattern detected the first three faults of the 6-tuple and did not detect the other three. For transistor N2, the entry is $(26)_8 = (010110)$. Therefore, F_{bd}, F_{ds}, and F_{gd} are detected. No faults were detected for P1 since the corresponding entry in the matrix is 0. The illegal states are removed and the matrix is used as a lookup table in test simulation.

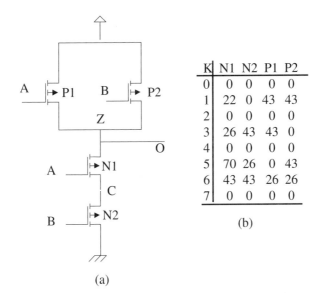

K	N1	N2	P1	P2
0	0	0	0	0
1	22	0	43	43
2	0	0	0	0
3	26	43	43	0
4	0	0	0	0
5	70	26	0	43
6	43	43	26	26
7	0	0	0	0

(b)

(a)

Figure 7.15   Test pattern generation using leakage fault model [Mao 1990 © IEEE. Reprinted with permission].

The full circuit is then logic simulated on the gate level using patterns from the functional test vectors. The values of the I/O of all gates are captured. The patterns for current testing are then selected according to the lookup table. The methodology has the advantage of being hierarchical and requiring only logic simulation. Experiments have shown that only 1% of the functional test patterns are needed for current testing. The approach thus streamlines functional testing with current testing.

7.7 IMPACT OF DEEP-SUBMICRON TECHNOLOGY

Current testing is based on the fact that the quiescent current of the fault-free circuit (defect-free I_{DDQ}) is much smaller than the quiescent current of a defective circuit. However, as indicated in Fig. 7.6, the transistors are still conducting in the off state, $V_{gs} < V_t$. The subthreshold current is a function of the threshold voltage. As the technology feature size decreased, this off-state current increases. Also, the circuits become denser and larger, which contributes to the increase defect-free current. Table 7.4 lists the off-state current per unit length for different feature sizes. The entries for 0.24 and 0.18 are only estimates.

As the off-state current is increased, the resolution between the fault-free and defective circuit currents, $M_g - M_d$, of Fig. 7.1 will be reduced. It is even predicted that change of this quantity normalized to M_g will be negligible for CMOS circuit by 2007 [Williams 1996]. Toward improving the discrimination between good and faulty responses, the notion of current signature has been proposed [Gattiker 1998]. This technique does not rely on a single threshold to determine a faulty IC. On the contrary, a *signature* is created by ordering all the measurements by magnitude. The signature is based on the state dependence of the off-state current. Another approach has also been recommended by [Maxwell 1998].

Different physical phenomena contribute to the increase in off-state current

TABLE 7.4 I_{DDQ} **Change with Technology Feature Size**

Technology (μm)	V_{dd} (V)	T_{ox} (A)	V_t (V)	L_{eff} (μm)	I_{off} (pA/μm)
1.0	5	200	—	0.8	4.1×10^{-4}
0.8	5	150	0.60	0.55	5.8×10^{-4}
0.6	3.3	80	0.58	0.35	0.15
0.35	2.5	60	0.47	0.25	8.9
0.25	1.8	45	0.43	0.15	24
0.18	1.6	30	0.40	0.10	86

Source: Data from [Keshavarzi 1997 © IEEE. Reprinted with permission].

[Soden 1996, Ferre 1997]. For technology features larger than 1 μm, the dominant mechanism is reverse-biased pn-junction leakage current. For submicron technologies, below 0.5 μm, the dominant mechanism is subthreshold leakage current. The decrease in channel length required a reduction in the threshold voltage, V_t, which in turn caused bloating in the subthreshold current, as illustrated in Fig. 7.6. The 50% decrease in V_t resulted in current three orders of magnitude higher.

In the past, efforts have concentrated on estimating the current in defective chips [Segura 1992, Soden 1996]. Nowadays, the concern is becoming more critical to estimate I_{DDQ} for the fault-free circuits [Ferre 1997, Maxwell 1998]. This analysis is needed to obtain the upper limit of defect-free I_{DDQ} current to be used as the comparison level for current test (I_{DDQ} test limit). To keep current testing effective for future technology, it is important to control the off-state current. Some solutions have been recommended [Williams 1996]:

1. Reverse biasing the substrate during test could give the effect of raising the threshold voltages of the devices and hence reduce the quiescent current [Keshavarzi 1997].

2. Devices could be cooled during testing to contain the leakage current. At lower temperatures the threshold voltages can be scaled with the power supply voltage and without the leakage problem [Gaensslen 1977].

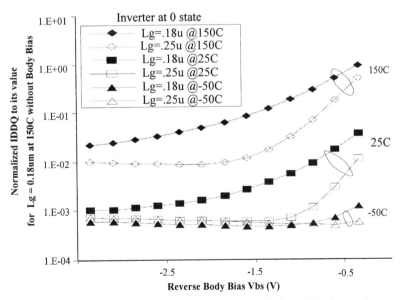

Figure 7.16 Change of the off-state current with back-body biasing and temperature.

3. Architectural changes such as using dual threshold devices where possible to minimize I_{DDQ} could temporarily delay the effect of the calculation in this work. Low-V_t devices can be used in parts of the logic where speed is critical, and high-V_t devices can be used for other noncritical parts of the design [Taur 1995].

4. The network could be partitioned and I_{DDQ} test performed on individual segments of the design. Individually, selectable foot-switch devices with a high V_t value could control the leakage current for the entire network that is implemented with low-V_t devices. Power supply buses could be split to reduce leakage current for I_{DDQ} test [Burr 1991].

The effects of the temperature and of reverse-body biasing on the off-state current of 0.18 μm devices have been investigated in an extensive simulation and measurement study. Some of the results are shown in Fig. 7.16 for a CMOS inverter (input = 0). Two channel lengths are used: 0.18 μm and 0.24 μm. The current is definitely reduced when the temperature is lowered. There is an optimal value of the bias voltage that reduces the off-state current. However, there are no conclusive results on the effectiveness of this approach.

REFERENCES

Acken, J. (1983), Testing for bridging faults (shorts) in CMOS circuits, *Proc. Design Automation Conference*, pp. 717–718.

Bollinger, S. W., and J. Figueras (1994), Test generation for I_{DDQ} testing of bridging faults in CMOS circuits, *IEEE Trans. Comput.-Aided Des.*, Vol. CAD-13, No. 11, pp. 1413–1418.

Burr, J., and A. Peterson (1991), Ultra low power CMOS technology, *Proc. 3rd NASA Symposium on VLSI Design*, pp. 1–11.

Daniels, R. G. (1990), The changing demands of microprocessor testing, Keynote address, *Proc. IEEE International Test Conference*, pp. 1–2.

Ferguson, F. J., and T. Larrabee (1991), Test pattern generation for realistic bridge faults in CMOS ICs, *Proc. IEEE International Test Conference*, pp. 492–499.

Ferre, A., and J. Figueras (1997), I_{DDQ} characterization in submicron CMOS, *Proc. IEEE International Test Conference*, p. 136.

Fritzemeier, R. R., C. F. Hawkins, and J. M. Soden (1991), CMOS IC fault models, physical defect coverage and I_{DDQ} testing, *Proc. Custom Integrated Circuits Conference*, pp. 13.1.1–13.1.8.

Gaensslen, F. H., et al. (1977), Very small MOSFET's for low-temperature operation, *IEEE Trans. Electron Devices*, Vol. ED-24, No. 3, pp. 218–229.

Gattiker, A. E., and W. Maly (1998), Current signatures: application, *Proc. IEEE International Test Conference*, pp. 1168–1177.

Hawkins, C. G., and J. M. Soden (1985), Electrical characteristics and testing considerations for gate oxide shorts in CMOS ICs, *Proc. IEEE International Test Conference*, pp. 544–555.

Hawkins, C. F., et al. (1994), Defect-classes: an overdue paradigm for CMOS IC testing, *Proc. IEEE International Test Conference*, pp. 413–425.

Hurst, J. P. (1997), A differential built-in current sensor design for high-speed I_{DDQ} testing, *IEEE J. Solid-State Circuits*, Vol. SC-32, No. 1, pp. 122–125.

Isawa, K., and Y. Hashimoto (1997), High-speed measurement circuit, *Proc. IEEE International Test Conference*, pp. 111–117.

Isern, E., and J. Figueras (1993), Test generation with high coverages for quiescent test of bridging faults in combinational circuits, *Proc. IEEE International Test Conference*, pp. 73–82.

Keating, M. (1987), A new approach to dynamic IDD testing, *Proc. IEEE International Test Conference*, pp. 316–321.

Keshavarzi, A., K. Roy, and C. F. Hawkins (1997), Intrinsic leakage in low power deep submicron CMOS ICs, *Proc. IEEE International Test Conference*, pp. 146–155.

Levi, L. (1981), CMOS is most testable, *Proc. IEEE International Test Conference*, pp. 217–220.

Liu, M., and S. Mourad (1999), Evaluation of body biasing engineering on performance of scaled CMOS devices, *Proc. SPIE Symposium on Solid State Technology*, pp. 138–145.

Malaiya, Y. K., and S. Y. H. Su (1982), A new fault model and testing techniques for CMOS IC defects, *Proc. IEEE International Test Conference*, pp. 25–34.

Maly, W., and P. Nigh (1988), Built-in current testing, a feasibility study, *Proc. International Conference on Computer-Aided Design*, pp. 340–343.

Maly, W., and M. Patyra (1992), Built-in current testing, *IEEE J. Solid-State Circuits*, Vol. , No. 3, pp. 425–428.

Mao, W., et al. (1990), Quietest: a quiescent current testing methodology for detecting leakage faults, *Proc. IEEE International Test Conference*, pp. 280–283.

Maxwell, P. C., and J. R. Rearick (1998), Estimation of defect-free I_{DDQ} in submicron circuits using switch level simulation, *Proc. IEEE International Test Conference*, p. 882.

McCluskey, E. J., and F. Buelow (1989), IC quality and test transparency, *IEEE Trans. Ind. Electron.*, Vol. IE-36, No. 2, pp. 197–202.

Millman, S. D., and E. J. McCluskey (1988), *Proc. IEEE International Test Conference*, pp. 773–783.

Miura, Y., and K. Konishita (1992), Circuit design for built-in current testing, *Proc. IEEE International Test Conference*, pp. 873–881.

Nelson, G. F., and W. F. Boggs (1975), Parametric tests meet the challenge of high-density ICs, *Electronics*, No. 12, pp. 108–111.

Nigh, P., and W. Maly (1990), Test generation for current testing, *IEEE Des. Test Comput.*, Vol. 7, No. 2, pp. 26–38.

Noyce, R. N. and M. E. Hoff, Jr. (1991), A history of microprocessor development at Intel, *IEEE Micro*, pp. 8–12.

Rajsuman, R. (1992), *Digital Hardware Testing: Transistor-Level Fault Modeling and Testing*, Artech House, Norwood, MA.

Riezenman, M. J. (1991), Wanlass's CMOS circuit, *IEEE Spectrum*, Vol. 28, No. 5, p. 44.

Righter, A. W., et al. (1996), High resolution I_{DDQ} characterization and testing: practical issues, *Proc. IEEE International Test Conference*, pp. 259–268.

Rius, J., and J. Figueras (1992), Proportional BIC sensor for current testing, *J. Electron. Test. Theory Appl.*, Vol. 3, No. 4, pp. 387–396.

Roca, M., and A. Rubio (1995), Current testability analysis of feedback bridging faults in CMOS circuits, *IEEE Trans. Comput.-Aided Des.*, Vol. CAD-14, No. 10, pp. 1299–1305.

Rodriguez-Montanes, R., J. Figueras, and A. Rubio (1993), Current vs. logic testability of bridges in scan chains, *Proc. European Test Conference*, pp. 329–396.

Sachdev, M. (1995), I_{ddq} test and diagnosis in deep sub-micron, *Proc. IEEE International Test Conference*, special session on I_{ddq}, pp. 84–89.

Segura, J. A., et al. (1992), Quiescent current analysis and experimentation of defective CMOS circuits, *J. Electron. Test. Theory Appl.*, Vol. 3, No. 4, pp. 337–348.

Shen, T., and J. C. Daly (1992), On-chip current sensing circuit for CMOS VLSI, *Proc. IEEE VLSI Test Symposium*, pp. 309–314.

Soden, J. M., et al. (1992), I_{DDQ} testing: a review, *J. Electron. Test. Theory Appl.*, Vol. 3, No. 4, pp. 291–304.

Soden, J. M., R. Gulati, and C. G. Hawkins (1993), Correct monitoring for efficient detection of CMOS defect and faults, Tutorial Notes, IEEE Computer Society, Baltimore, MD.

Soden, J. M., C. F. Hawkins, and A. C. Miller (1996), Identifying defects in deep-submicron CMOS ICs, *IEEE Spectrum*, Vol. 38, No. 9, pp. 66–71.

Storey, T. M., and W. Maly (1990), CMOS bridging fault detection, *Proc. IEEE International Test Conference*, pp. 842–850.

Taur, Y., et al. (1995), CMOS scaling into the 21st century: 0.1 μm and beyond, *IBM J. Res. Dev.*, Vol. 39, No. 1/2, pp. 245–260.

Wanlass, F. M., and C. T. Sah (1961), Nanowatt logic using field-effect metal-effect metal-oxide semiconductor triodes, *Proc. Solid State Circuits Conference*, pp. 32–33.

Williams, T. W., et al. (1996), I_{DDQ} test: sensitivity analysis of scaling, *Proc. IEEE International Test Conference*, pp. 786–792.

Xu, S., and S. Y. H. Su (1985), Detecting I/O and internal feedback bridging faults, *IEEE Trans. Comput.*, Vol. C-14, No. 10, pp. 553–557.

PROBLEMS

7.1. Calculate the peak current for a two-input CMOS NOR gate using minimum feature transistors for 1.2-μm technology and ignore channel-length modulation. The transconductance for the n- and p-channels are 80 and 27 μA/V^2; $L_n = L_p = 1.2$ μm, $W_n = 1.8$ μm, and $W_p = 5.4$ μm.

7.2. For the nor gate of Problem 7.1, calculate the fall and rise times.

7.3. The I_{DDQ} of a circuit is measured using an external monitor under the following conditions: The transient current has a peak of 2 A and lasts for 10 ns. The bypass capacitance is 5000 pF and the current is sampled every

200 ns. If the voltage across the transistor after it is turned off is 0.8 V, estimate the value of the quiescent current. Is the circuit faulty?

7.4. Using a leakage fault model, develop test patterns for a two-input NOR gate for current testing all intratransistor bridging faults.

7.5. Use the switch-level transistor model to determine the patterns to detect the shorts on all the transistors of the circuit in Fig. 7.14.

7.6. For the transistor-level circuit of Fig. P7.6, estimate the quiescent current of the circuit when $A = 1$, $B = C = 1$, and $A = B = C = 1$.

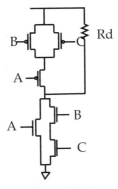

Figure P7.6

PART III

DESIGN FOR TESTABILITY

8

AD HOC TECHNIQUES

8.1 INTRODUCTION

In the first seven chapters we established the foundations of digital testing. In Part I, we discussed physical defects and their manifestation on the circuit and logic level. In Chapters 3 and 4 we placed testing within the design cycle and examined the various design representations that are appropriate for processing by different CAD tools. In Chapter 6, we found out that the process of test pattern generation is a NP-complete problem. In addition, we realized that the complexity of the test pattern is even more for sequential circuits. Because of the complexity, design approaches that facilitate testing have been developed and are very widely used at present. These approaches advocate introducing, into the circuit, constructs to make testing easier at the manufacturing level. The ease here extends to the processes, test pattern generation, and test application.

The design for test (DFT) discipline started with the insertion of test points into a design using an adhoc approach. Systematic techniques, such as internal scan, built-in self-test (BIST) and boundary scan, are detailed in Chapters 9 to 11. They are all approaches to increase *testability* by embedding test structures in the circuit under test. There is no formal definition for testability. An interesting attempt was given as: "A digital IC is testable if test patterns can be generated, applied, and evaluated in such a way as to satisfy predefined levels of performance (e.g., detection, location, application) within a predefined cost budget and time scale" [Bennetts 1984]. One of the key words is "cost." It is probably the cost of testing that deters semiconductor manufacturers from doing as much testing as is really needed to ensure reliable products. Assessing the ease of testing used to be done for design described on the gate level. However, testability measures on higher levels of abstraction have been formulated. The benefit of attempting to assess testability of a circuit at such an abstract level is to correct for problems early in the design cycle.

In this chapter we first discuss why DFT structures are very particularly relevant in present technology. We discuss testability measures in more detail than we discussed in Chapter 1. In Section 8.4 we review ad hoc techniques that were followed before the systematic DFT methods. In addition, we revisit pseudoexhaustive testing. Section 8.6 is devoted to regularly structured combinational designs that are tested either with a constant number of test patterns (C-testable) or with a test that depends on the number of repeated identical cells in the structure (scalable test).

8.2   CASE FOR DFT

There are several reasons why it is becoming highly desirable, if not mandatory, to adopt DFT constructs in present VLSI circuits. These reasons pertain to the testing process itself and the nature of modern circuits as they affect testing cost.

8.2.1   Test Generation and Application

Unless we use exhaustive testing, test pattern generation is a complex problem. However, we realized in Chapter 2 that exhaustive testing is not realistic except in very few instances. Thus what was saved in test generation time is offset by the test application time. Random test pattern generation is also a feasible low-cost process; it yields long test sets but does not guarantee 100% SAF coverage. Fault-based test pattern generation, also called deterministic test pattern generation, has proven to be NP-complete [Ibarra 1975]. This means, as we mentioned in Chapter 4, that the problem of test pattern generation cannot be solved with a polynomial time algorithm. "To illustrate this," in [Fujiwara 1986]'s words, "suppose that we have three algorithms whose time complexities are n, n^3, and 2^n. Assuming that time complexity expresses execution time in microseconds, the n, n^3, and 2^n algorithm can solve a problem of size 10 instantly in 0.00001, 0.001, and 0.001 seconds, respectively. . . . To solve a problem of only size 60, the 2^n algorithm requires 366 centuries whereas the other two algorithms require only 0.00006 and 0.216 seconds, respectively." Some work was done to find circuits that can have a tractable solution [Fujiwara 1990, Chakradhar 1991]. However, since these works require more theoretical knowledge than we intend to present in this book, we leave it as a problem to explore further.

8.2.2   Characteristics of Present VLSI

Changes in technology have resulted in digital ICs that are characterized by:

1. Smaller devices that are capable of switching faster, thus allowing an increase in the operating frequency.

2. Higher device density that leads the designer to make more complex and larger circuits.

3. Lower power supply voltage: the supply voltage decreased from 5 V to 3.3 V and, very soon, 2 V. Thus the circuit is becoming less immune to noise.

4. Thinner and longer interconnect wires: The smaller wire cross sections have caused an increase in resistance and capacitance. Other than capacitance to ground, there are fringing capacitances and mutual capacitance between interconnects on the same layer and on adjacent layers.

5. Design paradigm of a system on a chip (SOC): This approach to integrating several designs on the same chip involves the concept of reuse and brings new challenges to testing.

The second and fifth characteristics of modern VLSI have resulted in designs that are too large and hence increase the complexity of test pattern generation. In addition, accessibility to the circuit is becoming more limited. The I/O pins recently have increased, but not at the same rate at which the size of the circuit has, to the extent that the pin-to-gate ratio is continuously decreasing. This is clear from Fig. 8.1. The increase in circuit size causes an increase in its processing time by various CAD tools. As stated in Chapter 5, new approaches to simulation, such as cycle-based simulation and static timing analysis, have

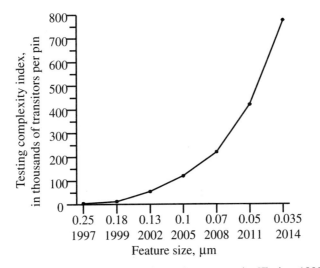

Figure 8.1 Testability index: number of transistors per pin [Zorian 1999 © IEEE. Reprinted with permission].

been adopted to cope with complex circuits without delaying time-to-market. Another issue is that synthesis tools have not yet matured, and the handling of a large circuit requires more attention and skillful designers. To meet the challenge of testing such circuits, more resources are needed—more powerful CAD tools, testers, and engineers trained in present technology. Next we revisit the testability measures that are still used as guides in test pattern generation.

8.3 TESTABILITY ANALYSIS

An attempt to quantify testability, in the sense of fault detection, was proposed by [Goldstein 1979] and [Grason 1979]. Two testability measures (TMs) were then defined: controllability and observability. *Controllability* reports the cost of placing a node in the circuit at a predetermined logic value. Placing a logic value on a primary input is "free." However, it is possible to assign a minimal cost, say 1, to any primary input. The cost increases as the node depth in the circuit increases. This cost will also depend on the type of gate, the logic value to be imposed on the line, and whether the circuit is combinational or sequential. For example, controlling the output of a multi-input AND gate to 1 requires the control of all its inputs to 1, whereas for an OR gate, it is sufficient to control only one of the inputs to 1. Thus controlling the output to 1 on an AND gate costs more than an OR gate. The most popular testability measures, SCOAP, list four controllability measures [Goldstein 1980]:

> CC0   combinational 0-controllability
> CC1   combinational 1-controllability
> SC0   sequential 0-controllability
> SC0   sequential 1-controllability

Here we concentrate on the combinational measures to illustrate the concepts and the use of the measures. The controllability measures of the outputs of elementary gates are given in Fig. 8.2, the gate within the circuit. If, however, they are primary inputs to the circuit, they are easily controllable and we can assume the cost to be 0 or 1. The cost at the output is increased by 1 to account for stepping one level over A and B. We can use these TMs to evaluate all nodes in a circuit based on such gates.

Example 8.1. As an example, let us calculate the TM of the nodes of circuit C1 shown in Fig. 8.3 as functions of the controllability of the input nodes. Although testability measure programs do not necessarily distinguish the branches from the stem, in this example we will, to allow us to demonstrate how redundant logic that creates an undetectable fault will yield finite TM. Branches of the primary input, A and B, will be indexed with the number of the gate in which they fan-in. This we have branches, $A1$, $A2$, $B1$, $B2$, $C1$, $C3$, $H4$, and $H5$. Also,

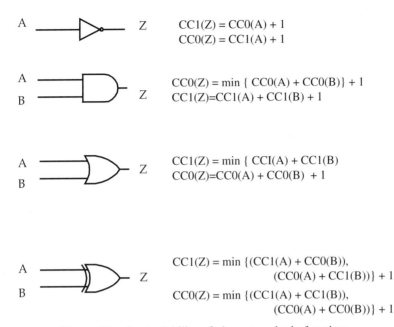

$$CC1(Z) = CC0(A) + 1$$
$$CC0(Z) = CC1(A) + 1$$

$$CC0(Z) = \min \{ CC0(A) + CC0(B)\} + 1$$
$$CC1(Z) = CC1(A) + CC1(B) + 1$$

$$CC1(Z) = \min \{ CCI(A) + CC1(B)$$
$$CC0(Z) = CC0(A) + CC0(B) + 1$$

$$CC1(Z) = \min \{(CC1(A) + CC0(B)),$$
$$(CC0(A) + CC1(B))\} + 1$$
$$CC0(Z) = \min \{(CC1(A) + CC1(B)),$$
$$(CC0(A) + CC0(B))\} + 1$$

Figure 8.2 Controllability of elementary logic functions.

since the circuit has more than one primary output, we calculate observability measures for both outputs, Y and Z.

To calculate the controllability, we need to proceed from the inputs to the outputs. Each logic level depends on the preceding levels. Here we assume that the controllability measures of the primary inputs, including branches, are all equal to 1. Then we calculate the numbers for the observability, this time starting from the most observable nodes, the primary outputs, and proceeding backward to the primary inputs. All results are summarized in Table 8.1. Initially, all TM values are set to infinity (∞). When the calculations are completed, the entries that remain with ∞ indicate that the corresponding nodes are either uncontrollable or unobservable at the corresponding primary output.

We calculate the TM for the nodes on the second level, F, G, and H. The controllability of $H4$ and $H5$ are equal to that of their stem, H.

$$CC1(F) = CC1(A) + CC1(B) + CC1(C) + 1 = 4$$
$$CC0(F) = \min\{CC0(A), CC0(B), CC0(C)\} + 1 = 2$$
$$CC1(H) = \min\{CC0(A), CC0(B)\} + 1 = 2$$
$$CC0(H) = CC1(A) + CC1(B) + 1 = 3$$
$$CC1(G) = CC0(C) + 1 = 2$$
$$CC0(G) = CC1(C) + 1 = 2$$

TABLE 8.1 Testability Measures of Circuit C1 in Fig. 8.3

Node	CC0	CC1	CO/Y	CO/Z	S1/Y	S1/Z	Patterns
A	1	1	6,8	7	7,9	8	2
$A1$	1	1	8	∞	9	∞	0
$A2$	1	1	6	7	7	8	2
B	1	1	6,8	7	7,9	8	2
$B1$	1	1	8	∞	9	∞	0
$B2$	1	1	6	7	7	8	2
C	1	1	8	5	9	6	3
$C1$	1	1	8	∞	9	∞	1
$C3$	1	1	∞	5	∞	6	2
G	2	2	∞	5	∞	7	2
H	3	2	4	4	7	7	1
$H4$	3	2	4	∞	7	∞	1
$H5$	3	2	∞	4	∞	7	1
F	2	4	5	∞	7	∞	1
Y	6	3	1	∞	7	∞	1
Z	5	3	∞	1	∞	6	3

These TM values are then used in calculating the primary output controllability:

$$CC1(Y) = \min\{CC1(F), CC1(H)\} + 1 = 3$$
$$CC0(Y) = CC0(F) + CC0(H) + 1 = 6$$
$$CC1(Z) = \min\{CC0(H), CC0(G)\} + 1 = 3$$
$$CC0(Z) = CC1(H) + CC1(G) + 1 = 5$$

The *observability* of a node indicates the effort needed to observe the logic value on the node at a primary output. Again, we assume that the cost associated with observing a primary output is 1. To observe an internal node, for example, node F of the circuit $C1$ in Fig. 8.3, it is necessary to control H to 0 in order to sensitize F to the primary output, Y. Thus the observability of F at Y is given by

$$CO_Y(F) = CO(Y) + CC0(H) + 1 = 5$$

Similarly, the observability of G is

$$CO_Z(G) = CO(Z) + CC1(H) + 1 = 4$$

For H we can observe it as well as its branch $H4$ through Y. Also, we can observe it and its branch $H5$ through Z; consequently, we can find two observability numbers:

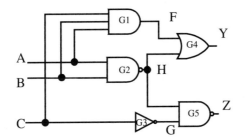

Figure 8.3 Circuit C1 to illustrate SCOAP testability measures.

$$CO_Y(H) = CO(Y) + CC0(F) + 1 = 4$$
$$CO_Z(H) = CO(Z) + CC1(G) + 1 = 4$$

The observability measures of the primary inputs are calculated next. Since the primary inputs may be observable at either Y or Z, we have different values to evaluate. Let us first consider line C. Through Z, the observability of this node as well as its branch, $C3$, is

$$CO_Z(C) = CO_z(G) + 1 = [CO(Z) + CC1(H) + 1] + 1 = 5$$

The observability of the same node and its branch, $C1$, through Y is given by

$$CO_Y(C) = CO_Y(F) + CC1(A) + CC1(B) + 1$$
$$= [CO(Y) + CC0(H) + 1] + CC1(A) + CC1(B) + 1 = 8$$

This result indicates that it is easier to observe C at node Z. In a similar manner, we can calculate the observability numbers for the other primary inputs, A and B. Actually, since the functions Y and Z are symmetrical in A and B, it is sufficient to calculate these numbers for one of these primary inputs. Notice also that the signals on these nodes may be sensitized to Y through gate G1 or G2. Sensitizing through G1 would yield the same observability number as for C through the same output, Y. Then $CO_{YF}(A) = CO_{YF}(B) = CO_Y(C) = 8$. Through G2 and observing at Y, the measure is

$$CO_{YH}(A) = CO_Y(H) + CC1(B) + 1 = 6$$

Observing through Z would yield

$$CO_Z(A) = CO_Z(H) + CC1(G) + 1 = 7$$

The TM numbers for all nodes in the circuit are summarized in Table 8.1. There are some entries in the table that still contain their initial value, infinity. This is an indication that the corresponding nodes are not observable at the corresponding primary output. Although the SA1 faults on the branches of $A1$ and $B1$ feeding into gate $G1$ are not detectable because of redundancy, they have finite TMs that are comparable to other detectable faults, for example C.

The results indicate that the TMs vary with the depth of the circuit. The controllability numbers are higher for the primary outputs, while the observability numbers are higher at the primary inputs. Both measures are equally important since an easily controllable node is not necessarily observable. Similarly, the ease of observability is useless unless we can control the node. Several attempts have been made to combine the two numbers to correlate with the ease of test pattern generation. For example, one can add (or multiply) the controllability 0 to the observability and use the resulting number as a measure of the testability of stuck-at-1 faults. Such values are listed in the sixth and seventh columns of Table 8.1. To assess these measures, it is possible to generate an exhaustive test set for the circuit and determine the number of the patterns that detect each fault. These numbers are shown in the seventh column of the table.

The merit of testability measures in testing was assessed in several works [Agrawal 1982, Savir 1983]. SCOAP testability measures are a quick estimate of the degree of difficulty in generating test patterns without actually generating them. Calculating TM is not as time consuming as test pattern generation. The complexity is only $O(n)$, where n is the circuit size (number of lines). There are several variations of testability measures [Kovijanic 1979, Bennetts 1981, Ratiu 1982]. In addition, statistical testability measures have also been developed [Agrawal 1985]. They have also been generalized to the design described on RTL and functional levels [Jamoussi 1991, Gu 1995]. As we learned in Chapter 6, these numbers have been used successfully as an aid in test pattern generation algorithms such as FAN and SOCRATES [Fujiwara 1982, Schultz 1988] and in determining test insertion points to facilitate controllability and observability for built-in self-test (BIST) [Lin 1993].

8.4 INITIALIZATION AND TEST POINTS

Ad hoc techniques tend to be heuristic rather than applicable to all circuits and low cost. On the other hand, structured techniques tend to solve general problems. Before DFT had been established formally, test engineers have always approached testing in a commonsense way. For example, they initialized a sequential circuit before testing it, included observation and control points to increase testability, and then divided the circuit into smaller partitions to make each part more manageable. These various ad hoc techniques are the topic of this section. Partitioning for testability is discussed in Section 8.5.

8.4.1 Initialization

Initialization of a circuit is not always necessary for proper operation. However, it is necessary to place the circuit in a known position in order to test it [McCluskey 1986]. We learned in Chapter 3 how to generate an initializing or homing sequence. Such sequences may be too long and may be computationally costly to develop. In addition, not every sequential circuit may have a synchronizing or homing sequence.

8.4.2 Observation Points

From the knowledge acquired on test pattern generation in previous chapters and the concepts of controllability and observability just discussed, it is clear that increasing the number of observation points facilitates the testing. This is illustrated with the circuit construct shown in Fig. 8.4, which consists of three main components: W, a NOR gate, and U. For example, if the node P is difficult to observe through U, then an observation point, OP, can be placed at this location as shown in Fig. 8.4b.

Example 8.2. Let us consider circuit C2 in Fig. 8.5a. This circuit requires five test patterns to detect all stuck-at faults. However, if we insert an observation point at F, only four patterns are necessary (or required) to detect the faults. If this shorter test set were used while observing only through the primary output, Z, the fault $B/1$ would not be observable. Thus adding this observation point improves testing by shortening the test length. However, these points cannot be added in a haphazard manner [Hayes 1975]. For example, an observation point at H would not have helped in reducing the test set.

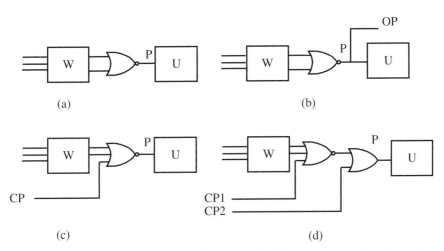

Figure 8.4 Test point insertion: (*a*) original circuit; (*b*) observation point; (*c*) controlling P to 0; (*d*) controlling P to 0 and 1.

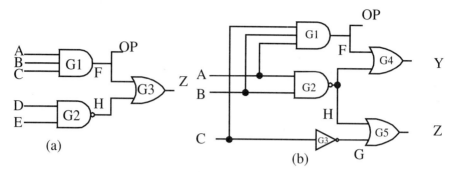

Figure 8.5   Examples to illustrate observation points.

Example 8.3. In circuit C1 of Fig. 8.3, the branches of *A* and *B* feeding into gate G1 have undetectable SA1 faults because of redundancy. These faults are observable at node *F*, but they are masked by *H*. Using *F* as an observation point, as illustrated in Fig. 8.5*b*, would make these faults detectable. Often, it is assumed that since a fault is redundant, there is no need to uncover it. However, this might cause unpredictable behavior by the circuit.

8.4.3   Control Points

As effective as inserting an observation point is the insertion of control points to ease forcing nodes to specific logic values. If, say, a zero is not obtainable from the NOR gate in Fig. 8.4*a*, it is possible to introduce the control point, CP, as an additional input to the NOR gate and hold it high, as shown in Fig. 8.4*c*. If *P* is also to be controlled to 1, the structure in Fig. 8.4*d* would achieve this goal whenever CP2 is held high. It is worth noticing, however, that the insertion of a control point affects the controllability of the nodes influenced by the controlled point (all nodes of *U*). But the observability of the nodes whose signals go through the controlled point will also be affected. Thus CP1 and CP2, which ease the control of *P* in Fig. 8.4*d*, affect the observability of the subcircuit, *W*. The observability of this subcircuit has more constraints—holding CP1 and CP2 low.

Test point insertion is not limited to combinational logic; it is actually needed more in sequential circuits [Gundhach 1990]. Sometimes, the reset of some storage devices is controlled by a complex logic, and this makes initialization of the devices difficult. Using a control point can facilitate this task, as illustrated in Fig. 8.6*a*. Test points can be very helpful in enhancing the testability of long counters. As was pointed out in Section 1.7, to test a SA0 on the ripple carry-out of a 32-bit counter, 2^{32} patterns must be applied. Inserting control points as shown in Fig. 8.6*b* makes it possible to test the two parts, *M* and *N*, independently. Whenever possible, it is preferable to design modular counters. For example, a Mod 24 counter may be designed as two freely running Mod 3

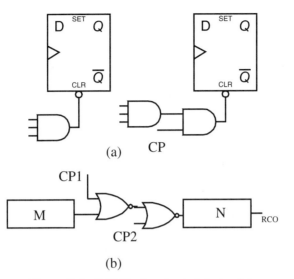

(a)

(b)

Figure 8.6   Initialization through test point.

and Mod 8 counters. In this fashion the test length can be reduced from 2^{24} to 2^8. Also, the ultimate of test point insertion in sequential logic is the introduction of scan path, which is discussed in Chapter 9 [Williams 1973].

Test point insertion for BIST application is also very important. We alluded to random pattern resistance (RPR) faults in Chapter 1. Those are the faults that are not easily detectable by PR patterns. To facilitate their testing, control points are inserted in the circuit and they are considered as primary inputs to the circuit. This topic is discussed in Chapters 11 and 14.

8.5   PARTITIONING FOR TESTABILITY

The testability measures of a circuit are functions of its depth. Smaller circuits are easier to test. If a large circuit is broken down into smaller subcircuits, it is possible to include some controllability and observability points, as illustrated earlier. For this, the circuit needs to be modified by inserting some hardware to facilitate insertion of these points. A general hardware partitioning technique using multiplexers is illustrated in Fig. 8.7. The subcircuits, G1 and G2, are not two independent partitions. They are two segments that have common circuitry. Using the multiplexers, it is possible to isolate each segment and improve its controllability and observability. The added multiplexers are transparent during normal operation, as illustrated in Fig. 8.7*b*. To test one of the segments, say G1, the circuit is configured as shown in Fig. 8.7*c*. Some input to G2 will be used to supply the test data directly to G1. Therefore, the lines from G2 feeding into G1 become test points that are easily controllable. Similarly, the lines from

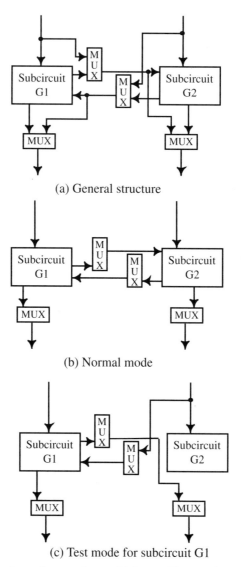

(a) General structure

(b) Normal mode

(c) Test mode for subcircuit G1

Figure 8.7 Partitioning scheme using multiplexers [Bozorqui-Nesbat 1980 © IEEE. Reprinted with permission].

G1 feeding into G2 are diverted through the multiplexer to the outputs of the circuit. They become easily observable at these outputs. We apply the concept to a concrete example, circuit, C3, in Fig. 8.8, which has been divided into three subsections, α, β, and γ. For example, to test the β partition of the same circuit, it is possible to insert two 2-to-1 MUXs as shown in Fig. 8.9 and to observe this partition directly ($S = 0$) or to synthesize the two other segments

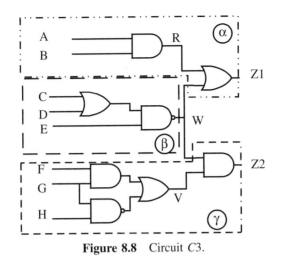

Figure 8.8   Circuit *C*3.

through their respective outputs ($S = 1$). The appropriate value is applied to I for testing each of these segments.

It is also possible to do this segmentation of the circuit by sensitization of the test data, and it is known as *sensitized partitioning*. For this we will revisit an approach that was proposed as an alternative to exhaustive testing [McCluskey 1981]. An N-input circuit is partitioned into M segments each of which has K_i primary inputs such that $K_i < N$, for all values of i. Testing all partitions exhaustively would require a test set of length: $\sum_{i=1}^{i=M} 2^{K_i} \leq 2^N$. Thus by testing each partition, rather than the whole circuit exhaustively, the complexity of the test is reduced while guaranteeing 100% fault coverage. We illustrate this approach with the circuit, C3, in Fig. 8.8.

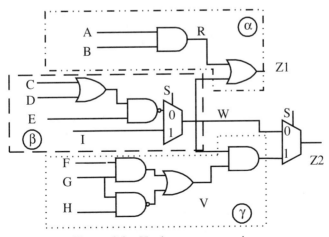

Figure 8.9   Hardware-segmentation.

Example 8.4. The circuit has eight primary inputs. The length of an exhaustive test is 256 patterns. Instead, the circuit is partitioned into three partitions: α, β, and γ. It is possible to test the first partition exhaustively using eight patterns and observing through the primary output, Z_1. The second partition, β, also depends on three of the inputs. It can be sensitized through Z_1 or Z_2. Partition γ can be tested exhaustively with 16 patterns since it depends on four inputs. The total length of the test is the sum of the three individual tests, 32 patterns instead of 256 patterns. To sensitize the test for partition α, we need to keep $W = 0$. But we have to have $W = 1$ to observe the response to the test for γ. For partition β, any time a pattern for α that makes $R = 0$ or a pattern for γ that makes $V = 0$, the response for β will be observed through Z_1 or Z_2, respectively.

It is possible to minimize the test set by testing the partitions concurrently. For example, patterns for α that result in $R = 0$ can be combined with patterns for β that make $W = 0$. Similarly, patterns for β that make $W = 1$ can be applied concurrently with those patterns for γ that make $V = 1$. Considering these facts, we can compact the test from 32 to 16 patterns. All patterns are listed in Table 8.2. It might be possible to compact the test further by taking advantage of the *don't care*, *x*, signals.

Another special case of exhaustive testing is known as *verification testing* [McCluskey 1984]. It is applicable to circuits where each primary output is a function of only a subset of primary inputs. Thus the circuit is divided into segments with overlapping components. We illustrate the concepts with three different circuits.

Example 8.5. Consider first the simple example, circuit C4, shown in Fig. 8.10. This circuit has five inputs ($n = 5$) and three outputs, $f(a, b, c) = a(b + c)$, $g(b, c, d) = d \oplus (b + c)$, and $h(e, d) = e + d$. Each output is a function of a subset of the primary inputs set. The three outputs have 3, 3, and 2 inputs, respectively. The largest number of input will be denoted by w and here $w = 3$. An exhaustive test set would consist of 32 patterns. If we exhaustively test one output at a time, the combined test will consist of 8, 8, and 4 patterns, a total of 20 patterns. Although this is shorter than the exhaustive test, it can still be reduced. It has been proven that if two inputs never appear in the same output function, they can have the same test signal applied to both [McCluskey 1984]. This results in reducing the number of required test signals, a scheme called *maximum test concurrency* (MTC). We follow a technique formulated by McCluskey. For circuit C4 we develop the *dependency matrix* shown in Table 8.3. This matrix consists of m rows and n columns, where m is the number of primary outputs and n is the number of primary inputs of the circuits. For this example, $m = 3$ and $n = 5$.

The first column lists the various functions. In this matrix, for each function, a 1 entry in any of the subsequent columns indicates the dependency of this function on the corresponding variable; otherwise, the entry in the matrix is 0.

TABLE 8.2 Pseudoexhaustive Testing for Circuit in Fig. 8.6

	A	B	R	C	D	E	W	F	GA	H	V
α	0	0	0				0				
	0	1	0				0				
	1	0	0				0				
	1	1	1				0				
	0	0	0				1				
	0	1	0				1				
	1	0	0				1				
	1	1	1				1				
β			0	0	0	0	1				
			1	0	0	1	1				
			0	0	1	0	1				
			1	0	1	1	0				
				1	0	0	1				
				1	0	1	0				
				1	1	0	1				
				1	1	1	1				
γ							0	0	0	0	1
							0	0	0	1	1
							0	0	1	0	1
							0	0	1	1	0
							0	1	0	0	1
							0	1	0	1	1
							0	1	1	0	1
							0	1	1	1	1
							1	0	0	0	1
							1	0	0	1	1
							1	0	1	0	1
							1	0	1	1	0
							1	1	0	0	1
							1	1	0	1	1
							1	1	1	0	1
							1	1	1	1	1

Next, we partition the primary inputs so that any partition is such that two or more of the inputs do not affect the same output. In other words, no partition would have more than one 1 in a row. For example, we combined a and d to form one partition since F depends on a and not d, and vice versa for G and H. Similarly, we combined c (or b) and e. This partitioning is not unique since we could have grouped a and e instead. Actually, there are three different ways of partitioning. In general, such partitioning is a NP-complete problem. The information provided by the partitioned matrix determines the total number of test patterns, 2^K, where K is the maximum number of 1's in a row. This is called the *maximum dependency*. Also, variables in the same partition can assume the

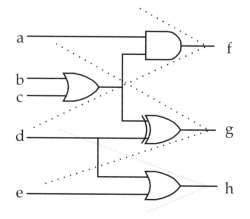

Figure 8.10   Circuit C4.

TABLE 8.3   Verification Testing for Circuit C4: (a) Dependency Matrix; (b) Two Different Input Partitionings

	a	b	c	d	e	a	d	b	c	e	a	e	b	c	d
f	1	1	1	0	0	1	0	1	1	0	1	0	1	1	0
g	0	1	1	1	0	0	1	1	1	0	0	0	1	1	1
h	0	0	0	1	1	0	1	0	0	1	0	1	0	0	1

| (a) | (b) |

same values in the test. For this example, $K = 3$ and the total number of patterns is only 2^3. This implies that only eight patterns are needed to test the entire circuit pseudoexhaustively! This is a 75% saving over the exhaustive test. The test is shown in Table 8.4. Notice that values of a and d are identical since the two inputs are in the same partition. The same is true for c and e.

TABLE 8.4   Test Set for Circuit C4

a	b	c	d	e
0	0	0	0	0
0	0	1	0	1
0	1	0	0	0
0	1	1	0	1
1	0	0	1	0
1	0	1	1	1
1	1	0	1	0
1	1	1	1	1

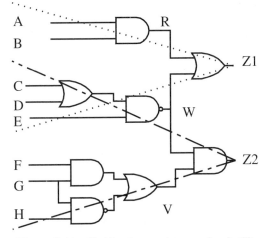

Figure 8.11   Verification testing on circuit C3.

Example 8.6. Circuit C3 has eight primary inputs and two primary outputs, Z_1 and Z_2. Each output is dependent on a subset of the primary inputs, 5 and 6, respectively, as can be determined from Fig. 8.11.

Exhaustive test sets for the partitions are of lengths 32 and 64; the total length of the test for the circuit is then 96, compared to the 256 patterns for an exhaustive test for the entire circuit. However, here again we can compact the test set because some of the signals on the shared inputs may be common to patterns for both partitions and therefore these patterns may be applied simultaneously.

To partition the matrix we combined A and F to form one partition, as shown in Table 8.5. Similarly, we combined B and G. This partitioning is not unique since we could have grouped A and H instead. Actually, there are six different ways of partitioning the matrix, since A and B can be combined with any of the inputs G, H, and F. The maximum dependency is $K = 6$. This implies that the number of test patterns needed is only $2^6 = 64$ instead of 256.

TABLE 8.5   Verification Testing for C3: (*a*) Dependency Matrix; (*b*) Input Partitioning

	A	B	C	D	E	F	G	H	A	F	B	G	C	D	E	H
Z_1	1	1	1	1	1	0	0	0	1	0	1	0	1	1	1	0
Z_2	0	0	1	1	1	1	1	1	0	1	0	1	1	1	1	1
				(*a*)									(*b*)			

8.6   EASILY TESTABLE CIRCUITS

Regularly structured circuits consist of an array of identical cells. They may be
arranged in one- or two-dimensional arrays as illustrated in Fig. 8.12. Examples
of such circuits are RAMs, FPGAs, array multipliers, or circuits that are deliberately designed in a modular fashion to facilitate layout on a chip. Although
these circuits may be combinational or sequential, it is on the first category that
we focus our attention in this section.

Combinational regular structures are usually referred to as iterative logic
arrays (ILAs). Often n-bit comparators are organized in a one-dimension array
and each compares the corresponding bit from two numbers. Another example
may be a parity tree where each cell is a two-input XOR. Array multipliers consist of some type of full adders that are usually arranged in a two-dimensional
array. Some of these structures are easily testable with a constant number of test
patterns that is independent of the size of the circuit—the number of identical
cells in the circuit [Friedman 1973]. Friedman called these ILAs C-testable. We
examine two examples: first, a parity tree, and second, an array multiplier that
is redesigned to make it C-testable. Other C-testable structures have a scalable
test length. That is, the test length grows proportionately to size. In this section
we examine the two types.

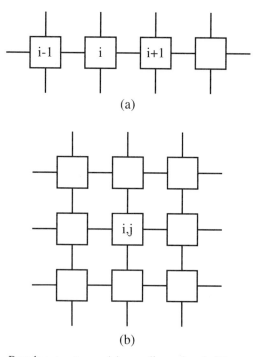

(a)

(b)

Figure 8.12   Regular structures: (*a*) one-dimensional; (*b*) two-dimensional.

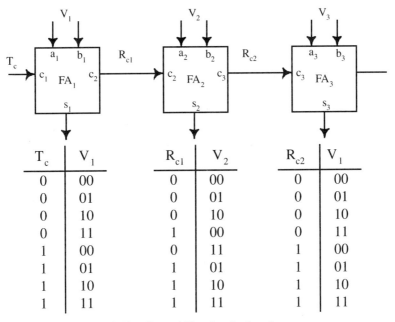

Figure 8.13   C-testability developing the test set.

Testing array structure will be encountered again in Chapters 12 and 13 when we study the testing of memories and FPGAs, respectively. C-testability and scalable testing are applicable to some of these structures.

8.6.1   C-Testability

Consider a one-dimensional array that is organized as shown in Fig. 8.13. Let T_c be an exhaustive test set for any cell j in the array, and let R_c be the response to this test. This test will detect all faults that keep the circuit combinational [Friedman 1973]. That is, the faults in cell j are sensitized to its outputs. If $R_c \subseteq T_c$, R_c will be an exhaustive test for cell $j + 1$. Notice that the order of the patterns in R_c is not necessarily the same as T_c. The test set for any cell j is the concatenation of T_c and V_j. If again $R_c \subseteq T_c$, then R_c concatenated with vector V_{j+1} forms the test for cell $(j + 1)$. Suppose that the cells are full adders, then an eight-pattern exhaustive test applied is sufficient to test the array as indicated in the figure. We will show two more example circuits.

Example 8.7. An *N*-input parity tree is easily testable with four patterns [Bossen 1970]. A parity tree is a fan-out free circuit. A test set for such a circuit consists of those patterns detecting faults on the primary inputs. The four patterns for an exhaustive test for each two-input XOR gate is {00,01,10,11}.

Consider gate 1 of the circuit shown in Fig. 8.14a. We apply an exhaustive

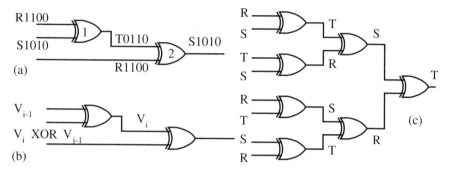

Figure 8.14 Parity tree: (*a*) labeling scheme; (*b*) labeling process; (*c*) labeled eight-input tree [Mourad 1989 © IEEE. Reprinted with permission].

test on it and we call the two vectors at the primary pins R and S. The result of the application of this test at the output of the gate is vector T. To observe the result of the test at the primary output, the output of gate 2, we can apply any signal, 1 or 0, at the other input of this gate. However, we apply the sequence R, which together with T, form an exhaustive test for gate 2. The output sequence is S. The three vectors have the following property: $V_i \oplus V_j = V_k$, where V_i, V_j, and V_k are selected from the set of vectors R, S, and T, in any order, provided that they are three different vectors. This property allows us to generalize the application of the test on any N-input tree. The labeling procedure illustrated in Fig. 8.14*b* of an eight-input tree is shown in Fig. 8.14*c*. You might be interested to verify the fault coverage of this test using a fault simulator. The test set for the eight-input parity tree is then:

| | Primary Inputs | | | | | | | |
Pattern	R	S	T	S	R	T	S	R
1	1	1	0	1	1	0	1	1
2	1	0	1	0	1	1	0	1
3	0	1	1	1	0	1	1	0
4	0	0	0	0	0	0	0	0

This approach was used successfully to determine the testability of parity checkers [Mourad 1989].

Example 8.8. Some circuits need just a slight modification to make them C-testable. An interesting example is the modification of the cell of an array multiplier that made it testable with only 16 patterns [Shen 1984]. Figure 8.15 shows a carry-save array multiplier and the basic unit, a full adder cell. This cell has four inputs, *a*, *b*, *c*, and *d*, and two outputs *x* and *y*:

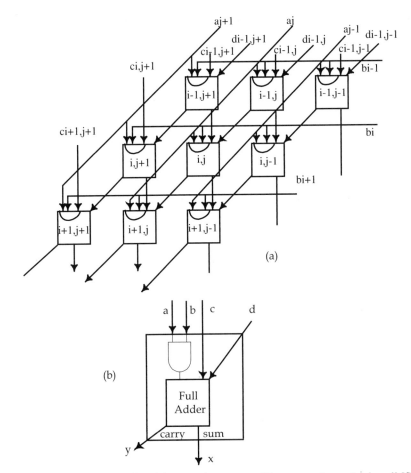

Figure 8.15 Array multiplier: (*a*) general structure; (*b*) carry and save basic cell [Shen 1984 © IEEE. Reprinted with permission].

$$x = (ab) \oplus c \oplus d$$
$$y = (ab)c + (ab)d + cd$$

As in the case of the parity tree, the aim is to use an exhaustive test for every cell and test all cells simultaneously.

Let $\{a, b, c, d\}$ be a test pattern applied to any of the cells. To be detected by the test, the fault has to be observable on either x, y, or both. A fault will cause any or both of the two outputs to be inverted. We notice also that the test patterns (a, b, c', d) and (a, b, c, d') yield the same results since x is a party of (ab), c, and d. Thus a fault appearing at c or d will be propagated to the output of the cell.

TABLE 8.6 **Truth Tables for the (a) Original and (b) Modified Array Multiplier Cell**[a]

	cd					cd			
ab	00	01	11	10	ab	00	01	11	10
00	0	0	1	0	00	0	1*	1	0
01	0	0	1	0	01	0	1*	1	0
11	0	1	1	0	11	0	1	1	1
10	0	0	1	0	10	0	0	1	0
		(a)					(b)		

Source: Data from [Shen 1984 © IEEE. Reprinted with permission].
[a]Asterisks denote proposed changes.

Next, we need the values on x and y to replicate a test pattern of the cell. That is, if $\{abcd\}$ is a pattern applied on cell (j, k) and $\{a_1 b_1 c_1 d_1\}$ is a pattern applied on cell $(j, k + 1)$, the $\{ab_2 x_1 y\}$ is a test for cell $(j + 1, k + 1)$. This is possible only if the main carry-save cell is slightly modified as indicated by the truth tables in Table 8.6. The proposed change, which is indicated by an asterisk, does not alter the functionality of the cell as a multiplier since, during normal operation, c and d on the first row of the array are both 0. Consequently, the patterns $\{abcd\} = \{0001\}$ and $\{0101\}$ can never appear on the input of any cell in the array. The array may thus be tested by 16 patterns regardless of its size. For details on constructing the test, you are referred to Shen's work [Shen 1984].

8.6.2 Scalable Testing

For arrays that are not C-testable the response, R_c, of cell j to the test, T_c, is such *that* $T_c \cap R_c \subset T_c$. For the response from cell j to be a test for cell $j + 1$, we need to apply on cell j an augmented test set, $T_a = T_c + T_p$, such *that* $T_c \subset R_a$, where R_a is the response of the cell to T_a. We can then write $R_a = T_c + T_q$. This is the test for $j + 1$ and hence it has to be augmented by T_p. In this fashion, when we reach the nth cell in the array, the test set will be $T_c + p(n - 1)$, where p is the length of the augmented patterns.

A scalable test can be obtained for the array comparator whose cell is shown in Fig. 8.16. The outputs, S and G, of this cell are defined in Table 8.7a. The minimal test set and the augmented set are shown in Table 8.7b and c. For any number of bits N, the test length is $K + p(N + 1)$, where K is the minimal test for one unit ($K = 6$) and p the number of added patterns ($p = 2$). Test scalability is becoming more relevant for core-based design [Al-Assad 1998] and will be revisited in a later chapter.

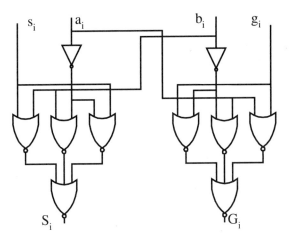

Figure 8.16   Two-bit magnitude comparator.

TABLE 8.7   Array Comparator: (*a*) **Definition;** (*b*) **Test Set** (T_c); (*c*) **Augmented Test** (T_a)

	GS								Cell *i*			Cell *i* + 1		
gs	*ab*	00	01	11	10		*gs*	*ab*	*GS*		*gs*	*ab*	*GS*	*gs*	*ab*	*GS*
00		00	01	00	10	1	11	00	11	1	11	00	11	11	00	11
01		01	01	01	10	2	00	00	00	2	00	00	00	00	00	00
11		*xx*	*xx*	*xx*	*xx*	3	10	11	10	3	10	11	10	10	10	10
10		10	01	10	10	4	01	11	01	4	01	11	01	01	01	01
						5	10	01	01	5	10	01	01	01	10	10
						6	01	10	10	6	01	10	10	10	01	01
										7	10	01	01	01	01	01
										8	01	10	10	10	10	10

| (*a*) | (*b*) | (*c*) |

REFERENCES

Agrawal, V. D., and R. M. Mercer (1982), Testability measures: what do they tell us, *Proc. IEEE Semiconductor Test Conference*, pp. 391–396.

Agrawal, V. D., and S. C. Seth (1985), Probabilistic testability, *Proc. IEEE International Conference on Computer Design*, pp. 562–565.

Al-Assad, J. P. Hayes, and B. T. Murray (1998), Scalable test generators for high-speed datapath circuits, Kluwer Academic, Norwell, MA.

Bennetts, R. G., C. M. Maunder, and G. D. Robinson (1981), CAMELOT: a computer-aided measure for logic testability, *IEEE Proc.*, Vol. 128, Part E, No. 5, pp. 177–189.

Bennetts, R. G. (1984), *Design of Testable Logic Circuits*, Addison-Wesley, Reading, MA, p. 164.

Bossen, D. C., D. L. Ostapko, and A. M. Patel (1970), Optimum test patterns for parity network, *Proc. AFIPS Fall 1970 Joint Computer Conference*, Vol. 37, pp. 63–83.

Bozorgui-Nesbat, S., and E. J. McCluskey (1980), Structured design for testability to eliminate test pattern generation. *Proc. International Symp. Fault-Tolerant Computing*, pp. 158–163.

Breuer, M. A., and A. D. Friedman (1976), *Diagnosis and Reliable Design of Digital Systems*, Computer Science Press, New York.

Chakradhar, S. T., et al. (1991), *Neural Models and Algorithms for Digital Testing*, Kluwer Academic, Norwell, MA.

Friedman, A. D. (1973), Easily testable iterative systems, *IEEE Trans. Comput.*, Vol. C-22, No. 12, pp. 1061–1064.

Fujiwara, H., and S. Toida (1982), The complexity of fault detection problem for combination circuits, *IEEE Trans. Comput.*, Vol. C-31, No. 6, pp. 555–560.

Fujiwara, H. (1986), *Logic Testing and Design for Testability*, MIT Press, Cambridge, MA.

Fujiwara, H. (1990), Computational complexity of controllability/observability problems for combination circuits, *IEEE Trans. Comput.*, Vol. C-39, No. 7, pp. 762–767.

Goldstein, L. H. (1979), Controllability/observability analysis of digital circuits, *IEEE Trans. Circuits Syst.*, Vol. CAS-26, No. 9, pp. 683–693.

Goldstein, L. H., and E. L. Thigpen (1980), SCOAP: Sandia controllability/observability analysis program, *Proc. Design Automation Conference*, pp. 190–194.

Grason, J. (1979), TEMAS: a testability measure program, *Proc. Design Automation Conference* pp. 156–161.

Gu, X., K. Kuchcinski, and Z. Peng (1995), An efficient and economic partitioning approach for testability, *Proc. IEEE International Test Conference*, pp. 403–412.

Gundhach, H. H. S., and K. D. Muller-Glaser (1990), On automatic test-point insertion in sequential circuits, *Proc. IEEE International Test Conference*, Philadelphia, PA, Oct., pp. 1072–1079.

Hayes, J. P. (1974), On modifying logic networks to improve their diagnosability, *IEEE Trans. Comput.*, Vol. C-22, No. 1, pp. 56–63.

Hayes, J. P., and A. D. Friedman (1975), Test points placement to simplify fault detection, *IEEE Trans. Comput.*, Vol. C-23, No. 7, pp. 727–735.

Ibarra, O. H., and S. K. Sahni (1975), Polynomially complete fault detection problems, *IEEE Trans. Comput.*, Vol. C-24, No. 3, pp. 242–249.

Jamoussi, M., B. Kaminska, and D. Mukhekar (1991), A new variable testability measure: a concept for data-flow testability evaluation, *Proc. IEEE International Test Conference*, pp. 239–243.

Kovijanic, P. G. (1979), Computer aided testability analysis, *Proc. IEEE Automatic Test Conference*, pp. 292–294.

Lin, C.-J., Y. Zorian, and S. Bhawmik (1993), PSBIST: a partial-scan based BIST scheme, *Proc. IEEE International Test Conference*, pp. 507–516.

McCluskey, E. J. (1984), Verification testing: a pseudo-exhaustive test technique, *IEEE Trans. Comput.*, Vol. C-33, No. 6, pp. 541–546.

McCluskey, E. J. (1986), *Principles of Logic Design with Emphasis on Semicustom Circuits*, Prentice Hall, Upper Saddle River, NJ, 1986.

Mourad, S., and E. J. McCluskey (1989), Testability of parity checkers, *IEEE Trans. Ind. Electron.*, Vol. IE-36, No. 2, pp. 254–260.

Ratiu, I. M., A. Sangiovanni-Vincentelli, and D. O. Peterson (1982), VICTOR: a fast VLSI testability analysis program, *Proc. IEEE International Test Conference*, pp. 397–401.

Savir, J. (1983), Good controllability and observability do not guarantee good testability, *IEEE Trans. Comput.*, Vol. C-32, No. , pp. 1198–1200.

Schultz, M. H., et al. (1988), SOCRATES: a highly efficient automatic test pattern generation system, *IEEE Trans. Comput.-Aided Des.*, Vol. CAD-8, No. 1, pp. 126–137.

Shen, J. P., and F. J. Ferguson (1984), The design of easily testable VLSI array multipliers, *IEEE Trans. Comput.*, Vol. C-33, No. 6, pp. 554–560.

Sridhar, T., and J. P. Hayes (1981), Design of easily testable bit-sliced systems, *IEEE Trans. Comput.*, Vol. C-30, No. 11, pp. 842–854.

Williams, M. J. Y., and B. Angell (1973), Enhancing testability of large scale integrated circuit via test points and additional logic, *IEEE Trans. Comput.*, Vol. C-22, No. 1, pp. 46–60.

Wulderlich, H.-J., and S. Hellebrand (1989), The pseudo-exhaustive test of sequential circuits, *Proc. IEEE International Test Conference*, pp. 19–27.

PROBLEMS

8.1. The two circuits in Fig. P8.1 are functionally equivalent. Compare their SCOAP testability measures.

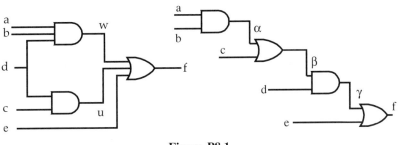

Figure P8.1

8.2. Verify the information presented in Example 8.2 about circuit C2 in Fig. 8.5*b* by developing a SA test for the circuit using **(a)** only Z as the output; **(b)** F and Z as outputs; and **(c)** H and Z as outputs.

8.3. Consider the logic block of ACT1 and generate a pseudoexhaustive test with a minimal number of patterns. For this circuit, see Fig. P3.6.

8.4. Use the labeling technique recommended in Section 8.6 and develop a four-pattern test set for a 16-input parity tree and verify the fault coverage using a fault simulator.

8.5. Develop a maximal concurrency test for **(a)** the parity generator network, part SN74630, and **(b)** the dual multiplexer network, part SN74153.

9

SCAN-PATH DESIGN

9.1 INTRODUCTION

Testing sequential circuits is a very complex process [Miczo 83]. Therefore, it has been important to find a scheme to facilitate their testing. The complexity of testing is a function of the number of feedback loops and their length. The longer a feedback loop, the more clock cycles are needed to initialize and sensitize patterns. In 1973, Williams and Angell proposed a design that increases the ease of testing of synchronous sequential circuits [Williams 1973]. Their scheme facilitates the initialization, controllability, and observability of the circuit. The suggested approach reduces the process of test pattern generation to that for combinational circuits. According to this scheme, a synchronous sequential circuit works in two modes, normal and test. The *normal mode* is depicted as shown in Fig. 9.1a, where the circuit is represented by the Huffman model introduced in Chapter 3. In the *test mode*, all the flip-flops are disconnected from the circuit and configured as a shift register, as illustrated in Fig. 9.1b. In this shift mode the flip-flops can be easily initialized. In this fashion, all the outputs of the flip-flops become pseudo primary inputs to the circuit. After initializing the flip-flops, a test pattern is applied to the primary inputs, the results are latched at the flip-flops, and then they are propagated to the output by placing the circuit in a test mode and clocking enough times to capture the results. This arrangement makes the input to a flip-flop an observation point. All such observation points are pseudo primary outputs to the circuit. To switch between normal operation and shift modes, each flip-flop needs additional circuitry to perform the switch. There are different implementations of storage devices and they are discussed in Section 9.6.

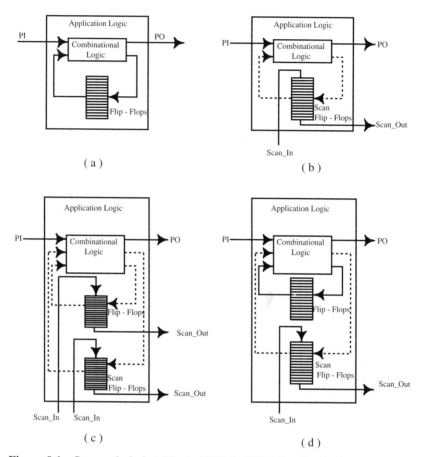

Figure 9.1 Scan-path design [Nagle 1989 © IEEE. Reprinted with permission].

Prior to the publication of scan-path methodology in 1973, the scheme was reported at a Japanese conference [Kobayashi 1963]. Also, at IBM, William C. Carter employed a very similar scheme for the purpose of design debugging [Carter 1964]. Scan-path design has evolved since Williams first proposed it. For example, modifications have been suggested for the storage devices. Instead of multiplexed flip-flops, two-port flip-flops or latches may be used. Also, different architectures have been recommended. It is possible to use a scan architecture that allows selective observation at any of the flip-flops in the chain. Another important approach is to organize the flip-flops in more than one scan chain, as illustrated in Fig. 9.1c. Instead of full-scan, it may be advantageous to scan some of the flip-flops. This approach, known as partial scan, is illustrated in Fig. 9.1d. Because of routing congestion in modern complex ICs, the placement order of the flip-flops present several problems that we discuss in Section 9.11.

9.2 SCAN-PATH DESIGN

Any synchronous sequential circuit may be modeled as shown in Fig. 9.2a, which is a detailed version of Fig. 9.1a. Here we assume that only D flip-flops are used. This assumption is very realistic and does not limit the applicability of the scheme to other flip-flop types. The combinational part represents the next-state forming logic and the output forming logic. For scan design, each flop-flop is preceded by a 2-to-1 MUX. The select lines of all flip-flops are connected together as shown in Fig. 9.2b. As this select line, scan enable (SE), is 0, the circuit works under normal operation. Since the other input of the MUX is connected to the output of the previous flip-flop, for SE = 1, all flip-flops are connected as a shift register. For the first flip-flop, the second input to the MUX is labeled as SI. Also, the output of the last flip-flop is labeled as SO.

This arrangement indicates that the scan-path approach requires three extra

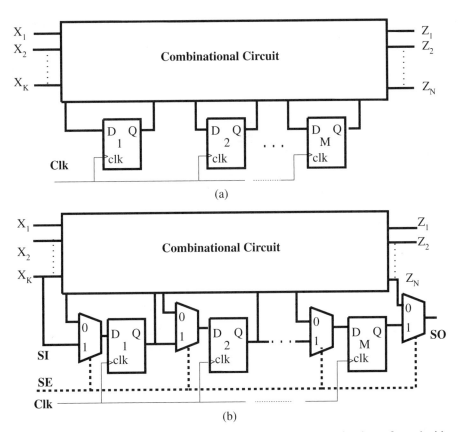

Figure 9.2 (a) Generic for a synchronous sequential circuit; (b) circuit configured with scan path.

I/O pins for the signals scan enable (SE), scan-in (SI), and scan-out (SO). However, instead of connecting SI to an extra pin, this signal is multiplexed with
one of the inputs. Similarly, the SO is multiplexed with one of the outputs. In
Fig. 9.2*b* we multiplexed SI with X3 and SO with Z_3. The different alternatives
of connecting these signals do not affect the operation of scan path.

9.3 TEST PATTERN GENERATION

Test patterns are generated only for the combinational circuit based on the following assumptions:

1. No asynchronous signals are in the circuit, including set and reset of flip-
   flops.
2. Latches are controlled by nonoverlapping clocks.
3. For test pattern generation:
   3.1. Propagation ends at an input of a flip-flop, which is considered as
        an output of the circuit.
   3.2. Justification ends at an output of a flip-flop, which is considered an
        input to the circuit.

Thus a circuit configured in scan path has $N + M$ primary inputs, where N is
the original primary inputs of the circuit and M is the number of the flip-flops
to be part of the scan chain. The inputs of the flip-flops are dispersed within
the circuit and thus allow improved controllability of the nodes. The increased
number of the outputs also improves the observability of the circuit and, in
effect, partitions the circuit into smaller components.

The example in Fig. 9.3 shows how the combinational part is actually partitioned. It consists of two parts. The first part consists only of gate G1; the
second comprises gates G2 and G3. Each partition has one primary input, F,
but while the first partition has one pseudoinput, Y_2, the other partition has two
pseudoinputs, Y_1 and Y_2.

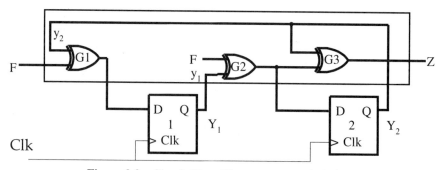

Figure 9.3 Circuit S1 to illustrate scan-path design.

9.4 TEST PATTERN APPLICATION

Test application consists of two stages. The flip-flops are tested, followed by the combinational part of the circuit. Testing the flip-flops is done first since they will be used in controlling and observing the testing of the combinational part of the circuit.

9.4.1 Testing the Flip-Flops

The flip-flops are configured as a shift register, SE = 1, and patterns are applied through the SI input. The output is observed at the SO pin. The pattern applied may consist of M 0's followed by M 1s, where M is the number of flip-flops. Because of the risk of pattern sensitivity, which will be discussed in conjunction with RAM testing in Chapter 12, other variations of patterns may be used. It is possible, for example, to alternate 0's and 1's.

9.4.2 Testing the Combinational Part of the Circuit

Test patterns have to be generated for each partition in the circuit, and the patterns need to be organized in the proper order of application. For stuck-at faults, the order is irrelevant; however, in the case of fault models for which the order of the sequence is pertinent, this has to be taken into consideration. After the application of each pattern, the response is first latched in the flop-flops and then it is scanned out of the circuit by configuring the flip-flops in a shift register. This scheme is summarized in the following protocol:

Repeat until all patterns are applied.
   a. Set SE = 1; shift in the initial values on the flip-flops. (These are the signals at the output of the latches for the first test pattern.)
   b. SE = 0; apply a pattern at the primary inputs.
   c. Clock the circuit once and observe the results at the primary outputs.
   d. Clock the circuit M times.
End repeat.

Since each pattern includes signals on the output of the flip-flops, we entered these signals prior to the application of the other signals of the patterns at the primary inputs. However, the initialization of the flip-flops may be performed as the previous test is scanned out—step d of the test application protocol given above. We can modify the test application protocol as follows:

Initialize.
Set SE = 1; shift in the initial values on the flip-flops for the first pattern.
Repeat until all patterns are applied.

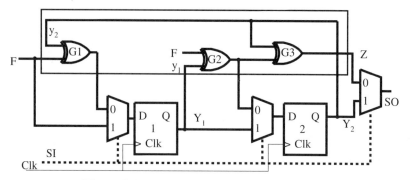

Figure 9.4 Circuit S1 with scan path inserted.

a. Set SE = 0; apply a pattern at the primary inputs.

b. Clock the circuit once and observe the primary outputs.

c. Set SE = 1.

- Apply the initialization for the next pattern at SI.
- Clock the circuit *M* times.
- Observe at the primary outputs and the SO pins.

End repeat.

The protocol is used to apply test patterns on the circuit shown in Fig. 9.4 and is described in the next section.

9.5 EXAMPLE FOR SCAN-PATH TESTING

Originally, the circuit had one input (excluding the clock) and one output. With scan-path structure, the PI and PO have been increased by the number of flip-flops. Also, the combinational part of the circuit has been partitioned into two components, as illustrated in Fig. 9.5. XOR 1 has inputs F and y_2, and the second component consists of the two other XOR gates with inputs F, y_1, and y_2, and outputs Z and D_2.

The first step in testing this sequential circuit will be to generate the patterns for the combinational parts of the circuit. Here we have three XOR gates that can be tested simultaneously. We learned from previous chapters that three patterns are sufficient to test this gate. However, assuming any type of implementation, we will use an exhaustive test, four patterns. Let us assume that we will use the test in the following sequence: $Fy_i = 00, 10, 01, 11$, where $i = 1$ or 2, depending on the gate tested. Thus the patterns on the second partition are $FY_1Y_2 = \{000, 100, 011, 111\}$.

We next apply the patterns. We summarize the steps taken in the application of the test in Table 9.1. We chose to apply the all-0's pattern followed by an

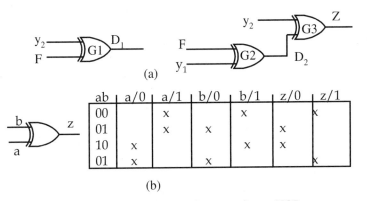

ab	a/0	a/1	b/0	b/1	z/0	z/1
00		x		x		x
01		x	x		x	
10	x			x	x	
01	x		x			x

(b)

Figure 9.5 Partitions of S1 and test set for an XOR gate.

all-1's pattern to check the flip-flops. This is given in the first four lines of the table. The first four columns of this table list the inputs to the circuit: the clk, SI, F, and SE. The fifth and sixth columns give the present state (PS) and the next state (NS) of the flip-flops. The present state for each clock is, of course, the next state from the previous clock.

To apply pattern 00 on G1 and G2 means that $Fy_1 = 00 = Fy_2 = 00$. That is, the pattern is $Fy_1y_2 = 000$. Thus for $y_1y_2 = 00$, we clock the circuit twice with

TABLE 9.1   Summary of Test Application to the Circuit of Fig. 9.4

clk	SE	F	SI	PS y_1y_2	NS Y_1Y_2	Z	SO
1	1	x	1	xx	1x	x	x
2	1	x	1	1x	11	x	xx
3	1	x	0	11	01	0	1xx
4	1	x	0	01	00	1	11xx
5	1	x	0	00	00	0	011xx
6	0	0	x	00	D_1D_2	0	Z011xx
7	1	x	0	D_1D_2	$0D_1$	x	$D_2$0011xx
8	1	x	0	$0D_1$	00	x	$D_1D_2$0011xx
9	0	1	x	00	D_1D_2	1	$ZD_1D_2$0011xx
10	1	x	1	D_1D_2	$1D_1$	x	$D_2ZD_1D_2$0011xx
11	1	x	1	$1D_1$	11	x	$D_1D_2ZD_1D_2$0011xx
12	0	0	x	11	D_1D_2	0	$ZD_1D_2ZD_1D_2$0011xx
13	1	x	1	D_1D_2	$1D_1$	x	$D_2ZD_1D_2ZD_1D_2$0011xx
14	1	x	1	$1D_1$	11	x	$D_1D_2ZD_1D_2ZD_1D_2$0011xx
15	0	1	x	11	D_1D_2	1	$ZD_1D_2ZD_1D_2ZD_1D_2$0011xx
16	1	x	a	D_1D_2	aD_1	x	$D_2ZD_1D_2ZD_1D_2ZD_1D_2$0011xx
17	1	x	b	aD_1	ba	x	$D_1D_2ZD_1D_2ZD_1D_2ZD_1D_1$0011xx

SE = 1 and SI = 0. However, since in testing the flip-flops we have already entered 0s on the flip-flops, the circuit was ready for application of the first test. Thus we apply $F = 1$ while SE = 0 and clock the circuit once. Then while SE = 1, we clock twice while applying 0 on SI. These two clock cycles will allow the observability of the response to the first pattern and the initialization of the flip-flops for the second test pattern, $Fy_1y_2 = 100$. To apply this pattern, we simply apply 0 on F and clock once. To observe the pattern, we clock twice while SE = 1. Again, as the results of the second pattern are scanned out, the initialization for the third pattern is performed. Such initialization is not needed when the last pattern is applied. Therefore, at the last clock cycles, 16 and 17, no particular signals are needed for subsequent test patterns. We deliberately indicated the data as a and b to emphasize that no particular signal is required.

9.6 STORAGE DEVICES

The most important requirement of storage device is the ability to work individually in normal mode or to be configured as a shift register in testing mode. In addition to multiplexed flip-flops, two other types are used: two-port flip-flops and level-sensitive latches. For a partial scan, the devices that are not scanned have an added requirement and that is of holding its value during shift mode.

9.6.1 Two-Port Flip-Flop

To configure the flip-flops in a shift register, we preceded each one by a 2-to-1 multiplexer. We can obtain the same functionality by substituting a two-port storage device that was originally proposed to minimize the gate size and its delays [Funatsu 1978]. A possible version of this device is the scan flip-flop, FD3S2, shown in Fig. 9.6a [LSI 1992]. It has two data lines, D and SI, and two clocking systems, CP and the combined SCK1 and SCK2 signals. The input at pin D is latched at the output at the positive edge of the system clock CP, whereas transition input at pin SI is captured after the pulses at SCK1 and SCK2

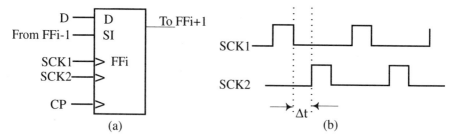

Figure 9.6 (*a*) Two-port D flip-flop; (*b*) nonoverlapping clocks.

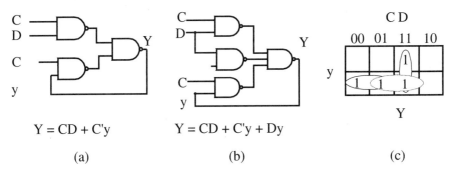

$$Y = CD + C'y \qquad\qquad Y = CD + C'y + Dy$$

(a)   (b)   (c)

Figure 9.7   Hazard-free clocked latch [Eichelberger 1985 © IEEE. Reprinted with permission].

are applied. SCK1 and SCK2 are used for shift mode in a non-overlapping fashion shown in Fig. 9.6b to minimize the hazards. After the application of each test pattern, CP is clocked once. Then SCK1 and SCK2 are clocked M times, where M is the number of two-port flip-flops in the scan chain.

9.6.2   Clocked Latch

The storage units in a sequential design do not have to be edge triggered. They may be latches. The latter devices are level sensitive and require a specific order on signal application. Figure 9.7 shows a clocked D-latch. Its next-state equation is $Y = CD + yC'$. However, to eliminate static hazard, the latch is redesigned as shown in Fig. 9.7b [Eichelberger 1965]. The governing equation is then $Y = CD + yC' + Dy$. As $C = 0$, the latch's output is unchanged, whereas when $C = 1$, any change on the input, D, is reflected immediately on the output Y. This is known as the transparency property of latches. The clock changes to 1 only when the input, D, is stabilized. But the slew rate of the clock does not alter the final value of the latch. It may cause delays, but no change results in the level of the output. This property is illustrated in Fig. 9.8, where the edge-triggered flip-flop and the clocked latch responses to the same input waveform are shown. Notice that when the clock, CK, is high, the input on D is seen immediately on Q for the clocked latch. This is known as the transparency property [McCluskey 1996].

Because of the transparency property, latches cannot be connected as a shift register while controlled by the same clock. When the clock is high, the value on the first latch will be passed immediately to all the next latches in a domino effect. For this, two latches are concatenated to perform the task of a flip-flop, as illustrated in Fig. 9.8. The circuit in Fig. 9.9 shows latches connected in a master–slave arrangement. However, to avoid essential hazards, the second latch is operated with clock CK2, which is nonoverlapping with clock CK1.

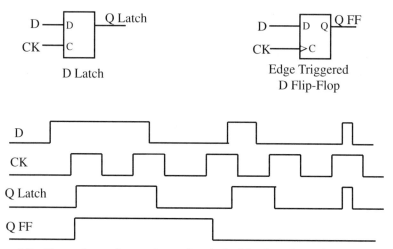

Figure 9.8 Comparison of operations of a clocked latch and an edge-triggered flip-flop.

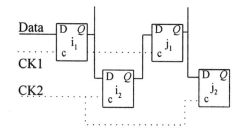

Figure 9.9 Formation of a shift register with clocked latches.

9.7 SCAN ARCHITECTURES

Scan-path design can be organized in various architectures [McCluskey 1986] and may be mixed with other DFT structures, such as built-in self-test (BIST), described in Chapter 11. In this section we examine only two architectures: level-sensitive scan design (LSSD) and Scan–Set. The first is a design that replaces flip-flops by latches and the second has the particular feature of being able to monitor the circuit on-line.

9.7.1 Level-Sensitive Scan Design

As described at the end of Section 9.6, the latch cannot be used in a shift register because when the clock is high, the value on one latch will change immediately because of the transparency property. For this, two latches are concatenated to perform the task of a flip-flop. The circuit in Fig. 9.10 shows two latches con-

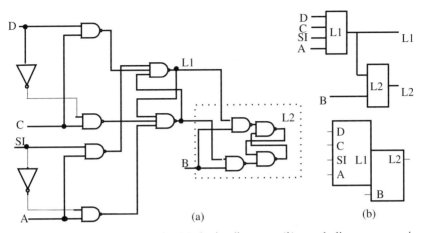

Figure 9.10 Polarity hold latch: (*a*) logic diagram; (*b*) symbolic representations [Eichelberger 1977 © IEEE. Reprinted with permission].

nected in a master–slave arrangement [Eichelberger 1977]. However, to avoid essential hazard, the second latch is operated with a clock *B*, which is nonoverlapping with clock *C*. This latch may then be used in a shift register as shown in Fig. 9.9, where the operation of the register is also illustrated.

To use the latch in a scan-path design, the design is augmented as shown in Fig. 9.7. This latch was proposed and patented by Edward Eichelberger and is the basic of IBM's level-sensitive scan design. Two clocks, *C* and *B*, are needed to perform the shift. To multiplex between normal operation and shift modes, a third clock, *A*, is used. The normal operation of the latch implies the use of clocks *C* and *B*. For shift mode, clocks *A* and *B* are used.

Each flip-flop in Fig. 9.2*a* is substituted for by a level-sensitive latch, and the circuit is then transformed to the one shown in Fig. 9.11. The test is performed as described in Section 2.2.2, with the exception that every time a clock signal is applied, it is replaced by the sequence clock, *C*, followed by clock *B*, or by clock *A* followed by clock *B* for the shift mode. We list the steps here for convenience.

Test the latches.

    Set *A* = *B* = 1.

    Apply 0 and 1 alternately at SI.

    Clock *A*, then clock *B*, *N* times.

Initialize.
Shift in the initial values on the flip-flops. (These are the signals at the output of the latches for the first test pattern.)
Repeat until all patterns are applied.

    a. Apply a pattern at the primary inputs.

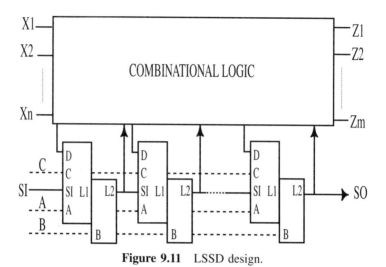

Figure 9.11   LSSD design.

b. Clock *C*, then clock *B* once and observe the results at the primary outputs.

c. Shift out the response
   • Apply the initialization for the next pattern at SI.
   • Clock *A*, then clock *B*, *M* times.
   • Observe at the primary outputs and the SO pins.

9.7.2   Scan-Set Architecture

An interpretation of the Scan-Set architecture is shown in Fig. 9.12. The circuit storage devices are not configured as a shift register and a shift register is added to the circuit [Stewart 1978]. The initialization data are introduced serially in the shift register, using clock TCK and SI port. The data are then transferred in parallel to all the latches of the circuit using the clock SCK. At this stage a test pattern is applied to the primary input and the system clock, SCK, is applied once to transfer the test response to the output of the latches. The shift register is then clocked *N* times with TCK, to scan the data to SO. The observation of the response is *on-line*. That is, it is possible to observe the circuit while it is working under normal operation. For both architectures described in this section, the storage devices may be arranged in one or multiple chains, as described next.

9.8   MULTIPLE SCAN CHAINS

Test application time is a function of the number of flip-flops scanned; the more flip-flips in the circuit, the longer the testing time will be. It is possible to reduce

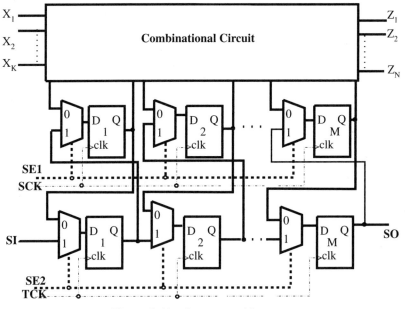

Figure 9.12 Scan-set architecture.

the time by forming more than one scan chain that can be operated in parallel. This arrangement is particularly effective for BIST design, which employs very long pseudo-random test sets. Also, some circuits use multiple clocks to control different parts of the circuit. It is easier in this case to configure the storage devices controlled by each clock separately. For externally applied test patterns, multiple chains require more pins or a more complex pin-multiplexing scheme. However, if BIST is used, there is no pin penalty.

In general, selecting the appropriate devices in each scan chain is far from trivial. In addition to having to balance the number of storage devices in the parallel chains, it is important to select the devices such that the interconnect wires are minimized. This topic is discussed in Section 11.10. The topology of the circuit in Fig. 9.13 is, for example, amenable to two-scan chain organization.

9.9 COST OF SCAN-PATH DESIGN

Scan-path design has greatly facilitated the testing of sequential circuits and has been widely accepted in industry. The ease of testing did not come free. The cost of scan-path design involves five principal issues: extra area, higher pin count, circuit performance, test application time, and power dissipation at speed testing.

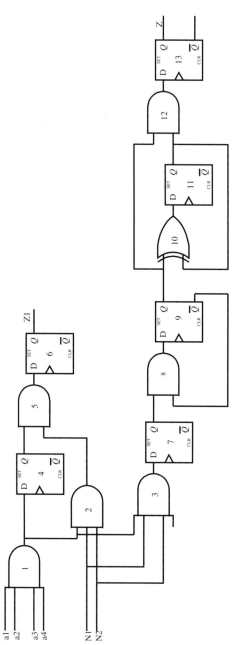

Figure 9.13 Example S2 for a sequential circuit.

9.9.1   Extra Area and Pins

The increase in silicon area is due to the complexity of the scan device, flip-flop, or latch. Although the overhead for one storage device may be high, the overall increase in hardware is of a lesser magnitude since all storage devices constitute a smaller percentage of the circuit. In addition, an efficient design of the scan cell can reduce the overhead. The most significant increase is due to the extra routing for connecting the flip-flops as a shift register. Efforts to order the device in the chain such that the area occupied by the interconnect wires is minimized can be done at placement and routing time. The extra pins are SI, SO, SE, or the second and third clocks that are used in multiple-port storage devices as described earlier in this chapter. SI and SO can possibly be multiplexed with input and output pins.

9.9.2   Performance

The major concern with performance is due to the added circuitry. In the multiplexed flip-flops, there is delay though the multiplexer. Two-port flip-flops (or latches) do not cause delay since the data can be available at the two ports simultaneously. The drawbacks of such types of devices is the need for a two-phase clock.

9.9.3   Test Application Time

As we have seen in Section 9.4.2, test application requires shifting response data for each test pattern. Thus test application time has been increased M times, where M is the number of flip-flops in the scan chain.

9.9.4   Heat Dissipation

Test application is usually done at a speed lower than the operating frequency of the circuit. However, when scan is used with built-in self-test (BIST), which is described in Chapter 11, testing is often done at speed, and in such a case the heat dissipation due to clocking all flip-flops is so excessive that it may burn the chip. At present, this problem is under investigation to minimize its impact.

9.10   PARTIAL SCAN TESTING

The penalties of full scan methods can be partially alleviated by scanning some but not all flip-flips of the circuit [Trischler 1980]. Thus the circuit remains sequential and requires a sequential test pattern generator. It is assumed, however, that such an approach will minimize the cost of the DFT circuitry: overhead area and delays. A trade-off is made between added hardware and fault coverage in terms of K, the longest loop in the circuit. Such a trade-off is shown in Fig. 9.14 for two circuits taken from ISCAS benchmark suite [Brglez 89].

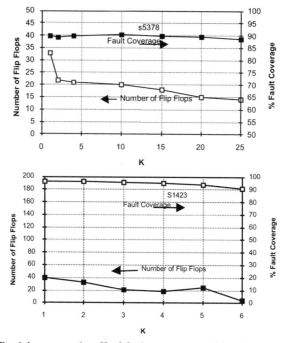

Figure 9.14 Partial scan, trade-off of fault coverage and hardware overhead [Greene 1998 © IEEE. Reprinted with permission].

The main issue in partial scan is the selection of the flip-flops to be included in the scan chain. Several studies have been conducted to make the best selection. By *best* is meant a selection that minimizes the cost of such a DFT structure:

- The circuit is easier to test by the sequential ATPG.
- The area overhead is minimized.
- The placement of the flip-flops is such that the interconnects are minimized.
- The delays are shortened.

Easing test pattern generation seems to be the most important factor. It has been shown that the complexity of test pattern generation is a function of the number of flip-flops in feedback loops [Agrawal 1988, Cheng 1990]. It is also affected by the number of flip-flops between two scanned flip-flops. Approaches for the selection of the appropriate flip-flops to include in the scan chain are based on one of the following techniques: testability analysis [Trischler 1980, Abramovici 1991]; structural analysis [Cheng 1990, Lee 1990, Park 1992]; and test pattern generation [Ma 1988, Chickermane 1991, Parikh 1993].

Techniques based totally on testability analysis are not reliable since these measures are not necessarily good predictors of problems faced during test pattern generation [Savir 1983]. It is possible to check the contribution of each flipflop by calculating the fault coverage when one flip-flop at a time is scanned. This approach will probably give the best results, but it is computationally expensive and it is better to use a less optimal approach that can be accomplished faster.

A more feasible approach is combining structural analysis with TM or test pattern generation. In structural analysis, the circuit is represented by a directed graph, the nodes of which are the flip-flops, with the edges being the combinational logic between the flip-flops. We will use some graph terminology that we have defined in Chapter 3 and list them here for ease of reference.

9.10.1 Definitions

- An *S graph* is a cyclic graph, $G(V,E)$, that is associated with a sequential circuit. A vertex, v_i, represents a flip-flop, and the edge, (v_i, v_j), is a path between the output of one flip-flop (v_i) and the input of another (v_j).
- A *path* is a sequence of edges from v_i to v_n, and its length is $n - i$.
- A *cycle* is a path in which the starting vertex is also the final vertex. These cycles represent the feedback loops in the circuit.
- The *sequential depth* of a circuit is the longest cycle in the circuit. This length causes difficulty in test pattern generation. Self-loops, feedback loops consisting of one flip-flop, are cycles of unit length. They usually do not cause the main problems in test pattern generation [Cheng 1990].

9.10.2 Selecting Scan Flip-Flops

Using the graph for the circuit, the idea is to select the minimal number of flipflops that change the S-graph to an acyclic graph. This minimal set of flip-flops is called the *minimal feedback vertex set* (MFVS). Under these conditions, it is possible to use, with some modifications, a combinational ATPG. Determining the MFVS is an NP-complete problem; therefore, heuristics are used to accomplish the task. The circuit used to illustrate scan-path test pattern application is shown in Fig. 9.3 and consists of two flip-flops, where the output of the second flip-flop feeds back into the first. Thus we have a cycle of depth 2. It is sufficient to scan one flip-flop and break the cycle. The circuit is so small that it really does not matter which flip-flop is scanned. But, in general, selecting the appropriate flip-flop is not as simple as it may seem.

We explain the approach using circuit S3, shown in Fig. 9.15. The figure also shows the S-graph of the circuit. The main idea is to select some of the flip-flops so that the graph becomes acyclic. The graph shows four cycles, of lengths, 1 (self-loop), 2, 3, and 4. The question is, then: What is the minimum number of flip-flops to scan (where to cut the cycles) such that the circuit becomes acyclic?

Table 9.2a gives the adjacency matrix representing the S-graph in Fig. 9.15.

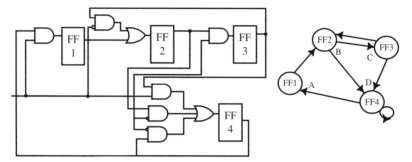

Figure 9.15   Sequential circuit S3 and its S-graph representation [Park 1992 © IEEE. Reprinted with permission].

An entry, a_{ij}, in the matrix indicates that flip-flop i feeds into flip-flop j if it is equal to 1; otherwise, it is 0. There are several approaches on how to use this matrix to search the graph for appropriate flip-flips to scan. One such approach uses the sum of indegree and outdegree for each vertex as an estimate of the number of possible loops through the vertex [Lee 1990]. This can be translated as adding the 1s in the corresponding flip-flop row and column and then selecting the largest number. The last column of Table 9.2a lists all the outdegrees for the flip-flops, and the last row gives the indegrees. For the example shown above, flip-flop 4 has the highest estimate, 5; this is followed by flip-flop 2. If we discount self-loops, both flip-flops would have the same number, 4. If we scan FF4, we cut all loops except that of depth 2, which is formed by FF2 and FF3. We can now update the matrix and calculate our cost function again. The updated values are shown in Table 9.2b, and to make the circuit fully acyclic, we can then scan FF2. Notice that we reach the same result by scanning FF3. However, remember that we are using a heuristic. There is another factor that we need to enter in selecting the appropriate flip-flops to scan: the proximity of the scannable flip-flops to each other.

TABLE 9.2   Adjacency Matrix for the S-Graph in Fig. 9.15

	FF1	FF2	FF3	FF4	Total		FF1	FF2	FF3	FF2T	Total
FF1	0	1	0	0	1		0	1	0	0	1
FF2	0	0	1	1	2		0	0	1	1	2
FF3	0	1	0	1	2		0	1	0	1	2
FF4	1	0	0	1	2		0	0	0	0	0
Total	1	2	1	3			1	2	1	2	

                          (a)   (b)

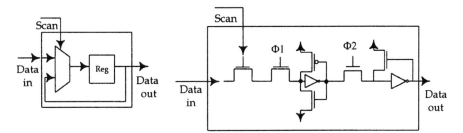

Figure 9.16   Hold mode for nonscannable flip-flop.

9.10.3   Test Application

Let us consider the most common approach to scan design using one clock and multiplexed flip-flop. After the test application, the system is clocked to latch the response on the flip-flops as usual. As the response is scanned out while clocking all flip-flops, the nonscanned flip-flops will latch to the scanned response. In other words, the circuit will not remain in the same state it was in when the test pattern was applied. Thus some provision must be made to keep the nonscanable flip-flops in the same state. This can be accomplished using separate clocks for the scanable and nonscanable flip-flops. Another alternative is to add a hold-mode feature in the nonscanable flip-flops, as illustrated in Fig. 9.16 [Steensma 1993, Debaney 1994]. It is, of course, possible to use Scan-Set architecture described in Section 9.7.2 in conjunction with partial scan. In this case, no special provision has to be made to maintain the state of the circuit at scan-out time since this process is done independent of the normal operation of the circuit.

9.11   ORDERING SCAN CHAIN FLIP-FLOPS

The order of the flip-flops in the scan chain affects two of the scan-path costs discussed in the preceding section: additional length of the interconnect wires and increased testing application time. It seems that an optimal approach to solving this problem would be during circuit synthesis. We discuss this approach in Chapter 14. Here we assume that after the circuit has been synthesized, the resulting netlist can then be optimized for the order of the chain, then placed and routed. Another approach is to actually place the netlist, then find a way "to stitch" the storage device for optimal order. It is important to realize that optimizing the test application does not necessarily result in optimal routing, or vice versa.

9.11.1   Optimizing for Test Application

The most general way for this optimization is multiple chains, but as we have mentioned before, this costs extra pins or multiplexers. In selecting the different

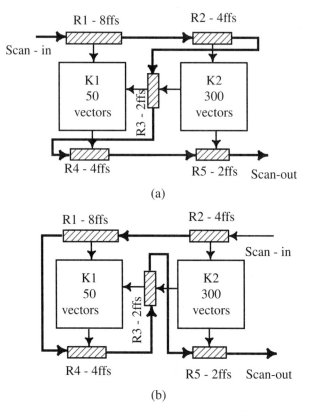

Figure 9.17 Ordering flip-flops, optimization for test application time [Narayanan 1992 © IEEE. Reprinted with permission].

devices in each parallel branch, it is better to do it after placement since the interconnect wires can be minimized. In some designs, such as datapath circuits, registers are used as output or input devices only; therefore, it is possible to take advantage of such structures to schedule multiple test sessions that minimize test application time. We illustrate this with an example taken from work done at University of Southern California [Narayanan 1992]. The circuit in Fig. 9.17 shows two arrangements of the registers for scan-path testing. There are five scan registers and two combinational partitions, K1 and K2, with test lengths of 50 and 300, respectively. The size of the registers is indicated in the figure and the total number of flip-flops used is 20. Assume that it is possible to merge the two tests and to apply only 300 patterns [Abramovici 1990]. Then, according to the test application process discussed in Section 9.4, the number of clock cycles required for test application is: 300(20 + 1) + 20 = 6320. This will involve all the flip-flops in the test results simultaneously in test application time that is independent of the register order in the scan chain.

It is possible to exploit the circuit topology to use an overlapped scheme

[Abramovici 1990]. Notice that registers R1 and R2 serve as input only to K1 and K2; they are called the *drivers*. Also, R4 and R5 serve as output only and they are called the *receivers*. Accordingly, R3 is a driver–receiver register. With this knowledge we rearrange the registers so that each of the 50 common patterns is shifted in the first three registers that are of combined length 14. At the same time, the result of the previous pattern is scanned-out through registers R3, R4, and R5, whose combined length is eight. The time taken for this test is then 50(14 + 1) = 750 cycles. The rest of the test, which consists of 250 patterns, is applied through driver R2. This takes 12 cycles for the arrangement shown in Fig. 9.17a. Meanwhile, the previous pattern is scanned out through R3 and R5 as receivers. The total testing time for this session is then 250(12 + 1) = 3250. The overall total testing time is then 750 + 3250 + 14 = 4014 cycles. In this fashion, the test application time has been reduced by 34%. Notice that this reduction of test application time is accompanied by lowering the heat dissipation during testing.

This saving can be even greater if the flip-flops in the scan chain are reordered as shown in Fig. 9.17b. The first session will take 50(18 + 1) = 950 cycles. The 18 cycles are required because to apply the pattern through drivers R1, R2, and R3, we also have to go through R4. For the other session, only R2 is used as a driver and both R3 and R5 are used as receivers. Then it takes only: 250(4 + 1) = 1250 cycles. The total test application time is 950 + 1250 + 18 = 2218 cycles. This is an improvement of 44.7% over the order shown in Fig. 9.17a, and of 66.9% over the traditional test application scheme. It is not realistic always to expect such a savings in test application time, because the savings is strongly dependent on the topology of the circuit and the length of test sets for the various partitions or segments of the combinational part of the circuit.

9.11.2   Optimizing Interconnect Wiring

Reordering the flip-flops in the scan chain to optimize on interconnect wire length is important for submicron technology chips. The delay due to the interconnect is becoming comparable to that of the logic gates. Also, interconnects occupy an appreciable part of the chip area. Determining the order of the flip-flops in the scan chain in such a way as to minimize interconnect length cannot be done on the logic level only. It is important to consider it on the physical level as well. The problem is analogous to placement and routing that we reviewed in Chapter 4. However, while in generic placement there is no emphasis on a particular class of gates, here the focus is on minimizing the interconnect between the scanned flip-flops. The interconnect wires in this case consist of:

- The control signal SE
- The direct connection from the output of one flip-flop to the input of the next flip-flop in the chain
- The clock signal

It is possible then to perform placement to evaluate the length of the inter-connect. This length can be used as a cost function in conjunction with a heuris-tic to reorder the position of the flip-flops. Such an approach may inadvertently cause an increase in the interconnects for the other gates of the circuit. To avoid this complication, it is also possible to decide about the order without chang-ing the physical locations of the flip-flops after placement is completed. The problem can be considered more as a routing problem: given the locations of the various flip-flops, how to find the shortest route such that each flip-flop is visited once and only once. This can be modeled by a graph $G(V, E, W)$, where $v_i \in V$ is any of the flip-flops, $(v_i, v_j) \in E$ is a connection from v_i to v_j, and $w_{ij} \in W$ is the length of the edge (v_i, v_j). The adjacency matrix for this graph is such that the entries represent the weights w_{ij}. We can use a modification of a greedy algorithm such as Krusal's algorithm that we used to find the minimal spanning tree in Chapter 4. We start with the smallest weight, w_{ij}, and then add other paths that are connected to either v_i or v_j. The vertices v_i and v_j are then removed from V so that they are no longer revisited. If the next flip-flop is v_k, the path is now v_i, v_j, v_k and we search for the next vertex that is nearest v_i or v_k. The process continues until all vertices have been visited.

According to [Makar 1998], finding the distance between all pairs of flip-flops is problematic. For example, for only 1000 flip-flops, 1 million distance computations are required. However, any flip-flop will be connected only to, at most, two flip-flops in its neighborhood. Hence it is less computationally intensive to find the distances from one flip-flop to others in its neighborhood. The designer can determine the size of the neighborhood, which is a factor of the size of the set V. Once the neighborhoods are determined, it is possible to find subchains as discussed above. At the end of the process some flip-flops may not belong to any of the subchain. However, the number of such isolated flip-flops is not expected to be too large to cause a serious problem. When all subchains are merged in one path, it is possible that some of the wire segments intersect. These segments may be routed at different metal layers. Changing the order of the flip-flop from the onset is more advantageous. This is illustrated by the example shown in Fig. 9.18. In this figure there are six flip-flops, a, b, c, d, e, and f, that have been placed. One shorter stitching pattern is shown in

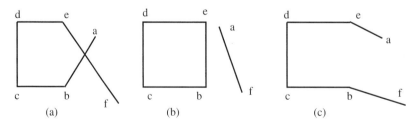

Figure 9.18   Rerouting for minimization of interconnect wires [Makar 1998 © IEEE. Reprinted with permission].

Fig. 9.18*a*. The segments, *ef* and *ab*, intersect. To break the intersection, it is possible to connect *b* to *e* and *a* to *f*, or *e* to *a* and *b* to *f*, as shown in Fig. 9.18*b* and *c*, respectively. The first solution is not acceptable since we obtain a loop. The second solution is acceptable, and in addition, it has been proven that it results in an even shorter path.

REFERENCES

Abramovici, M., M. A. Breuer, and A. D. Friedman (1990), *Digital Systems Testing and Testable Design*, IEEE Press, Piscataway, NJ.

Abramovici, M., J. Kulikowski, and R. K. Roy (1991), The best flip-flops to scan, *Proc. IEEE International Test Conference*, pp. 166–172.

Agrawal, V. D., K. Cheng, D. Johnson, and T. Lin (1988), Designing circuits with partial scan, *IEEE Des. Test Comput.*, Vol. 5, No. 4, pp. 8–15.

Brglez, F., et al. (1989), Combinational profiles of sequential benchmark circuits, *Proc. International Symposium on Circuits and Systems*, pp. 1929–1934.

Carter, W. C., et al. (1964), Design of serviceability features of the IBM System/360, *IBM J. Res. Dev.*, Vol. 8, No. 4, pp. 115–126.

Cheng, K., and V. D. Agrawal (1990), A partial scan method for sequential circuits with feedback, *IEEE Trans. Comput.*, Vol. C-39, No. 4, pp. 544–548.

Chickermane, V., and J. H. Patel (1991), A fault oriented partial scan design approach, *Proc. International Conference on Computer-Aided Design*, pp. 400–403.

Debaney, W. H. (1994), Quiescent scan design for testing digital logic circuits, *IEEE Dual Use Conference*, Session G4.

Eichelberger, E. B. (1965), Hazard detection in combinational and sequential circuits, *IBM J. Res. Dev.*, Vol. 9, No. 1, pp. 90–99.

Eichelberger, E. B., and T. W. Williams (1977), A logic design structure for LSI testability, *Proc. 14th Design Automation Conference*, pp. 462–468.

Eichelberger, E. B. (1985), Latch design using level-sensitive scan design, *Proc. Compcon.*, pp. 380–383.

Funatsu, S., N. Wakatsuki, and A. Yamada (1978), Designing digital circuits with easily testable considerations, *IEEE Semiconductor Test Conf.*, pp. 98–102.

Greene, B., and S. Mourad (1998), *IEEE Instrumentation and Measurement Technology Conference*, pp. 423–427.

Kobayashi, A., et al. (1963), A flip-flop circuit suitable for FLT (in Japanese), *Annual Meeting of the Institute of Electronics, Information and Communications Engineers*, Manuscript 892, p. 962.

Lee, D. H., and S. M. Reddy (1990), On determining scan flip-flops in partial-scan designs, *Proc. International Conference on Computer-Aided Design*, pp. 322–325.

LSI (1992), *Chip-Level Full Scan Design Methodology Guide*, LSI Logic Corporation, Milpitas, CA.

Ma, H. K. T., S. Devadas, A. R. Newton, and A. Sangiovanni-Vincentelli (1988), An incomplete scan design approach to test generation for sequential machines, *Proc. IEEE International Test Conference*, pp. 730–734.

Makar, S. (1998), A layout based approach for ordering scan chain flip-flops, *Proc. IEEE International Test Conference.*

McCluskey, E. J. (1986), A survey of design for testability scan techniques, *Summer 1986 Semicustom Design Guide*, VLSI System Design, CMP Publications, Manhasset, NY.

Miczo, A. (1983), The sequential ATPG: a theoretical limit, *Proc. IEEE International Test Conference*, pp. 143–147.

Narayanan, S., C. Njinda, and M. Breuer (1992), Optimal sequencing of scan registers, *Proc. IEEE International Test Conference*, pp. 293–302.

Parikh, P. S., and M. Abramovici (1993), A cost based approach to partial scan, *Proc. Thirtieth ACM/IEEE Design Automation Conference*, pp. 255–259.

Park, S., and S. Akers (1992), A graph theoretic approach to partial scan design by K-cycle elimination, *Proc. IEEE International Test Conference*, pp. 303–311.

Savir, J. (1983), Good controllability and observability do not guarantee good testability, *IEEE Trans. Comput.*, Vol. C-32, No. 12, pp. 1198–1200.

Steensma, J., F. Catthoor, and H. De Man (1993), Partial scan at the register-transfer level, *Proc. IEEE International Test Conference*, pp. 488–497.

Stewart, J. H. (1978), Applicant of scan/set for error detection and diagnostics, *Proc. IEEE Semiconductor Test Conference*, pp. 152–158.

Trischler, E. (1980), Incomplete scan path with an automatic test generation methodology, *Proc. IEEE International Test Conference*, pp. 153–162.

Williams, M. J. Y., and J. B. Angell (1973), Enhancing testability of large scale integrated circuit via test points and additional logic, *IEEE Trans. Comput.*, Vol. C-22, No. 1, pp. 46–60.

PROBLEMS

9.1. Determine if the counter in Fig. P9.1 has to go through an entire cycle to detect all stuck-at faults in the combinational part. How many test patterns are needed?

9.2. Repeat Problem 9.1 for the circuit shown in Fig. P9.2. Notice that the combinational circuit is the same for both circuits.

9.3. For the circuit in Fig. 9.12, how many scan paths would you use to minimize test pattern application? Redesign the circuit using scan-path design and generate a test set to detect all stuck-at faults in the combinational part.

9.4. The nodes on the grid shown in Fig. P9.4 are flip-flops that are to be stitched to form a scan path. Show how they can be connected for minimal interconnect length. Write a short algorithm to find the minimal spanning tree.

9.5. Develop a test set for the example in Fig. 9.2 when only the first flip-flop is scanned. Describe in detail application of the test.

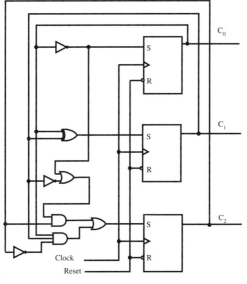

Figure P9.1

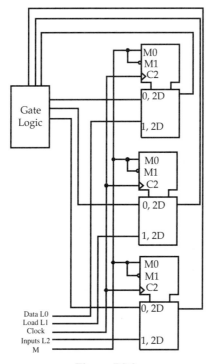

Figure P9.2

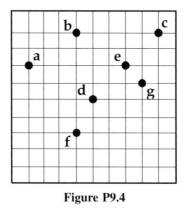

Figure P9.4

9.6. Select the appropriate flip-flops for partial scan of the circuit in Fig. 9.15 and justify your selection (*Hint*: Open the appropriate edges to make the S-graph acyclic.)

10

BOUNDARY-SCAN TESTING

10.1 INTRODUCTION

Although the emphasis of this book is on IC testing and designs for testability, it is important to include a chapter on Boundary-Scan, which was initially a DFT technique for a printed circuit boards (PCBs). This technique requires that ICs include extra hardware to facilitate communication between the board and the various ICs on which they are mounted during testing. Also with the present trend of a system on a chip (SOC), the ICs themselves consist of several embedded complex and diverse modules and start to beg for a DFT technique that facilitates their testing. Understanding Boundary-Scan can help in handling SOCs, as we describe in Chapter 15.

It has been customary through the late 1970s to access the various ICs on a board through a bed-of-nails fixture in an in-circuit tester. Because of the increase in device density, decrease in board size, and change to surface mounting in packaging, accessing the chips' test points became impractical. A small group of European electronics companies met in the Netherlands to discuss this problem in 1985. From the onset, consensus had been reached about the gravity of the problem and the need for a solution. The group which adopted the name *Joint Test Action Group* (JTAG), described the problems with board testing and proposed a methodology to resolve them. This proposal took the name JTAG Test Access Port and Boundary-Scan Architecture [JTAG 1988]. By 1988, the proposal gained widespread interest from industry and the support of the IEEE Computer Society. In 1990, the methodology became IEEE/ANSI Standard 1149.1 [IEEE 1990]. It was quickly adopted by the industry, to such an extent that it is now a standard feature in CAD tools and is supported by board test equipment. As early as 1990, it was embedded in some FPGAs, digital signal processors (DSPs), and other ASICs.

The use of Boundary Scan has been extended beyond boards to circuits [Hansen 1989]. For example, it is useful for internal testing and to run BIST. In addition to testing, Boundary Scan was adopted in debugging and diagnosis using in-circuit emulation protocols [Parker 1989]. Despite the proliferation of its use, the term *Boundary-Scan* as used in this chapter is synonymous with IEEE Standard 1149.1.

In this chapter we review briefly traditional board testing and determine the rationale for the adoption of a DFT technique to facilitate its testing. In Section 10.3 the architecture of Boundary-Scan is described, and its components are detailed in Section 10.5. The TAP controller is the topic of Section 10.4. With the knowledge acquired as to the architecture and controller, the modes of Boundary-Scan operation are given in Setion 10.7. Boundary-Scan description language (BSDL) is the topic of Section 10.8. As with any other DFT technique, Boundary-Scan has its cost, discussed briefly in Section 10.9.

10.2 TRADITIONAL BOARD TESTING

Through the late 1960s, PCBs included only discrete components and required a customized test bench. Testing could easily be completed in 3 to 4 minutes per board. As they became populated by ICs, their complexity has increased and their testing time has reached 3 to 4 hours per board [GenRad 1989]. As ATE equipment became driven by software, the testing throughput has increased.

Board testing used to consist of functional testing and in-circuit testing. The advantage of functional testing is that it can be done at speed. However, there are several disadvantages. First, it requires a long preparation time. Second, since it does not use a fault model, there is no way to verify its quality. In addition, hardly any diagnostics are possible. In-circuit testing allows testing individual components, discrete or ICs, on a fully assembled PCB. For this, all nodes must be accessible. In addition, all components need to be isolated during testing to avoid the device under test being affected by the other components on the board. This really amounts to partitioning the board and reducing testing complexity. It also allows diagnostic testing and isolating the faulty components.

This type of testing was practiced through the early 1980s using a *bed-of-nails* fixture. The nails on this fixture are small spring-loaded probes that touch the tracks and device leads on the unit under test (UUT). These nails are mounted on the fixture so that each nail lines up with a device pin or a via. The bottom of the spring-loaded probes is wire-wrapped to the socket as illustrated in Fig. 10.1. Connection to the tester is realized by wiring the sockets to the ATE system. Each custom-built bed-of-nails fixture typically has hundreds of test nails that are positioned to contact the solder pads on the bottom side of the UUT.

There are some problems with in-circuit testing. First, the fixture is custom

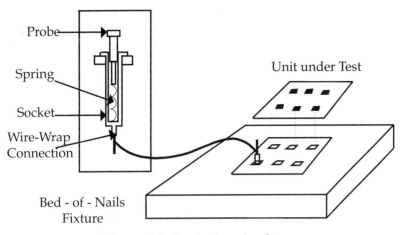

Figure 10.1 In-circuit testing fixture.

made and hence is expensive and requires a long lead time. The fixture cannot be built prior to finalizing the layout of the design. Yet, waiting until completion of the layout delays testing and therefore increases time to market. Second, the spring-loaded nails are easily damaged. Third, sometimes when testing an IC on the board, the appropriate input signals are not easily obtainable from the output of the driving IC; these signals have to be forced by the tester. This process may cause overheating of the test points and require long cooling periods; thus testing time can be prolonged. Also, the injected signals may produce noise on the board that results in wrong test response and damaging latch-up effects in CMOS circuits. These problems are exaggerated as the trace between the pins gets smaller, due to an explosion in the pin count as indicated by the trend shown in Fig. 10.2 for hand-held ICs [SIA 1997].

For modern boards, the challenge to in-circuit testing is the change in IC packaging along with rapid adoption of surface-mounted packages. The problem with these packages is pin density; that is, the pitch is getting too small. In addition, often the ICs are mounted on both sides of the board. Physical probing of test points on the board has become so difficult and costly that it motivated the search for a solution. Boundary scan is a promising alternative. According to [Parker 1998], it "actually helps one prolong the life of the in-circuit approach, because it allows the reduction of the number of nails needed to test a board while maintaining fault coverage."

10.3 BOUNDARY-SCAN ARCHITECTURE

The geneal IEEE Standard 1149.1 boundary-scan architecture is shown in Fig. 10.3. This configuration requires that the board and each IC that is part of the Boundary-Scan include the following principal hardware components:

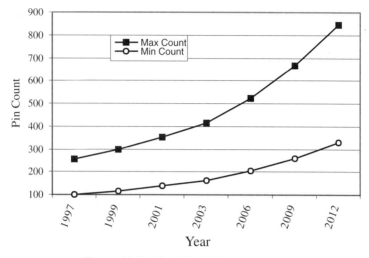

Figure 10.2   Hand-held ICs pin count.

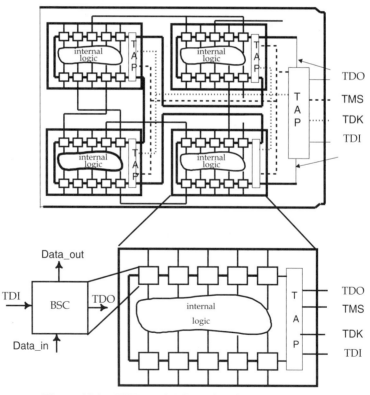

Figure 10.3   IEEE 1149.1 Boundary-Scan architecture.

- A test access port (TAP) with four to five pins
- A group fo registers: an instruction register (IR) and data registers (DRs)
- A TAP controller, a 16-state finite state machine

Four mandatory pins drive the TAP. They include two data pins, test data input (TDI) and test data output (TDO), and two control pins, test mode select (TMS) and test clock (TCK). It is important to mention that it is possible that some of the chips are not complying with the standard and their testing is not secured through the TAP.

The instruction register receives the instruction, then decodes it to perform the operations on the data registers. The instructions, which are described later in the chapter, include load, shift, or pause registers: IR or DR. There are two mandatory data registers, a bypass register and a boundary scan register (BSR), which consists of Boundary-Scan cells (BSC).

The I/O signals of all chips enter or leave the chip through the *Boundary-Scan cell* (BSC), described in Section 10.5.1. All such cells are provided with controls to allow them to be configured in a shift register and form a collar around the core logic of the chip. The TAP controller manages the exchange of data and instructions among the board and chips. Test data management is done according to a protocol dictated by the standard. As with any standard, this one evolves according to changes in technology and design practices. In addition, due to the emergence of mixed signal chips, a new standard, P1149.4, is being established to handle analog I/O [IEEE 1994]. The reader who plans to use the standard should consult the IEEE documentation [IEEE 1990, Maunder 1990].

During normal operation, the extra hardware on the chip or board is transparent. However, during testing mode, all input signals are scanned in from the in port, TDI, and scanned out through the out port, TDO, to the board. The information from all chips is passed to the outside of the chip through a shift register formed by flip-flops around the contour of the chip. That is, every I/O pin is registered, and all flip-flops may be organized as a shift register in a fashion similar to the scan-path technique that we studied in Chapter 9.

The Boundary-Scan architecture allows configuring the cells for the following testing modes:

- *External testing:* interconnects between the chips
- *Internal testing:* testing of the logic within the chip

10.4   TEST ACCESS PORT

In this section we describe the various signals used by the TAP controller. They are the signals applied to the four mandatory pins and the optional test reset (TRST). The latter pin can reset the test logic asynchronously. The four mandatory pins include two data pins, test data input (TDI) and test data output (TDO), and two control pins, test mode select (TMS) and test clock (TCK). As illustrated in Fig. 10.3, the TDI of the board is connected to that of the first chip. Also, the TDO of this chip is connected to the TDI of the next chip is the chain, and so on. The TDO of the last chip in the chain is connected to that of the board's TAP. The two control signals are, however, connected to the TAPs of all the chips.

TDI and TDO are to Boundary-Scan what scan-in and scan-out are to scan path design. However, unlike scan-path design the pins may not be one of the primary input/output pins on the chip that can be multiplexed for the two functions, as explained in Chapter 9. Both TDI and TDO are also connected to the TAP registers. Both TMS and TCK are distributed to all the chips on the board that are part of the scan design. It is very likely that the board will include logic that does not have Boundary-Scan cells.

- *TDI.* This is the test data input, which allows the introduction of test data in a fashion similar to the scan-in pin in traditional scan path. The TDI of the board is connected to its counterpart on the first chip in the scan chain. This signal is shifted in the registers at the positive edge of the TCK and, when not in use, is kept high.

- *TDO.* This is the test data output, which allows scanning out of the test data in a fashion similar to the scan-out pin in traditional scan path. The TDO of the board is connected to its counterpart on the last chip in the scan chain. Data are shifted out at the negative edge of TCK. When not in use, this input signal is kept in high impedance.

- *TCK.* This is the test clock, which operates the testing function synchronously and independent of the system clock. It controls the transfer to data and instructions among the TAP registers and shifting the data within any of the registers.

- *TMS.* This is the test mode select. The input stream to this pin is interpreted by the TAP controller and used to manage the various test operations. The signals are interpreted by the TAP controller, which is discussed in Section 10.6. When the TMS is not in use, it must be held high. For this, the standard requires that this be done automatically.

- *TRST.* This is an optional signal whose purpose it to reset all testing logic asynchronously and independent of TCK. It may also be used for reset at power-on.

10.5   REGISTERS

Figure 10.4 shows the various registers that support the testing of the board. Of these registers, only three are mandatory: the instructional register (IR), the bypass register, and the boundary-scan register (BSR), which consists of collections of the BSCs. To discuss the BSR, we first need to know about the

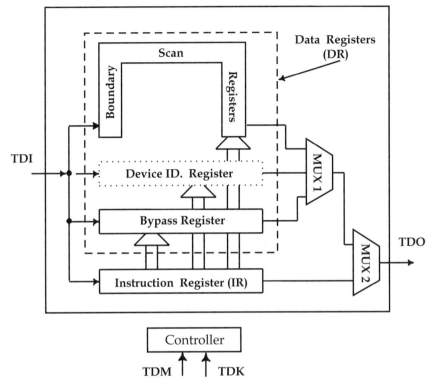

Figure 10.4 Test access port—the registers.

functionality of the BSC. A brief description of each of these registers is given in this section.

10.5.1 Boundary-Scan Cell

An example of a commonly used Boundary-Scan cell (BSC) is shown in Fig. 10.5. This cell may be used at the input or output pins. During normal operation the input signal is applied to the data-in pin and passes to the internal logic through MUX 2. Thus the value of Test/Normal should be 0, while the Shift/Load mode may be either 0 or 1. When the same cell is used as an output pin, the data-in are from the internal logic of the chip and pass, through MUX 2, to the output of the chip.

For test mode, the data are coming through TDI; thus the Shift/Load mode should be 1. The data are latched for internal testing or to be shifted to the next BSC when the clock is activated. For internal testing, the signal needs to pass through the second storage device by enabling the Update signal. It is then passed to the chip with Test/Normal being held high. The various configurations of the cell are also shown in Fig. 10.5.

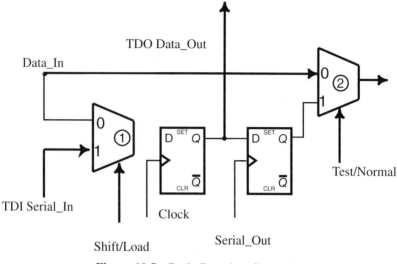

Figure 10.5   Basic Boundary-Scan cell.

10.5.2   Bypass Register

The bypass register is a single-stage register, as shown in Fig. 10.6. It is set to logic 0 at the rising edge of TCK when the TAP controller is in the Capture-DR state. Use of the bypass register allows the signal at the TDI to pass directly to the TDO of the chip, thus bypassing all the other BSCs of the chip. This approach is very useful. For example, if the board has, say, 30 chips each having 100 BSCs, there are 3000 stages in the Boundary-Scan register. Instead, when the bypass is used, there are only 129 stages in the BSR: 100 for the chips under test and one each for the other 29 chips on the board. Such an arrangement results in a reduction of both the testing time and the testing data.

10.5.3   Boundary-Scan Register

The Boundary-Scan register consists of all the BSC cells on the periphery of the chip. It is part of the testing of the interconnects and of any logic between the Boundary-Scan ICs on the board. This includes logic not configured in the Boundary-Scan as well as any ROM or RAM. In addition, it allows sampling and examination of the input and output signals without interfering with operation of the core logic.

10.5.4   Instruction Register

The instruction register is a serial-in, parallel-out register. Each flip-flop of the register is connected to an output latch as illustrated in Fig. 10.7. The latch holds the current instruction bits latched into it from the shift register in the

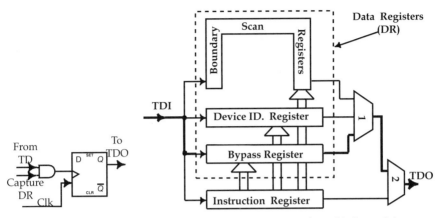

Figure 10.6 Bypass register: (*a*) functional representation; (*b*) flow of data.

Update-IR state. The IR must contain at least two shift-register cells that can hold instruction dtaa. These two mandatory cells are located nearest the serial output. They are the least significant bits (LSBs), as illustrated in Fig. 10.7.

10.5.5 Device Identificiaton Register

The device identification register is optional, but if included on the IC, it should comply with the standard. It must be a 32-bit-long parallel-in and serial-out.

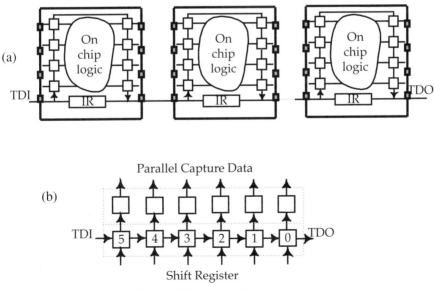

Figure 10.7 Instruction register.

It is intended to contain the manufacturer's number and the version number. This information facilitates verifying that the correct IC is mounted in a correct position and that it is the correct version of the chip. Unlike the other registers, the information is not passed to an input latch.

10.6 TAP CONTROLLER

The TAP controller responds to the TMS signals at the positive edge of TCK. Its main functions are (1) to load the instructions in the IR, (2) to provide control signal to load and shift the test data into TDI and out of TDO, and (3) to perform some test actions, such as capture, shift, and update test data. The TAP controller is a 16-state finite state machine that operates synchronously with TCK. Its state diagram is shown in Fig. 10.8 as defined by IEEE Standard 1149.1. Some of the states correspond to actual operations on the data (DR) or the instructions (IR), while others allow some flexibility in the flow of operations.

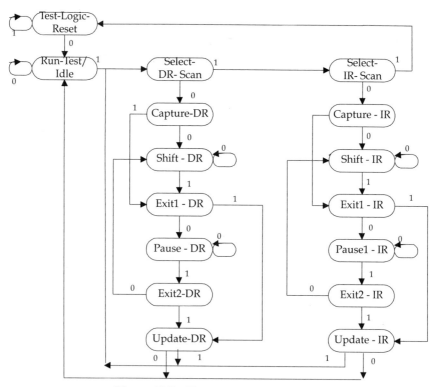

Figure 10.8   Test access port controller.

10.6.1 Controller's States

At power on, the controller is in the Test-Logic-Reset state. It remains in this state as long as TMS is high and the circuit is in normal operation mode. This state may also be reached if while TMS is high, TCK is clocked five times. As soon as TMS changes to logic 0, the controller is in the Run-Test-Idle state. In this state, either BIST may be initiated or the TAP idles between scan operations for other testing modes.

To start testing, the instruction needs to be loaded into IR. For this, TMS is held high and the TCK is clocked twice for the controller to reach the Select IR–Scan state. Now TDI and TDO are connected to IR, and all IR registers on the board are serially connected. Next, the controller passes to the Capture-IR state with TMS = 0. Once the instruction is loaded in the IRs, then, with TMS still low, the controller stays in the Shift-IR state for as many clock cycles as needed by the test mode. In this state, the previously captured data are shifted via TDI and via TDO, one shift register stage on each rising edge of the TCK pulse. If shifting is not needed, TMS = 1 and the controller bypasses Shift-IR and enters Exit1-IR state. The latter state as well as any of the other exit states is temporary. At the next positive edge of the clock, there is transition to another state. If TMS = 0, the next state is Pause-IR, and the control remains in this state until TMS = 1. The Pause-IR state is needed when the shift is done in a chain of different lengths. From this state, the control goes to Exit2-IR, then to Shift-IR if TMS = 0, or to Update-IR, if TMS = 1. The controller enters this state once the shifting process has been completed. The new data are latched into their parallel outputs of the selected data registers at the falling edge of the TCK. Depending on the value of TMS, the next state is either Run-Test-Idle or Select DR–Scan. When the controller is in the DR branch of the state diagram, it performs on the IR operations similar to those described above. Examples of controller timing diagrams are shown in Figs. 10.9 and 10.10 for instruction and data scan, respectively.

10.6.2 Instruction Set

The controller utilized only a few instructions. Only three of these are mandatory: BYPASS, EXTEST, and SAMPLE/PRELOAD. The most commonly used optional instructions are IDCODE, INTEST, and RUNBIST. The instruction set can also be extended with public and private instructions. Private instructions are those intended to be used only by the manufacturer of the device, and thus do not need to be documented. The instruction is loaded in the TDI and shifted in the IR during the Shift-IR state. It is then decoded and executed.

10.6.2.1 Bypass Instruction. This mandatory instruction permits bypassing of the current IC. It places the one-bit bypass register between TDI and TDO of the chip when another IC is being tested.

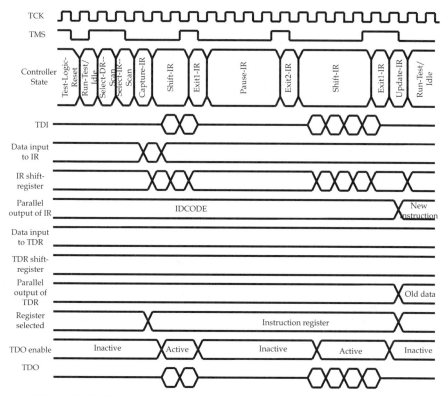

Figure 10.9   TAP controller: example of timing diagram, instruction scan.

10.6.2.2   EXTEST Instruction. This is a mandatory instruction, which is actually the primary reason for Boundary-Scan. It allows testing of the connectivities of the various pins of the ICs mounted on the board. The fault models used are stuck-at, bridging fault, and opens. For present deep-submicron technology circuits, noise faults should also be considered. The latter fault terminology was used in Chapter 2 to express noise failure, such as ground bounce and crosstalk.

10.6.2.3   Sample/Preload Instruction. This instruction is mandatory and is used to scan the BSR without interrupting the normal operation of the internal logic. It supports two functions: sampling the normal operations of the chip (taking snapshots) and preloading data for another test operation into the latched parallel outputs of the BSCs. This instruction is useful in debugging prototypes in the development phase of the board design.

10.6.2.4   IDCODE Instruction. Although this optional instruction does not involve testing of the board, it helps identifying misplaced ICs. Often, it is difficult to distinguish between similar devices, and discovering the reason for malfunction of the board may take unnecessarily a long time.

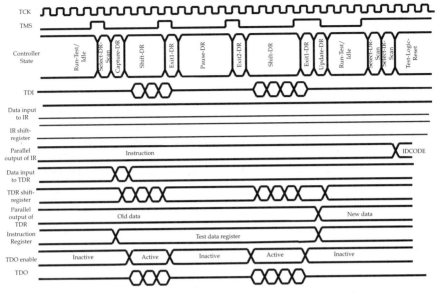

Figure 10.10 TAP controller: example of timing diagram, data scan.

10.6.2.5 INTEST Instruction. This is another optional instruction. It allows static testing of a particular IC using a bed-of-nails fixture and pin probing of ATE equipment. The test patterns are applied to the input of the chip and the response captured at the outputs. The test data are applied one at a time at the rate of TCK. For chips that need special operational speed such as those containing DRAMs that need to be refreshed, this instruction could be unusable.

10.6.2.6 RUNBIST Instruction. This instruction also is optional. It causes the execution of BIST test provided on the IC selected. It requires minimum data from outside the chip since, in BIST, the test patterns are generated internally on the chip. These patterns are applied dynamically. The instruction is run when the TAP controller is in its Run-Test-Idle state. The result of the test is captured when the controller is in Capture-DR state.

10.6.2.7 CLAMP Instruction. This optional instruction is used to control the output signal of a component to a constant level by means of a BSC. This is useful to hold values on some pins of the circuit, which are not involved in the test. These required signals are then loaded with other test patterns every time they are needed. This instruction, although useful, increases test application time.

10.6.2.8 HIGHZ Instruction. The HIGHZ instruction is optional and forces all outputs of a component to a high-impedance state. It is used, for example, when an in-circuit test is required for testing a non-BS compliant component.

In this way the other components are protected from any possible back drive signal by the tester.

10.7   MODES OF OPERATIONS

The TAP provides access to many test support functions. It consists of four input and one output connections. There is an optional input signal, which is used to reset the test logic synchronously. This reset function can also be performed by built-in logic in its controller. The controller receives input from TMS and TCK signals and passes the information to the IR register or the data register selected.

10.7.1   Normal Operation

During normal operation, the Boundary-Scan cell is transparent, allowing the input and output signals to pass freely through the test cells, thus enabling the device to perform its intended function. The select inputs to both multiplexers of the BSC, as discussed earlier and shown in Fig. 10.5, are held low. The TAP controller is in the Test-Logic-Reset state and TMS is kept high.

10.7.2   Test Mode Operation

10.7.2.1   External Testing.   Testing the interconnect between the ICs is one of the main objectives of Boundary-Scan testing. It is performed by the EXTEST instruction, which is one of the mandatory instructions of the IEEE Standard 1149.1. In addition to the interconnect circuitry, it is possible to test logic on the board: RAM, ROM, or glue logic that is not configured in Boundary-Scan design. In this mode the internal logic of the ICs is isolated from its input and output pins.

First, the EXTEST instruction is loaded into the IR register and decoded. Then the testing is performed according to the following steps, which are illustrated in Fig. 10.11. Notice that we use one cell to represent all input BSCs (to the left) and one cell to represent all output BSCs. In addition, the devices that are active in a certain step are highlighted.

1. *Shift-DR.* In this state the stimulus data are shifted in from the board's TDI through the BSR registers to the cells related with the output pins of the ICS.

2. *Update-DR.* When all the data are entered, the stimulus is latched at the output cells and is thus applied to the board interconnections.

3. *Capture-DR.* The data on the interconnect are then captured at the input cells of the input pins of the receiving ICs. This is indicated in Fig. 10.11 as Internal Logic 2.

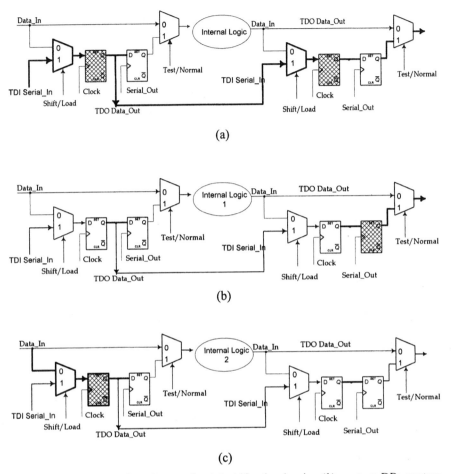

Figure 10.11 External testing mode: (*a*) shift stimulus in; (*b*) capture-DR: capture data at the outputs of internal logic 1; (*c*) capture-DR: capture data at the inputs of the receiving IC, internal logic 2.

4. *Shift-DR.* The results of the test are shifted through the BSR toward TDO for observation.

This procedure enables the testing of faults, such as open, stuck-at, or bridging faults. These faults are due to structural problems of the interconnect between the pins and with the pins themselves.

10.7.2.2. Testing Internal Logic. This mode may be performed by the INTEST instruction. It is one of two optional instructions that permit testing on-chip logic. This is performed in the same fashion as the external test, except that test data are applied to one of the ICs as illustrated in Fig. 10.12. The stimulus

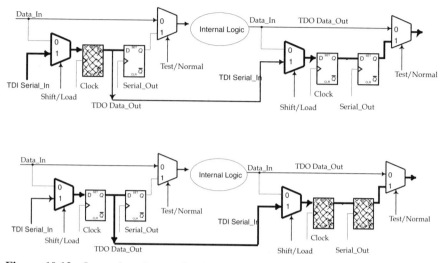

Figure 10.12 Internal testing mode: (*a*) capture-DR: simulus data: (*b*) capture-DR: capture the response data.

is entered in through TDI, applied to the internal logic, and then shifted out in the same way as the EXTEST:

1. *Shift-DR.* Shift stimulus data in from TDI through the registers of the cells related to the input pins of the IC.

2. *Update-DR.* Apply stimulus to the internal logic.

3. *Capture-DR.* Capture the status of the logic at the output pins of the same IC.

4. *Shift-DR.* Shift out the results through the BSR toward TDO for observation.

For Shift-DR states, the operations are done exactly as in the case of the external testing. Only the operations in the two other states are illustrated in Fig. 10.12. This testing allows the detection of only limited faults since it is not based on any internal DFT technique.

10.7.2.3 BIST Execution. This is an optional mode. It is performed by the RUNBIST instruction while the TAP controller is in Run-Test-Idle state. In this testing the data transfer is kept to a minimum since the instruction triggers the on-chip testing. The test response is prevented from appearing at the output until the test application is completed. The results are then captured by the BSR and shifted through TDO for observation.

10.7.3   Testing the Boundary-Scan Registers

All the testing modes have made use of the Boundary-Scan registers to scan-in and scan-out the data. Thus these registers should be tested thoroughly to validate the testing of the ICs and the interconnect. This is very similar to verifying the integrity of the scan-path chain prior to using it for scan-path testing. For details of this integrity testing of the Boundary-Scan register, the reader may consult references such as [Parker 1998].

10.8   BOUNDARY-SCAN LANGUAGES

From what we have learned so far about Boundary-Scan, clearly, the basic principles are relatively simple compared to managing the data when applying the test patterns. It was thus necessary to adopt use of the standard by creating software tools to support its automation in testing. For this, Boundary-Scan Description Language (BSDL) was developed to facilitate the description of the architecture of a specific board and the flow of the data for testing. According to IEEE Standard 1149.1-1990, it has been intended to be used by chip developers and ATE manufacturers to promote consistency throughout the industry.

This language, which was initiated by Hewlett-Packard in 1990, is based on a subset of VHDL [Parker 1990, 1991]. To facilitate data management, the language needs two attrbitues. First, it has to be easy to use. Secion, it has to be simple to make it unambiguously parasable. In virtue of its raison d'être, the BSDL is not a general-purpose language. On the contrary, it implies knowledge of the devices and the standard, that is, how the device captures, shifts, and updates data. Thus the elements of the design that are absolutely mandatory for compliance with the IEEE Standard are not part of the language. For example, the mandatory registers are not described; they are implied. In addition, the language does not describe the logic on the chip, simply the connectivities of the Boundary-Scan registers to the terminals of this design.

Like any other VHDL language, BSDL consists of an entity and an architecture, which comprise a description of the package and of the package body. Among other parameters, the entity includes the TAP description and the package pin mapping.

10.9   COST OF BOUNDARY-SCAN DESIGN

Initial reluctance to the use of Boundary-Scan design was due primarily to the hardware overhead, delays through different multiplexers, and the need for special design tools to incorporate the structure on the chip, board, and the tester. The cost of hardware overhead is really relative to the size of the chips and the board. Consider a small chip such as the 74H322, an octal D-latch SSI [TI 1990]; the overhead is really too large to be acceptable. For relatively larger chips, this overhead is dwarfed. However, the cost of adding four pins may be a larger factor to consider.

At present, many commercial and proprietary tools that insert Boundary-Scan structure in a design. Similar to other DFT constructs, adding Boundary-Scan may increae the design time. However, it can actually shorten the manufacturing test engineering and hence the time to market [Bleeker 1993, Parker 1998]. This is because:

- Many vendors provide tools for automatic Boundary-Scan insertion.
- More people have experience with the IEEE Standard 1149.1.
- The silicon cost is declining.
- Boundary-scan technology is more appropriate for concurrent engineering development, which has been shown to shorten the design cycle.
- In addition to helping in manufacturing testing, Boundary-Scan is helpful in prototype debugging.

10.10   TO EXPLORE FURTHER

It is important to follow the changes in the standards as they evolve. Monitoring the Web pages of the IEEE section on standards is an easy way to be informed. With the increase in mixed-signal circuits, it is also important to learn about the new standard for analog and mixed-signal circuits, IEEE Standard p1149.4 [Parker 1998]. How the board-level testers handle Boundary-Scan design is another important topic. This knowledge is critical so that the designers understand the impact of this DFT constructs on the testers [Lefevre 1990, Tinaztepe 1991].

REFERENCES

Bleeker, H., P. V. D. Eijnden, and F. de Jong (1993), *Boundary-Scan Test: A Practical Approach*, Kluwer Academic, Norwell, MA.

Feugate, R. J., et al. (1986), *Introduction to VLDI Testing*, Prentice Hall, Upper Saddle River, NJ.

GenRad (1989), *Introduction to Multi-strategy Testing*, GenRad, Inc., Concord, MA.

Hansen, P. (1989), Testing conventional logic and memory clusters using boundary scan devices as virtual ATE channels, *Proc. IEEE International Test Conference*, pp. 166–173.

IEEE (1990), *IEEE Standard Test Access Port and Boundary-Scan Architecture*, IEEE Standard 1149.1, IEEE Press, New York.

IEEE (1994), *IEEE Standard Module Test and Maintenance (MTM) Bus Protocol*, IEEE Standards Board, New York.

JTAG (1988), *JTAG Boundary-Scan Architecture Standard Proposal, Version 2.0*, Technical Sub-Committee of the Joint Test Action Group, Ipswitch, United Kingdom, March.

Lefevre, M. F. (1990), Functional test and diagnosis: a proposed JTAG sample mode scan tester, Paper 16.1, *Proc. International Test Conference.*

Maunder, C. M., et al., (1990), *The Test Access Port and Boundary-Scan Architecture,* Tutorial, IEEE Computer Society Press, Los Alamitos, CA.

Parker, K. P. (1989), The impact of boundary-scan on board test, *IEEE Des. Test Comput.,* Vol. 6, No. 4, pp. 18–30.

Paker, K. P., and S. Oresjo (1990), A language for describing boundary-scan devices, *Proc. IEEE International Test Conference,* pp. 222–234.

Parker, K. P., and S. Oresjo (1991), A language for describing boundar-scan devices, *J. Electron. Test. Theory Appl.,* Vol. 2, No. 1, pp. 43–74.

Parker, K. P. (1998), *The Boundary-Scan Handbook: Analog and Digital,* Kluwer Academic, Norwell, MA.

SIA (1997), *The National Technology Roadmap for Semiconductor,* Semiconductor Industry Association.

TI (1990), *SCOPE Testability Products: Application Guide,* Texas Instruments, Inc., Dallas, TX.

Tinaztepe, C. (1991), Device oriented test program generation for the sequence-per-pin test system architecture, *Proc. European Test Conference,* Munich.

PROBLEMS

10.1. Develop a synchronizing sequence for the TAP controller FSM, if any.

10.2. The IDCODE instruction is optional. Would you recommend using it? Why or why not?

10.3. You are to implement the execution of RUNBIST on a chip. Describe all the necessary signals and their sequence to execute the instruction. Indicate any extra circuitry needed to accomplish this task.

10.4. In a PCB, you decided to use a chip that follows IEEE Standard 1149.1. All the other ICs do not use include boundary-scan architecture. Would you implement this architecture on the board level? Explain why or why not. If the answer is affirmative, would it be necessary to use the bypass instruction?

11

BUILT-IN SELF-TEST

11.1 INTRODUCTION

In Chapter 8 we justified the need for DFT techniques. The motivation for the development of the built-in self-test (BIST) technique arose particularly from the cost of test pattern generation and the volume of data that keeps increasing with circuit size. In addition, BIST enables testing at speed. This aspect of testing is especially relevant for present technology, where the effects of the interconnection are causing delays that previously were ignored. Originally, the SAF model was used in conjunction with BIST. Recently, however, this testing is also used to uncover defects that cannot be represented by SAF.

Although there are many BIST schemes, in general, any of those techniques encompasses test pattern generation, test application, and response verification. There are several schemes to generate the test set and to verify the response. These schemes are customized for the type of circuit under test (CUT) and the testing environment. The methodology used for memory testing, *memory BIST*, is different from that used with random logic circuits, *logic BIST*. A combination of these methodologies may be used with RAM-based FPGAs. In this chapter we concentrate on logic BIST. BIST for memories and FPGAs is discussed in Chapters 12 and 13, respectively.

Logic BIST uses mostly pseudorandom (PR) tests. These tests are generated using a Linear Feedback Shift Register (LFSR) or cellular automata. They are usually much longer than deterministic tests, but are definitely less costly to generate. A control signal may trigger the test pattern genration at will. Because of the large volume of test patterns, some type of compactor is used to reduce the response data. There are several compacting schemes, but signature analyz-

IC

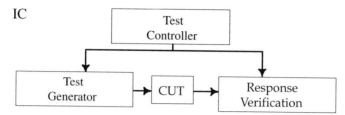

Figure 11.1 General BIST architecture.

ers are the most popular. Both the random test generator and the compactor are *built* on the chip, as shown in Fig. 11.1. A test controller is added to manage application of the test.

We first describe pseudorandom test generation—the mathematical foundation of LFSRs and the hardware implementations. This is followed by response compaction. Both time and space compaction are explained using LFSRs. One major drawback of compaction is the loss of information. Although the test applied has more than one pattern to detect the fault, the compaction yields results that are identical to the fault-free response. This property is known as *aliasing*. Faults that are hard to detect with PR tests are called *random pattern resistant* (RPR) and are addressed in Section 11.4. Section 11.5 is devoted to BIST architectures for general logic ICs. Some of these are configurations of BIST with other DFT constructs.

11.2 PSEUDORANDOM TEST PATTERN GENERATION

In random test pattern generation, a pattern may be repeated several times in the process. However, using pseudorandom yields random patterns without repetition. This is equivalent to selection without replacement. The length of the test generated in such a manner depends on the seed of the random number generator. It is conceivable to use some software tools to generate the random patterns and then store them in a ROM that is placed on the chip as illustrated in Fig. 11.1. Two main types of hardware may be used instead: a *cellular automata* [Hortensius 1989, Bardell 1990] or an LFSR [Bardell 1987].

11.2.1 Linear Feedback Shift Register

Examples of LFSRs are shown in Fig. 11.2. For simplicity we use only three- and four-long LFSRs without loss of generality. As its name implies, the LFSR is a shift register with feedback from the last stage and other stages. Besides the clock, it has no other inputs. This is why it is also referred to as *autonomous LFSR* (ALFSR). The outputs of the flip-flops form the test pattern. For this example, there are 3-bit test patterns. The number of unique test patterns is equal to the number of states of the circuit, which is determined, by the number and locations of the individual feedback tabs.

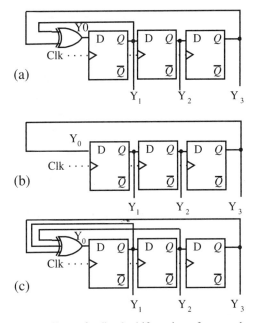

Figure 11.2   Autonomous linear feedback shift register for pseudorandom test pattern generation.

In Fig. 11.2a, the parity of all feedback leads is the input to the register. Thus we have $Y_0 = Y_1 \oplus Y_3$, $Y_1 = y_0$, $Y_2 = y_1$, and $Y_3 = y_2$, where $y_1 y_2 y_3$ represents the present state and $Y_1 Y_2 Y_3$ is the next state of the register. The LFSR was arbitrarily initialized to 001, which is the first pattern appearing on the LFSR. Next we clock the circuits as many times as it is necessary to reproduce this first pattern as illustrated in Table 11.1. In all, there are seven patterns, since the all-zeros pattern cannot be generated. This LFSR produces a *maximal cycle*.

TABLE 11.1   Pseudorandom Pattern Generated by LFSRs in Fig. 11.2

Clk	Y_0	Y_1	Y_2	Y_3	Clk	Y_0	Y_1	Y_2	Y_3	Clk	Y_0	Y_1	Y_2	Y_3
	1	0	0	1		1	0	0	1		1	0	0	1
1	1	1	0	0	1	0	1	0	0	1	1	1	0	0
2	1	1	1	0	2	0	0	1	0	2	0	1	1	0
3	0	1	1	1	3	1	0	0	1	3	0	0	1	1
4	1	0	1	1							1	0	0	1
5	0	1	0	1										
6	0	0	1	0										
7	1	0	0	1										
	(*a*)					(*b*)					(*c*)			

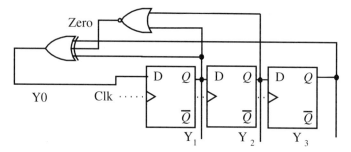

Figure 11.3 LFSR of Fig. 11.2a with adjustment for all-0's pattern.

TABLE 11.2 Pseudorandom Pattern Generated by LFSR of Fig. 11.3

Clk	Y_0	Zero	Y_1	Y_2	Y_3
	0	1	0	0	1
1	1	1	0	0	0
2	1	0	1	0	0
3	1	0	1	1	0
4	0	0	1	1	1
5	1	0	0	1	1
6	0	0	1	0	1
7	0	0	0	1	0
8	0	1	0	0	1

Had the feedback been only from the last stage, or from all stages as illustrated in Fig. 11.2b and c, it is easy to determine that the cycle length would have been reduced to only three and four patterns, respectively. The corresponding patterns are also shown in Table 11.1a.

The design has been modified by adding a NOR gate to the circuit in Fig. 11.3 to allow the inclusion of the all-zeros pattern. The results are shown in Table 11.2, where eight distinct patterns are listed. The output of the LFSR can also be tapped from the last flip-flop or any other flip-flop. The sequence obtained will then be one of the last three columns of Table 11.1 or 11.2. This sequence has some properties that are dependent on the connectivity of the LFSR.

11.2.2 LFSR Configurations

The circuit in Fig. 11.4 shows a generic representation of a *standard* LFSR. There is a feedback from the ith D flip-flop if $C_1 = 1$; otherwise, the output of this flip-flop is not tapped. There is another configuration of the LFSR in which the XOR gates are inserted between the flip-flops. This *modular* arrangement is shown in Fig. 11.5. The *C*'s values represent the same meaning as for the

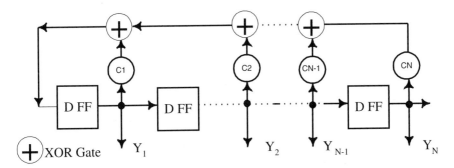

Figure 11.4   Generic standard LFSR with possible tabs from all flip-flops.

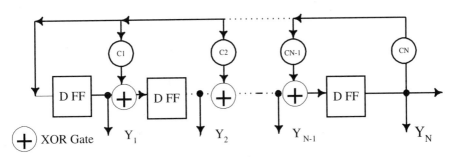

Figure 11.5   Generic modular LFSR with possible tabs from all flip-flops.

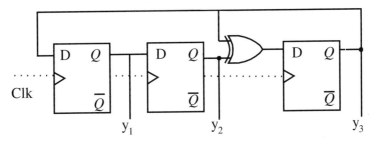

Figure 11.6   Modular configuration of the LFSR in Fig. 11.2a.

standard configuration. Also, the XOR gate at the input of flip-flop 1 is not necessary because there is only one input, the output of the nth flip-flop. The output sequence at the last stage is a function of the initial seed in the LFSR and the feedback coefficients, C_i. The modular LFSR equivalent to the LFSR in Fig. 11.2a is shown in Fig. 11.6. Another example for standard and modular LFSRs is shown in Fig. 11.7 for the polynomial $1 + X^3 + X^4$.

External Connections (STANDARD)

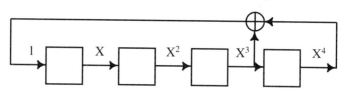

Internal Connections (MODULAR)

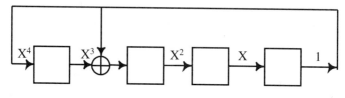

$$P(X) = 1 + X^3 + X^4$$

Figure 11.7 Standard and modular configurations for the same polynomial.

11.2.3 Mathematical Foundation of LFSR

First, we discuss the XOR operation. The truth table for this function is shown in Table 11.3. It represents the addition and difference, modulo 2, which are also defined in the same table. In these definitions we include the carry and the borrow, just for clarification, but since the operations are modulo 2, both terms are not part of the results of the operations. In the next derivation we use the plus sign to indicate an XOR operation, addition or subtraction, according to the context.

As a function of time, Y_j can be represented for the standard form as

$$Y_j(t) = Y_{j-1}(t-1) \qquad \text{for } j \neq 0 \qquad (11.1)$$

TABLE 11.3 Modulo 2 Operations

a	b	$a \oplus b$	$a+b$ Sum	$a+b$ Carry	$a-b$ Difference	$a-b$ Borrow
0	0	0	0	**0**	0	**0**
0	1	1	1	**0**	1	**1**
1	0	1	1	**0**	1	**0**
1	1	0	0	**1**	0	**0**

This is a translation in time of the value at the flip-flop preceding it. We can thus express Y_j in terms of Y_0 as

$$Y_j(t) = Y_0(t - j) \tag{11.2}$$

If we denote the translation operator as X^k, where k represents the time translation units, we can write the Eq. (11.2) in the form

$$Y_j(t) = Y_0(t)X^j \tag{11.3}$$

On the other hand,

$$Y_0(t) = \sum_{j=1}^{j=N} C_j Y_j(t) \tag{11.4}$$

where the summation is equivalent to an XOR operation. Then substituting Eq. (11.3) and Eq. (11.4), we get

$$Y_0(t) = \sum_{j=1}^{j=N} C_j Y_0(t)X^j \qquad \text{for } 1 \leq j \leq N \tag{11.5}$$

Because of the linearity property, we can rewrite Eq. (11.4) as

$$Y_0(t) = Y_0(t) \sum_{j=1}^{j=N} C_j X^j \tag{11.6}$$

and

$$Y_0(t)\left(\sum_{j=1}^{j=N} C_j X^j + 1 \right) = 0 \tag{11.7}$$

We can then write this expression as

$$Y(t)P_N(X) = 0 \tag{11.8}$$

For nontrivial solutions, $Y_0(t) \neq 0$, we must have

$$P_N(X) = 0 \tag{11.9}$$

where

$$P_N(X) = 1 + \sum_{j=1}^{j=N} C_j X^j \qquad (11.10)$$

$P_N(X)$ is called the *characteristic polynomial* of the LFSR.

We illustrate the use of this polynomial for the 3-bit LFSRs ($N = 3$) shown in Fig. 11.2. For the first one, we have $C_3 = 1$, $C_2 = 0$, and $C_1 = 1$. Thus the characteristic polynomial is

$$P_3(X) = X^3 + X + 1 \qquad (11.11)$$

For the second LFSR, $C_3 = 1$, $C_2 = 0$, and $C_1 = 0$, and the $P_3(X) = X^3 + 1$. Finally, for the third LFSR, $C_3 = 1$, $C_2 = 1$, and $C_1 = 0$, which results in $P_3(X) = X^3 + X^2 + X + 1$. Using similar analysis on the modular LFSR shown in Fig. 11.6 will result in the same polynomial given by Eq. (11.11).

11.2.3.1 Reciprocal Polynomial. The reciprocal polynomial of $P(X)$ is defined by

$$P_N^R(X) = \begin{cases} X^N P_N(1/X) = X^N \left(1 + \sum_{j=1}^{j=N} C_j X^j \right) & (11.12) \\ \\ X^N + \sum_{j=1}^{j=N} C_j X^{N-j} & \text{for } 1 \leq j \leq N \qquad (11.13) \end{cases}$$

Thus every coefficient C_i in $P(X)$ is replaced by C_{N-1}. For example, the reciprocal of polynomial given by Eq. (11.11) is $P_3^R(X) = 1 + X^2 + X^3$. The length of the LFSR sequence is determined by its characteristic polynomial. An *N-degree* polynomial yields a maximal length cycle $2^N - 1$ if it is a *primitive* polynomial. Only a *primitive* polynomial guarantees a maximal-length sequence. But first, let us define operations on polynomials.

11.2.3.2 Operations on Polynomials. All operations are modulo 2; thus addition and subtraction are identical according to the definitions in Table 11.3. We illustrate the multiplication and division by examples. For these operations we need to use the *modulo 2* addition property: $x^i + x^i = 0$. Consider the fourth-degree polynomial, $S(x) = x^4 + x^3 + 1$. We multiply it by the first degree, polynomial $x + 1$, and then divide by the second-degree polynomial, $D(x) = x^2 + 1$.

$$x^4 + x^3 \qquad + 1$$
$$x + 1$$

$$\overline{\qquad\qquad\qquad\qquad}$$

$$x^4 + x^3 + \quad + 1$$
$$x^5 + x^4 + \quad + x$$

$$\overline{\qquad\qquad\qquad\qquad}$$

$$x^5 + \quad + x^3 + x + 1 \qquad \text{since } x^4 + x^4 = 0$$

Division is of particular interest when LFSRs are used for response compaction:

$$x^2 + \ x \ + \ 1$$

$$x^2 + 1 \,\overline{)\,x^4 + x^3 + \qquad\qquad + 1}$$
$$x^4 + \qquad + x^2$$

$$\overline{\qquad\qquad\qquad\qquad}$$

$$x^3 + x^2 + \quad + 1$$
$$x^3 + \qquad + x$$

$$\overline{\qquad\qquad\qquad\qquad}$$

$$x^2 + x + 1$$
$$x^2 + \quad + 1$$

$$\overline{\qquad\qquad\qquad}$$

$$x$$

The result of the division of polynomials $S(x)$ by $D(x)$ is the quotient $Q(x) = x^2 + x + 1$ and the remainder $R(x) = x$.

11.2.3.3   *Properties of Polynomial*

1. An irreducible polynomial is that polynomial which cannot be factored, and it is divisible only by itself and 1.

2. An irreducible polynomial of degree n is characterized by:

   - An odd number of terms including the 1 term

   - Divisibility into $1 + x^k$, where $k = 2^n - 1$

3. Any polynomial with all even exponents can be factored and hence is reducible.

4. An irreducible polynomial is *primitive* if the smallest positive integer k that allows the polynomial to divide evenly into $1 + x^k$ occurs for $k = 2^n - 1$, where n is the degree of the polynomial.

All polynomials of degree 3 that also include the term 1 are

TABLE 11.4 Primitive Polynomials of Degree N

N	Polynomials
4	$1 + X + X^4$
5	$1 + X^2 + X^5$
8	$1 + X^2 + X^3 + X^4 + X^8$
	$1 + X + X^5 + X^6 + X^8$
10	$1 + X^3 + X^{10}$
12	$1 + X + X^3 + X^4 + X^{12}$
16	$1 + X^2 + X^3 + X^5 + X^{12}$
	$1 + X + X^3 + X^4 + X^{16}$
24	$1 + X + X^3 + X^4 + X^{24}$
32	$1 + X + X^2 + X^{22} + X^{32}$
48	$1 + X + X^{27} + X^{28} + X^{48}$

$$x^3 + 1 = 0$$
$$x^3 + x^2 + 1 = 0$$
$$x^3 + x + 1 = 0$$
$$x^3 + x^2 + x + 1 = 0$$

The second and third polynomials are primitive and it is not difficult to verify that they will divide evenly the polynomial $x^7 + 1$. The other two are reducible. As an exercise, you may verify that

$$x^3 + 1 = (x + 1)(x^2 + x + 1)$$
$$x^3 + x^2 + x + 1 = (x + 1)(x^2 + 1)$$

There are several primitive polynomial of degree N. However, we are interested in those that include the fewer terms since this means using less XOR gates in the LFSR.

Table 11.4 lists the primitive polynomials of degree N, where $1 \le N \le 48$ [Peterson 1972, Golumb 1982, Bardell 1987]. Notice that for a given N there may be more than one primitive polynomial with the same number of terms as indicated for $N = 8$ and $N = 16$. Any primitive polynomial includes the terms 1 and X^N. All these polynomials correspond to LFSRs with at most three tabs (XOR gates). This is true also for primitive polynomials of degrees up to 300 [Bardell 1987].

11.2.3.4 *Properties of Maximal-Length Sequence*

1. The maximal-length shift register sequence has a period of $2^N - 1$.
2. If the sequence associated with an n-stage LFSR is of length $2^n - 1$, the characteristic polynomial associated with the LFSR is primitive.

3. The numbers of 1's in a maximal-length sequence differ from the number of 0's by 1.
4. Any maximal sequence produces a run of n 1's and $n - 1$ of 0's.
5. For $n > 4$, the maximal sequence for a polynomial P_n is the reverse sequence of that associated with the reciprocal polynomial P_n^R.
6. In any such sequences, one-half the runs have length 1, one-fourth have length 2, and so on.
7. The sequence obtained at any stage j is one clock cycle behind the sequence at stage $j - 1$.

You can easily verify these properties for the polynomials discussed so far. We next use the example for the fifth-order polynomial: $X^5 + X^2 + 1$ with the initial seed of $\{00001\}$. The output of one of the last flip-flop gives the sequence:

$$S(t) = \{(10000 \ 10010 \ 11001 \ 11110 \ 00110 \ 11101 \ 0)1000 \ 01\ldots\}$$

For the reciprocal LFSR, $S^R(t) =$ includes several runs of different lengths: for example, one run of five and four 1's, two runs of length 3 and length 2, and five runs of length 1.

This distribution shows that the patterns produced have a predetermined distribution of grouping bits [Rajski 1998]. Also, the sequences from the different stages are autocorrelated. For example, the sequences for the second and third stages of the LFSR in Fig. 11.2 and given in Table 11.1 are

$$(1 \ 0 \ 0 \ 1 \ 1 \ 1 \ 0)1\ldots$$
$$(0 \ 0 \ 1 \ 1 \ 1 \ 0 \ 1)0\ldots$$

Because of these facts, pseudorandom patterns may fail to detect some faults. Such faults are called *random pattern-resistant* (RPR) faults and are discussed in Section 11.4.

11.3   RESPONSE COMPACTION

It is necessary to compress the responses in BIST since the interest is to check the response of the test on a chip. In such a manner it is possible to obtain a signature for the circuit at the end of the test application. Comparing the faulty and fault-free signatures will allow us to detect faults. Usually, we think of data compression as a process that preserves data integrity. This is why we give more attention here to data compaction, which may result in some losses. For example, if the signature of a faulty circuit, due to the compaction scheme, is equal to the fault-free signature, the fault escapes detection. This type of information loss is called *aliasing*. The probability of aliasing decreases as the length of the test increases.

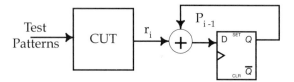

Figure 11.8 Parity testing.

There are several compaction testing techniques: (1) parity testing, (2) one counting, (3) transition counting, (4) syndrome calculation, and (5) signature analysis. We describe briefly the first three techniques and illustrate them using single-output circuits. Signature analysis is described in Section 11.3.4. Subsequently, we consider multiple-output circuits in the section on *space compaction*.

11.3.1 Parity Testing

This is the simplest of all techniques but also the most lossy. The parity of responses to the test patterns is calculated as $P = \sum_{i=1}^{i=L} r_i$, where L is the length of the test and r_i is the response for the ith test pattern. The summation here is modulo 2. The response of the circuit under test (CUT) to pattern i and the partial product P_{i-1} is illustrated in Fig. 11.8. The result on the storage device after the last test is applied is the parity to be compared to the fault-free circuit parity. If an odd number of test patterns detect the fault, the parity of the faulty circuit is different from that of the fault-free one and the fault is definitely detected. Otherwise, the number is even, and the parity is indistinguishable from the fault-free parity.

11.3.2 One Counting

In this scheme, which is shown in Fig. 11.9, the number of 1's in the response stream is calculated and compared to the number of 1's in the fault-free responses [Akers 1989]. We refer to this number as a 1-count. The responses to the test are applied to a counter that counts up each time, the response $r_i = 1$.

Consider the circuit in Fig. 11.10, on which we applied an exhaustive test consisting of eight patterns. For the fault-free circuit, 1-count = 5, notice that this number represents the number of minterms of the function. Why? For $a/0$ and $a/1$, the counts are 4 and 6, respectively. Both faults will thus be detected.

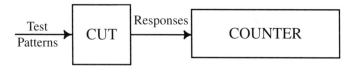

Figure 11.9 Counting compaction scheme.

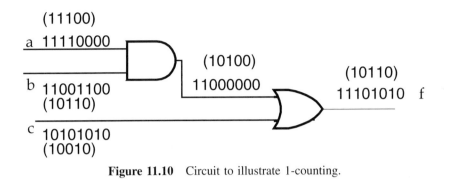

Figure 11.10   Circuit to illustrate 1-counting.

A fault that causes equal changes of 1's to 0's, and vice versa, will have the same number of 1's as in the fault-free count. If we have a test of length L and the fault-free count is m, the possibility of aliasing is $[C(L, m) - 1]$ patterns out of a total number of possible strings of length L, $(2^L - 1)$. Thus the upper limit of the aliasing probability is $\text{Pa}(m) = [C(L, m) - 1]/(2^L - 1)$. However, because the distribution of $C(L, m)$ is symmetrical and has a peak at $L/2$, this probability is lower for small and large values of m. Also, this probability is smaller the larger L is.

For the example in Fig. 11.10, where $m = 5$ and $L = 8$, this probability will be $\text{Pa}(m) = 55/255 \approx 0.2$. However, it will be left to the problems to show that this test will not cause any aliasing. Notice also that not all of the 255 strings of length 8 will actually be generated since there are only as many strings as there are faults. In this case we have only 10 faults.

Consider a shorter test, say, the pseudoexhaustive test shown in parentheses on the circuit. The good circuit count is 3 and the test length is 5. The aliasing probability is then $10/31 \approx 0.3$. Actually, aliasing happens only for one fault, $a/0$. For this fault as well as the fault-free circuit, 1-count $= 3$. Faults which without compaction are detected by an odd number of patterns will always be detected by 1-counting. Only some of those detected by an even number of patterns will have a 1-count that is different from the fault-free response. In addition, the count is independent of the order of pattern application for a combinational circuit, and unchanged if the output is inverted. The aliasing probability was estimated assuming that all possible strings will be generated by test application. However, this is exaggerated since the number of faults will probably be lower in numbers. The probability that the fault-free circuit will produce m 1's is only $P_m = C(L, m)/2^L$. Thus the aliasing probability is only $\text{Pa} = \text{Pa}(m)P_m = [C(L, m) - 1]/(2^L - 1) \cdot C(L, m)/2^L$.

11.3.3   Transition Counting

In transition counting compaction, it is only the number of transition $0 \rightarrow 1$ and $1 \rightarrow 0$ that are counted. Thus the signature is given by $\sum_{i=1}^{L-1} r_i \oplus r_{i+1}$, where

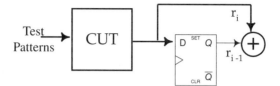

Figure 11.11 Transition count compaction scheme.

the summation is ordinary addition and $\oplus$ is XOR operation. The compaction scheme is shown in Fig. 11.11.

11.3.4 Signature Analysis

Signature analysis is the most popular technique for response compaction and it was originally called *cyclic code checker* [Benowitz 1975]. The technique was then applied for field testing of electronic equipment by Hewlett-Packard electronics [Frohwerk 1977] and became known as *signature analysis*. The compactor in this case is an LFSR to which the response string, $M(t)$, from the CUT is applied. At the end of the test application, the content of the LFSR is called the *signature* of the circuit. It reduces the response of all L test patterns to N-bit, where N is the length of the LFSR; that is, N is the number of flip-flops in the shift register. A fault in the circuit is expected to produce a signature that is different from that of the good circuit. The loss of information due to compaction may cause a faulty circuit to produce a good signature that results in an *aliasing*.

We will first take an example of a test response, a string $M(t)$, and generate the signature of this response. We will then show how the response is represented as a polynomial $M(X)$ and the signature is just the remainder of its division by the characteristic polynomial of the LFSR, $P(X)$ [Bhavsar 1981]. In addition, we show how aliasing may occur and estimate its probability. Assume that the application of an eight-pattern test to a circuit results in the data stream $M(t) = \{10110010\}$. The leftmost bit is the first signal entering any of the LFSRs shown in Fig. 11.12. The first is of the standard type and its polynomial is $X^3 + X^2 + 1$ and the second is of the modular type and characterized with the same polynomial. The initial state of the storage elements in the shift register is assumed to be $\{000\}$. As this stream of data is shifted in the LFSR, the storage elements change their states following the timing shown in Table 11.5.

We selected one standard LFSR, a modular one with the same characteristic polynomial and a modular one with the reciprocal polynomial as indicated in the figure. The three signatures are $\{100\}$, $\{100\}$, and $\{001\}$, respectively. The signatures on the modular LFSRs are the result of the polynomial divisions of $M(X)$, which represents the response string, $M(t)$, by the characteristic polynomials of the LFSRs.

The data stream, $M(t)$, can be represented by a polynomial in a fashion sim-

TABLE 11.5 Signature Analysis

T	M(t)	Standard $(X^3 + X^2 + 1)$				Modular $(X^3 + X^2 + 1)$					Modular $(X^3 + X + 1)$				
		y_0	y_1	y_2	y_3	y_0	y_1	D_2	y_2	y_3	y_0	y_1	y_2	D_2	y_3
0	1	1	0	0	0	1	0	0	0	0	1	0	0	0	0
1	0	0	1	0	0	0	1	1	0	0	0	1	0	0	0
2	1	0	0	1	0	1	0	0	1	0	1	0	1	1	0
3	1	0	0	0	1	0	1	0	0	1	0	1	0	1	1
4	0	0	0	0	0	0	0	0	0	0	1	0	1	0	1
5	0	0	0	0	0	0	0	0	0	0	0	1	0	0	0
6	0	0	0	0	0	0	1	0	0	0	0	0	1	1	0
7	1	1	0	0	0	1	1	1	0	0	0	0	0	1	1
8	x	**X**	**1**	**0**	**0**	x	**1**	**1**	**0**	**0**	x	**0**	**0**	**1**	**1**
							$R(X) = X^2$					$R(X) = 1$			

275

$$P(X) = 1 + X^2 + X^3$$

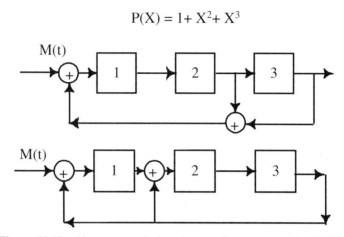

Figure 11.12   Signature analysis using standard and modular LFSR.

ilar to the representation of the output of an autonomous LFSR in Section 2.3. Each bit is translated in time by one unit. Again, Eq. (11.5) defines the translation operator. Thus the data stream, $M(t) = \{10110001\}$, is represented by the polynomial $M(X) = X^7 + X^5 + X^4 + 1$. The characteristic polynomial of the first modular LFSR in Fig. 11.12 is the same as the standard one, $P(X) = X^3 + X + 1$. Notice, however, that the taps are the reciprocal of those of the standard configuration. The division of $M(X)$ by $P(X)$ is performed in the same manner as presented in Section 11.2.3 and the results of the operation are a quotient $Q(X)$ and a remainder $R(X)$ such that $M(X) = P(X)Q(X) + R(X)$.

$$X^7 + X^5 + X^4 + 1 = (X^4 + X^3 + 1)(X^3 + X^2 + 1) + X^2$$

$$
\begin{array}{r}
Q(X) \qquad\qquad X^4 + X^3 + 1 \\[2pt]
\hline
X^3 + X^2 + 1 \overline{)\, X^7 + \qquad X^5 + X^4 + \qquad\quad + 1} \\
X^7 + X^6 + \qquad X^4 \\[2pt]
\hline
X^6 + X^5 \qquad\qquad\quad + 1 \\
X^6 + X^5 + \qquad + X^3 \\[2pt]
\hline
X^3 + \qquad\quad + 1 \\
X^3 + X^2 + 1 \\[2pt]
\hline
R(X) \qquad\qquad\qquad\qquad\qquad X^2
\end{array}
$$

The division can also be performed using the coefficients C_k that we introduced in Section 11.2.3.2. As we recall, $C_k = 1$ only when there is feedback from the last stage to the kth flip-flop; otherwise, it is 0 for the other modular LFSR. The taps are the same as for the standard LFSR, but the polynomial is the reciprocal of the latter's polynomial.

$$
\begin{array}{r}
1\ 1\ 0\ 0\ 1 \\[2pt]
\hline
1\ 1\ 0\ 1\ )\overline{1\ 0\ 1\ 1\ 0\ 0\ 0\ 1} \\
1\ 1\ 0\ 1 \\
\hline
1\ 1\ 0\ 0\ 0\ 0\ 1 \\
1\ 1\ 0\ 1 \\
\hline
1\ 0\ 0\ 1 \\
1\ 1\ 0\ 1 \\
\hline
0\ 1\ 0\ 0
\end{array}
$$

As another example, we divide $M(X)$ by $P(X) = X^3 + X^2 + 1$.

$$X^7 + X^5 + X^4 + 1 = X^4(X^3 + X + 1) + 1$$

$Q(X) \qquad\qquad\qquad X^4 \qquad\qquad\qquad\qquad\qquad 1$

$$
X^3 + X + 1\ )\overline{X^7 + X^5 + X^4 + 1} \qquad\quad 1\ 0\ 1\ 1\ )\overline{1\ 0\ 1\ 1\ 0\ 0\ 0\ 1}
$$
$$
X^7 + X^5 + X^4 \qquad\qquad\qquad\qquad 1\ 0\ 1\ 1
$$

$R(X) \qquad\qquad\qquad\qquad\qquad\quad 1 \qquad\qquad\qquad\qquad\qquad 0\ 0\ 0\ 1$

This remainder corresponds to the signature {001} left in the corresponding LFSR as indicated in Table 11.5.

11.3.4.1   Fault Detection.

Consider a CUT and a test set of length L such that there is at least one pattern that detects any fault, f. We will apply the response of the faulty circuit to an LFSR that we will call a *signature analyzer*, as we use it in compaction. This signature analyzer has N stages and a characteristic polynomial, $P(X)$. The fault f is detected through the signature analyzer if the faulty signature S_f is such that $S_f \oplus S = 1$, where S is the signature of the fault-free circuit; otherwise, the fault escaped detection. It is masked and we say that *aliasing* has occurred. In terms of polynomial division, aliasing implies that the division of the faulty and faulty-free responses, $M_f(X)$ and $M(X)$, yield the identical remainders. That is, $R_f(X) = R(X)$.

Instead of analyzing the response stream, we consider the error, $E(X)$.

$$M(X) = P(X)Q(X) + R(X)$$
$$M_f(X) = P(X)Q_f(X) + R_f(X)$$

Adding the two equations, we get

$$E(X) = P(X)q(X) + r(X)$$

The fault is detected if $r(X) \neq 0$.

For a given polynomial $P(X)$, there are only a finite number of remainders, and therefore it is not possible to eliminate aliasing but only to minimize it. For example, if the length of the test set is equal or smaller than the length of the LFSR's cycle, $L \leq N$, all faults will be detected. However, most pseudorandom test sets are very long and it is not practical to have a very long LFSR.

11.3.4.2 Aliasing Probability.

We calculate the aliasing probability for a test of length L and a signature analyzer of N stages. Thus any response of the circuit to the test consists of L bits. It is any of $k = 2^L$ strings of L. No aliasing is possible for those response strings with $k - N$ leading zeros since they are represented by polynomials of degree $N - 1$ that are not divisible by characteristic polynomial of the LFSR. There are 2^{L-N} such strings. Hence the probability of aliasing is Pa $= (2^{L-N} - 1)/(2^L - 1)$ since $L \gg 1$ and $L \gg N$; then Pa $= 2^{-N}$. This value of the aliasing probability is reached for test lengths is 50 patterns. Thus the larger the N, the smaller the aliasing probability. Notice that this probability is based on the assumption that the response streams are equally likely to occur. That is, the number of faults is actually $2^L - 1$. However, there are generally fewer faults.

Several approaches have been proposed to minimize aliasing—reversing the test sequence, using multiple MISR, taking multiple signatures. It is possible to double the test length by applying the reverse sequence. This can be achieved by changing the LFSR polynomial to its reciprocal. The aliasing probability will be reduced to 2^{-2N}; however, this requires some hardware overhead to reconfigure the LFSR and doubles the test application time. Another approach is to use several LFSRs with polynomials that are relatively prime. An orthogonal practice is to use the same LFSR and take signatures at several time intervals. For w signatures the probability will then decrease to 2^{-wN} More storage is then needed for the various signatures. An extreme case of this scheme is to take a signature every N cycles. This scheme would have an aliasing probability of zero. The proof is straightforward using polynomial division and is left as an exercise.

11.3.5 Space Compaction

So far we assumed that the CUT has one primary output. This is not realistic and we need to deal with compaction schemes for k-output circuits, as illustrated in Fig. 11.13. Different possible space compaction schemes are shown in Fig.

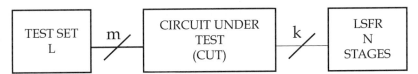

Figure 11.13   Compaction using signature analysis.

11.14 for a case where $N = 4$, $L = 5$, and $k = 5$. The characteristic polynomial of the signature analyzer is $P(X) = X^4 + X + 1$.

11.3.5.1   Single-Input Signature Analyzer (SSA). Responses from one primary output at a time are compacted through the signature analyzer using an *n*-to-1 multiplexer as illustrated in Fig. 11.14*a*. This process is slow, but it reduces

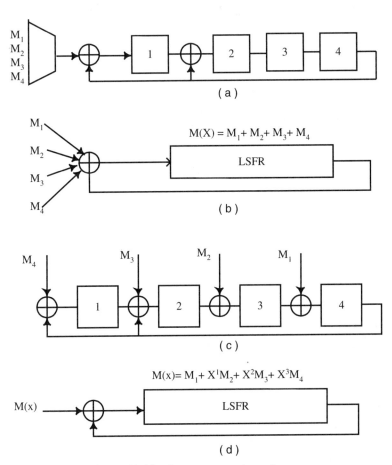

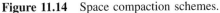

Figure 11.14   Space compaction schemes.

the aliasing probability since the scheme is equivalent to elongating the test by a factor of N. The probability of aliasing is still 2^{-N}; however, faults may be detected through more than one output and thus having reduced aliasing.

11.3.5.2 Bellmac. This approach, which is featured in Fig. 11.14b, takes its name from the product in which it was used for the first time. In this case parity and signature analyzer compaction techniques are used. All the responses from the primary outputs are compacted through a parity tree before a second compaction by the analyzer. In the example used in this section, the errors at the four outputs of a circuit, in response to five patterns, are shown in Fig. 11.15a. The leftmost column of Fig. 11.15b shows the parity of these errors for each pattern. The parity stream is then compacted in an LFSR. The polynomial representing this stream of data is $X^4 + 1$. Its division by the characteristic polynomial $P(X) = (X^4 + X + 1)$ results in the remainder $R(X) = X$.

11.3.5.3 Parallel Signature Analysis (PSA). This scheme is also known as multiple input signature register (MISR) and illustrated in Fig. 11.15c. According to this compaction, the responses from the different primary outputs are introduced at the different stages of the LFSR. Thus the different responses, $M_i(X)$ are staggered into the signature analyzer. This scheme is equivalent to compaction through a SSA with the input stream being shifted in time, $M(X) = M_0(X) + X_1(X) + \cdots + X^n M_n(X)$. This is illustrated in the example of Fig. 11.15c. The resulting error polynomial due to the four outputs is

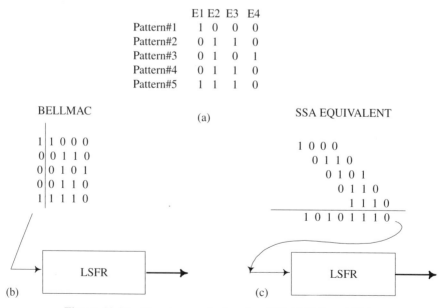

(a)

(b) (c)

Figure 11.15 Signature analysis using BELLMAC and PSA.

$$E(X) = E_1(X) + XE_2(X) + X^2E_3(X) + X^3E_4(X)$$

It is divided by the characteristic polynomial, $X^4 + X + 1$, to result in a remainder of $R(X) = X^3 + X + 1$. As a consequence of this equivalence, the aliasing probability for MISR is still 2^{-N}, where N is the number of stages of the signature analyzer.

In the example considered to illustrate MISR, the number of outputs is equal to the number of stages in the signature analyzer. However, we need to determine space compaction when $n \neq N$. If $n < N$, the outputs will be introduced in n of N flip-flops and the balance will not have any input from the circuit. In the other situation, $n > N$, it is possible to use additional LFSRs or to partition the outputs into N partitions. The parity of each partition is then fed to the different stages of the MISR.

11.4   RANDOM PATTERN RESISTANT FAULTS

The effectiveness of a test set is measured by its fault coverage, its length, and its hardware and data storage requirement. PR tests generated according to the method we described above are usually long and result in unacceptable fault coverage. Figure 11.16a shows a typical plot of the fault coverage versus the number of random test patterns. At first, a rapid increase in fault coverage is followed by saturation. The difference between this saturation level and 100% is denoted by ΔFC, and it represents faults that are hard to detect by random patterns. They are called random pattern resistant (RPR) faults. The fault coverage can be improved by reducing the aliasing probability. In Section 11.3.4.1 we briefly listed several approaches. However, the low coverage is mostly due to lack of patterns to detect some faults. There are faults that are detected by very few, if not by only one pattern. If this pattern is not part of the test set, the fault will remain undetected. We explain next why some of these patterns are hard to obtain.

PR patterns generation is based on the fact that all flip-flops have equal probability of producing a 1 or a 0. That is, all the inputs of the circuit under test have the same probability for a 0 or a 1. For example, if a six-input NOR gate is considered, SA0 faults on its inputs are detected by one and only one pattern. The probability of generating this pattern is only 1 out of 64 (2^6). In addition, the sequence generated is based on an equally likely probability for 0 and 1 to appear at each of the CUT's inputs. Consider the circuit in Fig. 11.16b, where the minimal SAF test set is shown for a three-input AND and a three-input OR. More 1's are needed on the AND gate and more 0's are required for the OR gate. For the circuit in Fig. 11.16c, six patterns out of a 32-long exhaustive test set guarantee 100% fault coverage. It is obvious from the test set that the 0's and 1's do not appear equally on all inputs. Also, inputs d and e need to be more exercised than the other primary inputs (PIs). Such an observation prompted the

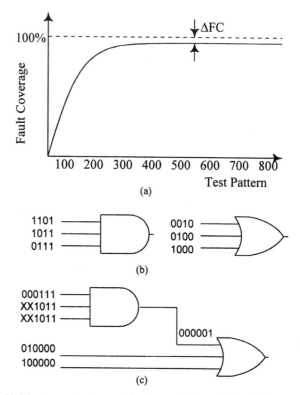

Figure 11.16 Example to explain the need for weighted PR test patterns.

introduction of weights for the various PIs [Schnurmann 1975, Bardell 1987]. The weights are the probabilities for the input signals to be equal to 1. This approach, which is called *weighed PR test pattern* generation, improved the fault coverage appreciably, but also added extra hardware.

Several techniques have been used to generate optimal weights. Some are based on circuit structure analysis [Waicukauski 1989], and others are based on fault detection probabilities [Lisanke 1987, Wunderlich 1987]. Despite the improvement resulting from this method, there were still some faults that are hard to detect. For example, going back to Fig. 11.16*b*, if the inputs of the AND and OR circuits are tied together, the weights selected will favor one structure or the other, and multiple weights are required. However, this multiple-weight approach requires considerable increase in the hardware.

Another approach to overcome the RPR fault problem is test point insertion. The location of the test points is determined by testability measures. As we discussed in Chapter 8, test point insertion requires knowledge of the circuit topology and extra hardware, with possible impact on performance [Schotten 1995]. Reseeding the LFSR is an alternative approach that has low-overhead hardware [Hellebrand 1992]. A similar technique used a multiple polynomial LFSR. However, these techniques generate test sets of excessive length.

Recently, a mixed-mode approach has been used. With fault simulation it is possible to determine the RPR faults and use a deterministic test approach to generate the patterns to detect them. BIST execution then combines the PR pattern test (R) and the deterministic test (T) according to a certain protocol. The test T has to be stored in a ROM on the chip or implemented as a hardware circuit that is activated at testing time. Bit fixing is one of these techniques. The output bits of the LFSR, at a certain position in the sequence, are changed to generate the RPR test patterns [Pomeranz 1993, Chatterjee 1995, Touba 1995, 1996, AlShaibi 1996]. An alternative to bit fixing is bit flipping [Wunderlich 1996]. Most of these variations of controlling the bits of the LFSR sequence have not yet solved the RPR problem and they were applied to circuits that combine BIST with scan path. These circuits are described in the next section as part of BIST architectures.

11.5 BIST ARCHITECTURES

Several testing schemes make use of pseudorandom test pattern generation (an LFSR) and response compaction (a signature analyzer). Some of these schemes include both an LFSR and a signature analyzer on chip for internal testing. Others use them off-chip for external testing. Typically, logic BIST approaches combine pseudo-random testing with other DFT constructs such as scan path and Boundary-Scan. The intent here is to optimize on the advantages of the various DFT constructs.

11.5.1 Built-In Self-Testing

The built-in self-testing architecture is shown in Fig. 11.17. In addition to the circuit under test (CUT), the chip includes an LFSR for test pattern generation and a MISR for response compaction. Some control circuitry is required to allow for configuring the circuit for normal and testing modes. The signature from the MISR is checked automatically versus the good circuit response, which is stored in a ROM on the chip. Under normal operation the signal is applied to the CUT and observed directly on the primary output. During test mode, the patterns are applied from the LFSR, the response of the circuit to these patterns is compacted through the MISR, and the signature is compared to the good one. The effectiveness of this testing for a given circuit depends on the appropriate choice of the LFSRs, their length, and configurations. It usually yields high fault coverage for combinational circuits.

11.5.2 Autonomous Test

In the autonomous test approach [McCluskey 1981], the circuit is partitioned into subcircuits to facilitate its testing, as explained in Chapter 8. The partition-

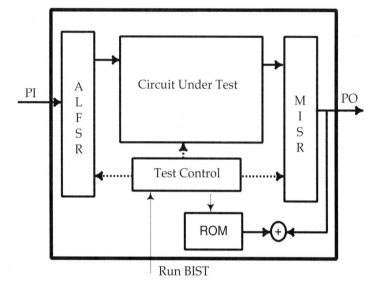

Figure 11.17   BIST architecture.

ing is acomplished through an added multiplexing scheme or by path sensitization. Each partition is tested independent of the others. The test may be supplied externally, but here we consider built-in testing. Thus the scheme, which is illustrated in Fig. 11.18, uses the same LFSR and the same MISR for test generation and response compaction. The solid lines illustrate the testing of module-1. If the multiplexers are too many for optimal timing of the circuit, a sensitization approach should be taken, although it may take some more effort to configure.

11.5.3   Circular BIST

Circular BIST is intended for register-based architecture [Krasniweski 1989]. In Fig. 11.19 the self-test shift registers (STSRs) are part of the circular self-test path. The other registers are not used for testing and they are denoted by SR. Each of the storage devices, STSRs, is designed as shown in Fig. 11.20. The MISR formed by all STRSs has a characteristic polynomial of $1 + X^m$, where m is the number of storage elements in the chain. All these cells should be initializable to put the CUT in a known state before testing. The test process requires three phases:

1. *Initialization.* All registers are placed in a known state.
2. *Testing of the CUT.* The circuit is run in test mode; registers that are not in the self-test path operate in their normal mode. During this phase the self-test path operates as both an LFSR and a MISR.

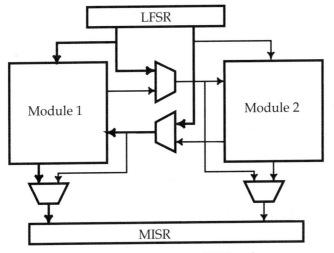

Figure 11.18 Autonomous BIST testing.

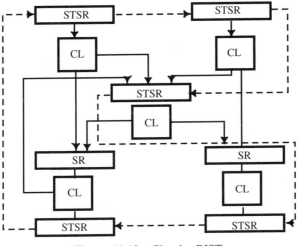

Figure 11.19 Circular BIST.

3. *Response evaluation.* During this phase the circuit is also in test mode, but now the sequences of outputs from one or more STSR are compared with a precomputed fault-free value.

11.5.4 BILBO

In this and the next subsections, we address the various schemes that combine BIST with scan path. The objective is to optimize on the strength of both tech-

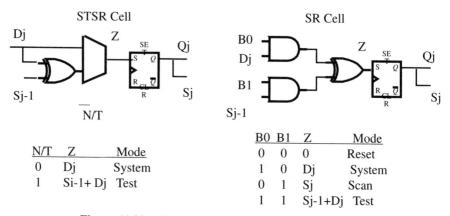

Figure 11.20   Circular BIST: storage cell configurations.

niques to facilitate IC testing. One of the early DFT constructs that combined BIST and scan path is the built-in logic blocks observer (BILBO) [Konemann 1979, 1980]. This architecture, which combines the functions of test pattern generation and response compression into a single unit, was developed for bus-oriented systems. It takes advantage of registers or storage devices on the chip to configure them as pseudorandom test pattern generators PRTPG and signature analyzers. In Fig. 11.21a the flip-flops can be configured to form any of the following structures: a shift register for scan design, an LFSR, or a MISR

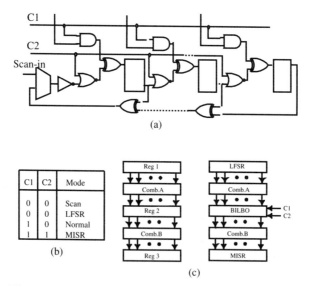

Figure 11.21   BILBO structure and modes of operations.

according to the mode of operation determined by the two control lines $C_1 C_2$. A circuit using BILBO is modeled as shown in Fig. 11.21b. The two combinational parts, Comb-1 and Comb-2, are tested, one at a time, using registers R1, R2, and R3. To test Comb-A, R1 is configured as a PRTPG and R2, which is a BILBO block, is used as a MISR. For the Comb-B, it is the BILBO block that is used as a PRTPG and R3 is used as the MISR. These different testing sessions need to be coordinated by a test manager.

11.5.5 Random Test Socket

This is a generic architecture that combines scan path and BIST [Bardell 1982] as illustrated in Fig. 11.22. A modification subsequently modified resulted in more efficient DFT architectures, such as STUMPS, discussed in Section 11.5.6. In random test socket architecture, all primary inputs (PIs) to the circuit are connected to the taps of a pseudorandom pattern generator (LFSR 1) and all primary outputs (POs) are connected to a MISR. The storage cells of the CUT are configured for scan-path operations and form the shift register, SR. The SI pin is connected to another pseudorandom pattern generator (LFSR 2) and the SO is connected to an SSA. The testing is applied as follows:

1. Load the SR with an initialization pattern by clocking the LFSR and SR simultaneously while the scan enable, SE = 1. On each cycle a new random bit is generated. Thus we need n clock cycles, where n is the number of storage units in SR.

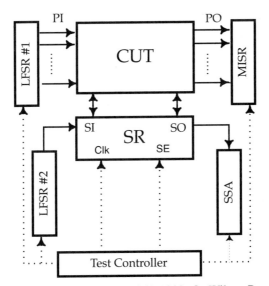

Figure 11.22 Random test socket [Bardell 1982 © Wiley. Reprinted with permission].

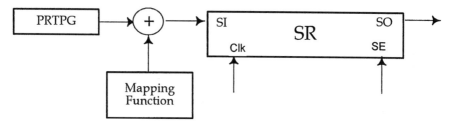

Figure 11.23 Mixed mode control for RPR faults.

2. Apply a test pattern from LFSR 1 to the CUT's PI.

3. Clock the system once with SE = 0 to latch the response to the pattern in SR.

4. Capture the results in MISR by clocking once while keeping SE = 0.

5. Unload the data of the SR into the SSA by making SE = 1 and clocking both the SSA and SR n times.

Steps 1 and 5 can be overlapped as we have done in traditional scan path (see Chapter 9). This approach to applying the test pattern in the scan-path registers is called *test per scan*, as opposed to *test per clock*, used so far in conjunction with BIST.

The advantages of this scheme are low-cost ATPG and improved observability. Also, instead of using two LFSRs for PRTPG, one will generally be used to initialize the scan chain and to stimulate the primary inputs. Similarly, one MISR can be used to compress the response data from the primary outputs and from the scan chain. The limitations are increased overhead and long test application time. In addition, the RPR faults are still a problem for fault coverage.

We illustrate next how to use mixed-mode technique to detect RPR faults. The concept is illustrated in Fig. 11.23. The output of the LFSR is applied to the scan resisters as described above; however, these output bits can be changed is such a way as to allow a pattern to detect the RPR faults. Fixing the bits is obtained from the mapping function, which was generated using knowledge about the circuit—the test patterns to detect RPR as obtained form a deterministic test pattern generator and are verified by a simulator.

11.5.6 STUMPS

Self-test using MISR and a parallel shift register sequence generator (STUMPS) architecture was originally used in multichips board in which each board has its SR, which allows the testing of all boards using the same DFT overhead. We discuss it here in term of one chip having multiple scan chains. The scheme is illustrated in Fig. 11.24. The multiplicity of the chains speeds up testing appli-

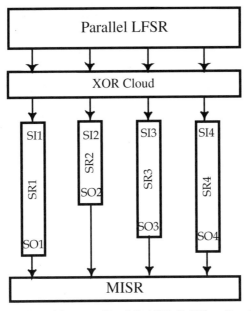

Figure 11.24 STUMPS architecture [Bardell 1982 © Wiley. Reprinted with permission].

cation. The SI pins of all chains are supplied from a multiple-output LFSR. Also all 50 pins are connected to the various stages of a MISR.

All SRs are loaded using the shift register sequence generator (SRSG). Both registers are clocked simultaneously. The number of cycles is equal to the longest scan chain of the SRs. The results are first captured back in the SRs, then scanned out while another test is loaded. In the example shown, we have $m = 4$ and nothing is shown about how the CUT receives test patterns. We can consider two alternatives: Use a parallel LFSR or use Boundary-Scan cells. The second solution, of course, make the test controller more complex, but it is feasible. This testing architecture requires only one additional I/O pin, the test mode signal. The use of multiple scan chains accelerates test application. The main limitation of the architecture is the need of addition interconnections, which makes routing more difficult.

The sequences obtained from adjacent bits of a parallel LFSR are not linearly independent; the neighboring scan chains contain test patterns that are highly correlated [Chen 1986]. This can affect fault coverage adversely since the patterns seen by the CUT do not really form a random sequence. To alleviate the consequence of this problem, phase shifters are placed between the LFSR and the SI pins of the scan chains [Bardell 1987]. Phase shifters are formed with an XOR network and form the XOR cloud box shown in the figure. An example of the shifters attached to the PRTPG is shown in Fig. 11.25.

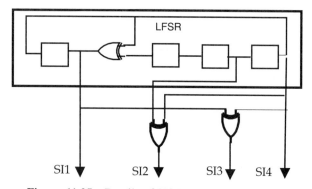

Figure 11.25 Details of XOR cloud in Fig. 11.24.

REFERENCES

Akers, S. B., and B. Krishnamurthy (1989), Test counting: a tool for VLSI testing, *IEEE Des. Test Comput.*, Vol. 6, No. 10, pp. 58–72.

AlShaibi, M. F., and C. R. Kime (1996), MFBIST: a BIST method for random pattern resistant circuits, *Proc. IEEE International Test Conference*, pp. 176–185.

Bardell, P. H. (1982), Self-testing of multi-chip logic modules, *Proc. IEEE International Test Conference*, pp. 200–204.

Bardell, P. H., et al. (1987), *Built in Test for VLSI: Pseudo-random Techniques*, Wiley, New York.

Bardell, P. H. (1990), Analysis of cellular automata used as pseudo-random pattern generators, *Proc. IEEE International Test Conference*, pp. 762–767.

Benowitz, N., et al. (1975), An advanced fault isolation system for digital logic, *IEEE Trans. Comput.*, Vol. C-24, No. 5, pp. 489–497.

Bhavsar, D. K., and R. W. Heckelman (1981), Self-testing by polynomial division, *Digest of Papers, 1981 International Test Conference*, 181–189 pp.

Chakrabarty, K., et al. (1997), Test with compression for built-in self testing, *Proc. IEEE International Test Conference*, pp. 328–337.

Chatterjee, M., and D. Pradham (1995), A novel pattern generator for near-perfect fault coverage, *Proc. IEEE VLSI Test Symposium*, pp. 417–425.

Chen, C. L. (1986), Linear dependencies in linear feedback shift registers, *IEEE Trans. Comput.*, Vol. C-35, No. 12, pp. 1086–1088.

Frohwerk, R. A. (1977), *Signature Analysis: A New Digital Field Service Method*, Hewlett-Packard Journal, Vol. 28, No. 9, pp. 2–8.

Golumb, S. W. (1982), *Shift Register Sequences*, Aegean Park Press, Laguna Hills, CA.

Hellebrand, S., et al. (1992), Generation of vector patterns through reseeding a multiple-polynomial linear feedback shift registers, *Proc. IEEE International Test Conference*, pp. 120–129.

Hortensius, P. D., et al. (1989), Cellular automata based pseudo-random number generators for built-in self-test, *IEEE Trans. Comput.-Aided Des.*, Vol. CAD-8, No. 8, pp. 842–859.

Koenemann, B. (1979), Built-in logic block observable technique, *Proc. IEEE International Test Conference*, pp. 37–41.

Koenemann, B. (1980), Built-in test for complex digital integrated circuits, *IEEE J. Solid State Circuits*, Vol. SC-15, No. 3, pp. 315–318.

Krasniweski, A., and S. Pilarski (1989), Circular self-test path: a low cost BIST technique for VLSI circuits, *IEEE Trans. Comput.-Aided Des.*, Vol. CAD-8, No. 1, pp. 46–55.

Lisanke, R., F. Brglez, and A. Degeus (1987), Testability-driven random test pattern generation, *IEEE Trans. Comput.-Aided Des.*, Vol. CAD-6, No. 6, pp. 1082–1087.

McCluskey, E. J., and S. Bozorgui-Nesbat (1981), Design for autonomous test, *IEEE Trans. Comput.*, Vol. C-30, No. 11, pp. 860–875.

McCluskey, E. J. (1986), *Logic Design Principles with Emphasis on Testable Semicustom Circuits*, Prentice Hall, Upper Saddle River, NJ.

Peterson, W. W., and E. J. Weldon (1972), *Error-Correcting Codes*, Colonial Press, Birmingham, AL.

Pomeranz, I., and S. M. Reddy (1993), 3-Weight pseudo-random test generation based on a deterministic test set for combinational and sequential circuits, *IEEE Trans. Comput.-Aided Des.*, Vol. CAD-12, No. 7, pp. 1050–1052.

Rajski, J., and J. Tyszer (1998), *Arithmetic Built-in Self-Test for Embedded Systems*, Prentice Hall, Upper Saddle River, NJ.

Schotten, C., and H. Meyr (1995), Test-point insertion for an area-efficient BIST, *Proc. IEEE International Test Conference*, pp. 515–523.

Schnurmann et al. (1975), The weighted random test-pattern generator, *IEEE Trans. on Comput.*, Vol. C-24, No. 7, pp. 695–700.

Touba, N. A., and E. J. McCluskey (1995), Transformed pseudo-random patterns for BIST, *Proc. IEEE VLSI Test Symposium*, pp. 2–8.

Touba, N. A., and E. J. McCluskey (1996), Altering a pseudo-random bit sequence for scan-based BIST, *Proc. IEEE International Test Conference*, pp. 167–175.

Waicukauski, J., et al. (1989), A method for generating weighted random test patterns, *IBM J. Res. Dev.*, Vol. 33, No. 2, pp. 149–161.

Wunderlich, H. J. (1987), On computing optimized input probabilities for random tests, *Proc. ACM Design Automation Conference*, pp. 392–398.

Wunderlich, H. J., and G. Kiefer (1996), Scan-based BIST with complete fault coverage and low hardware overhead, *Proc. IEEE European Test Workshop*, pp. 60–64.

PROBLEMS

11.1. Show that the polynomial $P(X) = 1 + X + X^4$ has a maximal test sequence.

11.2. Is the polynomial $P(X)$ $1 + X^5$ primitive? Explain.

11.3. Give, if any, the characteristic polynomial of the LFSR that causes aliasing when the following error sequence is

$$t = 0\ 1\ 2\ 3\ 4 \dots$$
$$E = 1\ 0\ 0\ 0\ 0 \dots 0\ 0\ 0\ 0\ 0$$

11.4. Suppose that a modular LFSR whose characteristic polynomial is $1 + X + X^4$ is used as a signature analyzer for the circuit and the test shown in Fig. P11.4.

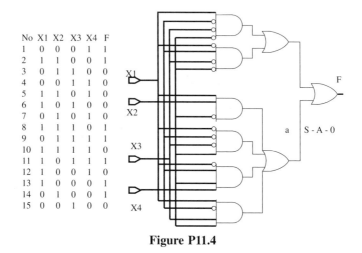

No	X1	X2	X3	X4	F
1	0	0	0	1	1
2	1	1	0	0	1
3	0	1	1	0	0
4	0	0	1	1	0
5	1	1	0	1	0
6	1	0	1	0	0
7	0	1	0	1	0
8	1	1	1	0	1
9	0	1	1	1	1
10	1	1	1	1	0
11	1	0	1	1	1
12	1	0	0	1	0
13	1	0	0	0	1
14	0	1	0	0	1
15	0	0	1	0	0

Figure P11.4

(a) Calculate the signature of the good circuit.

(b) Calculate the signature if the test sequence is 1, 2, 3, ... , 15.

(c) Repeat part (b) with the sequence reversed, 15, 14, 13, ... , 1.

11.5. For the circuit shown in Fig. P11.5, use the stuck-at test given and find the signature of the good circuit if the time compaction used **(a)** parity testing; **(b)** one count; and **(c)** transition count. For each scheme, list the undetected faults and explain why they were not detected.

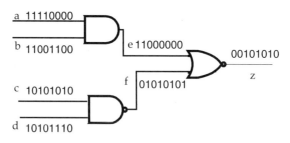

Figure P11.5

11.6. Using a single-input signature analyzer (SSA), we have shown that the aliasing probability is given by

$$p(\text{aliasing}) = \frac{2^{L-N} - 1}{2^N - 1}$$

where L is the test of length and N is the degree of the compactor primitive characteristic polynomial $P(X)$. Develop the aliasing probability if instead of an SSA we use an N-bit MISR with an N-output CUT.

11.7. Suppose that we record the signature every N clock cycles.

(a) How many signatures will be formed for the test?

(b) What will be the aliasing probability?

(c) What are the advantages and disadvantages of such a scheme?

PART IV

SPECIAL STRUCTURES

12

MEMORY TESTING

12.1 MOTIVATION

There are a few very good reasons why memory testing deserves special attention. First, memory is vital to electronic products. We hardly find a digital system that does not include a certain type of memory. Second, like other digital circuits, the device density and the circuit complexity of memory chips are continuously increasing. This trend is shown in Fig. 12.1. The dramatic rate of increase is even higher than that of microprocessors. Third, even though unlike most digital circuits, memory chips have regular structures that initially may imply an ease of testability, the regularity, together with sequential circuit property of the cells, in some ways causes testing problems that are not necessarily encountered in other regular combinational logic circuits, such as the array multipliers that we discussed in Chapter 8. The fourth reason for testing complexity is the proliferation of memory types. Volatile memories, also known as random access memories (RAMs), may be static (SRAMs) or dynamic (DRAMs). Some nonvolatile memory devices are read-only memories (ROMs), programmable read-only memories (PROMs), erasable (reprogrammable) memories (EPROMs), UV erasable memories (UVPROMs), and electrically erasable memories (EEPROMs). Flash memories are also being used whenever high-density nonvolatile memories are needed. In this chapter we concentrate on RAM testing.

In Section 12.2 we present RAM models and architectures. Fault models for these RAMs include traditional fault models encountered in combinational and sequential circuits such as stuck-at and bridging faults. In addition, there are faults that better represent failures in RAMs. These are coupling fault (CF)

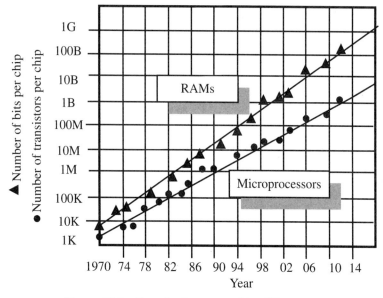

Figure 12.1   Growth of memory chips [SIA 1997].

and pattern-sensitive fault (PSF) models [Hayes 1975]. Fault models will be discussed in Section 12.3. RAM testing is covered in the next three sections. This includes functional testing and built-in self-test (memory BIST).

12.2   MEMORY MODELS

Circuit models are useful in simulation as well as development of algorithms for test pattern generation. Models for memory chips are available at different levels in each domain of design hierarchy: physical, logic, and system. Here we consider the logic and circuit levels to discuss failure modes in the cells and the functional level to generate test patterns.

12.2.1   Functional Model

A generic functional model of a RAM is shown in Fig. 12.2. The storage elements are usually organized to form the *array*. In this figure the array consists of $r \times c$-bit words. This array is surrounded with peripheral circuits. The address from the *memory address register* (MAR) is decoded by row and column decoders to select the appropriate row or column of cells to be accessed. Writing and reading is then performed using the *write drivers* and the *sense amplifiers*, respectively. The data to be written to the memory or obtained from the memory are usually placed in the *memory data register* (MDR). In addition

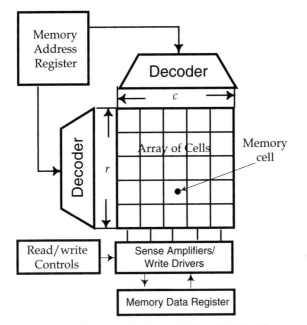

Figure 12.2   Functional model for a RAM chip.

to these different components, a refresh circuit is needed when a dynamic RAM is used.

The memory array consists of rows and columns of memory cells. The cells in each word are arranged in a row and selected by the word line (WL). The data are supplied (write operation) and collected (read operation) through the bit lines (BLs) to all cells in a column. However, one cell and only one cell of the column is affected by the operations at a time. In word-oriented memory all bits in the word are written or read simultaneously. In content-addressable memories (CAMs), the bits in a column are, instead, addressed simultaneously.

12.2.2   Memory Cell

On the logic level, the memory cell may be represented by a D flip-flop on which one can write or store a bit, 0 or 1. The information is held until it is overwritten. It is assumed that when the cell is not accessed, it retains the stored value. The retention depends on the type of cell implementations. Cells may be *static* or *dynamic*. The terminology refers to the method according to which the charge is stored in the cell. A very commonly used static RAM (SRAM) cell consists of two cross-coupled inverters, as shown in Fig. 12.3*b*. To *write* on the cell, the data and its complement are placed on bit lines BL and BL′ (respectively). Then the word line is held high and the pair of inverters will store the new information. To *read* from the cell, the bit lines are precharged

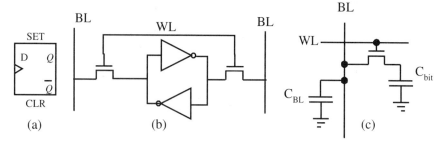

Figure 12.3   Memory cells: (*a*) logic; (*b*) static CMOS; (*c*) dynamic MOS.

to complementary values; then the word line is again held high and one of the inverters will discharge in one of the bit lines.

Other implementations are also possible using fewer transistors. However, this number can really be decreased using dynamic logic. The dynamic RAM (DRAM) cell may consist of three, two, and even one transistor. An example of a one-transistor cell is shown in Fig. 12.3*c*. This cell relies on storing the bit value in a MOS capacitor [Rideout 1989]. The write operation is done in a fashion similar to the static cell. The read operation is performed by charging the BL to V_t, then holding WL high. The storing capacitance, C_{bit}, discharges in the bit line, BL. The read operation is destructive and thus requires a write back.

Because of charge sharing, this type of cell is more difficult to design. In addition, because of the possibility of charge leakage, it is necessary to *refresh* it. Refresh circuitry performs periodical data update on a row basis. No bit-line selection is necessary. As the bits of the word are read, they are then rewritten. Despite all these extra operations, dynamic RAMs are still the most widely used because compared to static RAMs, they are more compact and can operate at higher speeds. They are becoming even more compact using stacked and trench capacitors [Watanabe 1995, Tsuruda 1997].

12.2.3   RAM Organization

Two important factors that affect test pattern generation are the architecture of the array as well as whether the memory is an independent IC or a macrocell inserted as one of modules in a system on a chip (SOC). As *N*-bit RAM may be organized as arrays of *N* by 1 or *r* by *c*, where $rc = N$. For example, a 16-kB RAM may consist of 128 rows and 128 columns or 256 rows and 64 columns, or yet, 64 rows and 256 columns. One of the criteria for selecting one array versus another is space availability on the floor of an SOC. The array may also be partitioned in subarrays that are organized to share the sense amplifiers.

With the rapid increase in device density, functions that used to be on a separate chip can now be combined in one IC. This approach minimizes off-chip interactions, increases the bandwidth, and therefore improves performance.

This integration is not limited to random logic but to a mix of subsystems, including RAMs. For example, memory-intensive ASICs such as DSP processors and microprocessors, in general, include several on-chip RAMs. These on-chip RAMs are called *embedded RAMs*. It is almost certain that any system on a chip (SOC) nowadays includes embedded RAMs. Because embedded RAMs cannot be directly accessed for controllability and observability, they are difficult to test. BIST seems a more feasible approach to test them [Zorian 1990]. Memory BIST is the topic of Section 12.6.

12.3 DEFECTS AND FAULT MODELS

12.3.1 Defects

RAM ICs are subject to the same physical defects encountered in logic ICs, which were described in Chapter 2. The defects include the presence and absence of material and of imperfection in the manufacturing process. Some defects, such as gate oxide breakdown, are more likely to cause failure in RAMs than other defects. Also, RAMs, particularly DRAMs, particularly DRAMs, seem to be more susceptible than other ICs to some interference noise and disturbances. For example, because of long, parallel access and sensing wires, there is a good chance for crosstalk to occur [Yang 1999]. In addition, the memory cells are subject to single-event upsets (SEUs), which are induced by charged particles penetrating an electronic circuit. Such upsets occur not only in space but may be triggered by particle emanating from the metal used in fabricating the circuit.

12.3.2 Array Fault Models

Detection of the faults depends on where the fault is located in the various parts of the RAM circuitry. It is important to distinguish between the faults in the memory cell array and those in the surrounding circuitry. The latter consist of the decoding circuit, the write drivers, the sense amplifiers, and the registers. Although it seems that the surrounding circuitry can be handled as any random logic, the only way to observe its response to a test is through the RAM cells. Thus testing the array should also include testing the surrounding circuitry.

It might seem plausible to use for RAM cells the same structural fault models as those used for random logic. However, memory arrays are far denser than random logic. Using structural SAF or bridging faults would not be practical because of the extremely large volume of test data. Also, testing time would be excessively long. Because of these problems, a functional fault model is more appropriate for RAM arrays. A properly functioning RAM is capable of storing 1's and 0's in every cell, having each cell transition from 1 to 0, and vice versa, and having each cell return to its value after reading. In addition, the cells are expected to retain the information. A functional test that is capable of detecting all faults in these operations, while possible, is hardly practical since it would be on the order of 2^N for an N-cell memory [Hayes 1975]. To make the test length manageable, the set of fault types is restricted.

The set of functional memory fault models used include stuck-at and bridging faults. However, these models are not sufficient to represent all types of failures in memory circuits. The sets also include faults that are very special to memory circuits and architecture. These are retention faults, transition faults, coupling faults, and pattern-sensitive faults. All the faults that we describe in the rest of this section are functional faults and not structural faults.

12.3.2.1 Stuck-at Faults. We are familiar with gate-level stuck-at faults (SAFs) from Chapter 2. For memory cells, a stuck-at fault is defined as a functional fault that makes the cells behave as if they have a 1 or a 0 stored permanently in them. This is illustrated in Fig. 12.4, where the cell is modeled as a flip-flop that is in either a state of 0 or a state of 1. To detect such functional SAF, it is necessary to write a 1 (and a 0) on the cell and read the result of the operation. This needs to be performed for all cells in the array of size N. The process can be described in an algorithmic form as follows: For all cells $C(j)$:

1. Write 0 on cell $C_j(C_j \Leftarrow 0)$.
2. Read C_j.
3. Write 1 on cell $C_j(C_j \Leftarrow 1)$.
4. Read C_j.

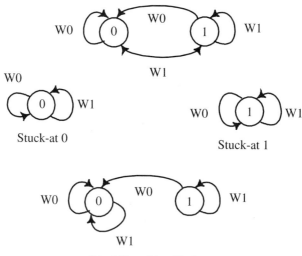

0 to 1 Transition Fault

Figure 12.4   Fault representation of SAF and transition fault.

The test then requires accessing each cell C_j four times—two write and two read operations. If the access time is τa, testing time is $4N\tau a$, where N is the number of all cells in the array. Thus the complexity of the test is $O(n)$. This testing approach will also allow detection of multiple stuck-at faults (MSAs).

12.3.2.2 *Transition Faults.* This is a special case of stuck-at fault. A transition fault (TF) occurs when one of the cells cannot make the transition 0 to 1 ($\uparrow$) or 1 to 0 ($\downarrow$). The up- and down-arrow notations are due to [van de Goor 1998]. A cell may have one type and not the other. Thus a cell may function properly from 0 to 1, and when it undergoes a 1-to-0 transition, the cell may remain at state 1 exhibiting a stuck-at-1 value. To detect these faults, it is important to make every cell undergo 0-to-1 and 1-to-0 transitions and to check the results.

12.3.2.3 *Coupling Faults.* Because of the regularity of its structure, a memory chip may experience a change in one cell due to an intended change in another cell. This is known as a coupling fault (CF). The coupling is due to shorts or parasitic effects such as stray capacitance. There are three CF types: inversion, indempotent, and bridging and state. *Inversion coupling* (CF_{ins}) occurs when one cell's transition causes inversion of another cell's value. That is, a 0-to-1 transition in cell i causes cell j to go from its logic value, a, to its complement, a'. An *idempotent* fault (CF_{ids}) occurs when a transition on cell i causes cell j to be in a particular logic value, 0 or 1. These faults are illustrated in Fig. 12.5.

The two-cell coupling used to define the faults is just a special case of k-way coupling, which can become too complicated because of the many ways these cells may interact. Here we concentrate on two-cell coupling. Techniques

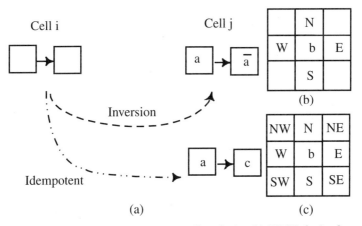

Figure 12.5 Fault representations: (*a*) coupling fault; (*b*) NPSF fault: five-cell, (*c*) nine-cell.

for detecting CF_{ids} can also detect CF_{ins}. We examine next how indempotent coupling can be detected. Assume that as cell C_i changes state it will affect cell C_j. We can detect the fault by performing the following operations:

1. Write 0 on all cells.
2. For all $C_j, j \neq i$:
   2.1. Change C_i.
   2.2. Read C_j to check if it has changed.
   2.3. Restore the value of C_i.

Also, they have to be repeated with the array initialized to 1. The time taken would then be

$$2[N + 3(N - 1)N] = 6N^2 - 4N$$

Bridging and *state coupling* faults (SCFs) are a special type of coupling fault. They occur when a certain state in one cell causes another specific state in another cell. They are caused by logic coupling rather than by transition. A bridging fault (BF) occurs when two or more lines are inadvertently shorted together. The shorted lines behave as if they are ANDed or ORed together, depending on the strengths of the drivers and the load transistors. Some of these faults may be detectable by stuck-at test patterns, but this is not always the case [Millman 1990]. Test algorithms have been developed for locating a bridging fault of multiplicity k [van de Goor 1991]. For a definition of multiplicity, see Section 2.7. An effective technique for detecting bridging faults is based on current testing (I_{DDQ}) monitoring. However, as the technology scales down to very deep submicron, current testing may lose its advantage, as we mentioned in Section 7.10.

12.3.2.4 *Pattern-Sensitive Faults.*

Another way a memory cell fails is due to the different activities of other cells in the array, leading to pattern-sensitive faults (PSFs) [Brown 1972]. These faults are caused primarily by unwanted interference between the cells due to the high packed density of the cells. It is not possible to examine the effects on each other of all cells in the array, since this would require a test set of length $(3N^2 + 2N) \cdot 2^N$ [Hayes 1975]. It is probably more practical to examine the fault due to cells in the proximity, the neighborhood, of the one under test. These are called neighborhood PSFs (NPSFs). The definition of a neighborhood is centered on a *base cell*. The neighbors may be identified as E, W, N, and S, as shown in Fig. 12.5b, and it is a five-cell neighborhood. It may also be extended to a nine-cell neighborhood, as illustrated in Fig. 12.5c.

NPSFs are classified into three categories: *active*, *passive*, and *static*. Given a certain pattern change in neighboring cells, the base cell may (1) change value (active fault), (2) remain at a fixed value regardless of the attempts to change it (passive fault), or change to a specific value (static fault). To detect an *active* NPSF, it is important to know the state of the base cell, to change a neighborhood cell, and to read the base cell. For a neighborhood of k cells, this will require 2^k changes, and for each change the other cells, $k - 1$, assume all possible values. This is a total of $(k - 1) \cdot 2^k$ [Suk 1980]. In the case of a *passive* fault, it is important to find out if for all possible patterns in the neighbor cells, the base cell will or will not change. Thus we perform two transitions for the base cell for all possible $2^{k - 1}$ neighborhood patterns. In all, 2^k are required. In a similar fashion, it can be shown that 2^k patterns are needed for static NPSF faults. For all NPSF faults, a total of $(k + 1) \cdot 2^k$ patterns per neighborhood.

Detecting NPSF faults requires the exercising of the memory in such a manner as to detect other fault models. The NPSF model therefore covers many other memory fault models. However, due to the fact that these faults are not common and require very high complexity test sets, PSFs are not usually tested in today's memories.

12.3.3 Surrounding Logic

In addition to the array, the surrounding logic also has to be tested. A functional fault is also used in the address decoding circuit. An n-to-k decoder is exercised by making all possible selections and ensuring that a 0 and a 1 are appearing in the corresponding output. For example, for a 2-to-4 decoder, all four outputs are checked for correct operation by applying an exhaustive test on the select lines, as shown in Fig. 12.6. If any of the outputs is stuck at 1, the decoder will work correctly in one case and select the corresponding bit; otherwise, for the other cases, two bits will be selected simultaneously. However, if any output is stuck at 0, the corresponding bit will never be selected. With this knowledge, it is possible to find out if the decoder is functioning correctly or by monitoring write and read operations on the RAM cells. The input, too, can be faulty, and this will result in selecting only half of the words. For example, if the first select line is SA0, it will read the first and second words correctly, but instead of the third and fourth, it will repeat words 1 and 2. This pattern can then be traced to the data written and read from the array cells. Having one of the input SA0 (or SA1) is equivalent to having the MAR's cell corresponding to this select line SA0 (SA1). All these faults are detectable by patterns that are generated to test the RAM array.

12.4 TYPES OF MEMORY TESTING

As for any other integrated circuit, memory circuits are subjected to *parametric tests* (DC and AC), *functional tests*, *dynamic tests*, and I_{DDQ} tests. Memory functional testing, which involves generation of test patterns and their application using ATE, may be performed on the IC level, array level, or board level. However, to expedite testing, more and more built-in self-testing is used, as

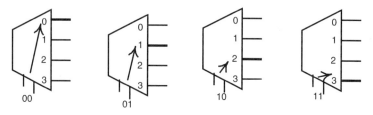

Figure 12.6   Testing the decoding circuitry.

described in Section 12.6. Parametric testing involves measurement of voltage, current, and frequency. Recently, I_{DDQ} testing has augmented it. Sometimes, the circuit might be performing its intended function but is not capable of driving a bus. This circuit will fail a parametric test and will immediately be considered defective. Also, characterization testing determines its operating limits. Parametric testing is used (1) to verify that the product meets the specifications and (2) to characterize the product. The first provides the operating points of the circuit in normal operation mode.

12.4.1   Specification Testing

Memories are designed according to specific values of the supply current, the output voltage range, and seek time. The fabricated product has to conform to these specifications. However, conforming to these specifications does not guarantee that the product will actually work since there might be changes in the ambient temperature or humidity, the power supply, or loading conditions. Both DC and AC parametric testing are usually performed.

12.4.2   Characterization Testing

Characterization testing involves a repetitive sequence of measurements used to locate the operating limits of the circuit. For example, the Shmoo plot in Fig. 12.7 distinguishes, in the V_{dd}–τp space, the region in which the circuit operates correctly, as indicated by white tiles. The black tiles specify incorrect operation.

12.4.3   Functional Testing

The main task in functional testing is to generate test patterns that can detect the types of faults discussed in previous sections of this chapter. Since RAMs have many bits, the complexity of the algorithm generating the test needs to be evaluated for efficient test pattern generation and effective fault coverage. Here, application of the patterns consists of writing and reading from the cells.

As we have learned from the onset, detecting a stuck-at fault requires that we place on the faulty node the complement of the logic value at which the node is stuck. For all but the memory cell, the test pattern generation is similar

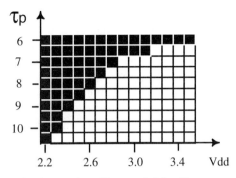

Figure 12.7 Characterization testing: Shmoo plot for V_{dd} versus τp [Poulton 1997 © IEEE. Reprinted with permission].

to that discussed in Chapters 1, 2, 6, and 7. For the memory cells, however, we need to write a 0 (and 1) on each cell and check to determine if a 0 (and 1) is actually written by reading it. As a result of applying these test sets to the array, we may also detect other faults on the auxiliary circuitry such as the decoders. Consider, for example, the case where one of the decoder outputs experiences a stuck-at fault that results in inhibiting the accessing of the corresponding cell (or cells). It will not be possible to write a 0 and a 1 on the cell and verify the result of the writing. The stuck-at fault on the decoder will then be detected as part of the functional test of the array.

For transition faults, a test must transition each cell from 0 to 1 and then read it immediately. The test must also be repeated for transition from 1 to 0. Detecting coupling faults is more involved since it necessitates the testing of neighboring cells in a particular order. Thus to sensitize and detect coupling faults, we must write on cell i and then read cell j. This two-step operation must be done in:

1. Ascending order since any number of cells with lower address may be affected

2. Descending order since any number of cells with higher addresses may be affected

NPSF faults are more complex and require a variety of methods. Test algorithms are available to perform such tests, but they are not as efficient for high fault coverage in a short testing time [Suk 1981, Saluja 1985, Sharma 1997]. Instead of applying a test set to detect every type of fault independently of the others, it is possible to devise one test set that detects several of these fault models, as described in the next section.

12.4.4   Current Testing

Current testing, discussed in Chapter 7, has been shown to be very useful in screening out potential failure of RAMs. The quiescent current for a normally functioning RAM may be on the order of tens of nanoamperes, while some failure may elevate this current several orders of magnitude [Hawkins 1985]. In a study, 1582 devices failed current testing only; 1490 also failed functional testing. This implies that the 92 devices escaped functional testing [Meershoek 1990]. While current testing is applicable to present technology-feature SRAMs, it is not recommended for DRAMs.

12.5   FUNCTIONAL TESTING SCHEMES

There are several test algorithms for RAMs [van de Goor 1990]. They are categorized into two classes: The first class comprises tests that require reading each cell after a specific change has been performed to other cells. For the other class of tests, only the cell that changed value is read. Commonly used algorithms are given next. Some of these algorithms are presented for historical and academic reasons. However, only those that are mostly effective are used for actual test application in production testing. An effective test is a test that detects as many faults as possible with the fewer possible patterns. The length of the test is measured in terms of the size N of an $N \times 1$ memory array. Of course, if the memory is arranged in $r \times c$ array, where c is the word width and $N = rc$, it is possible to read or write in all bits of the word simultaneously and shorten the test length by r.

12.5.1   MSCAN

The MSCAN memory scan sequence is sometimes also referred to as a solid *pattern*. It consists of writing an all-0's pattern (and an all-1's pattern), then reading all the cells. In addition to SAF testing, it serves in finding worst-case power dissipation. It also serves as a preparation for the application of other test sequences. Testing time is $T = 4N$; that is, the complexity is linear in the memory size.

12.5.2   GALPAT Algorithm

The GALloping PATtern algorithm is also known as the walking 1(0) or Ping-Pong test [Barraclough 1976]. The algorithm consists of the following steps:

1. A solid all 0 (1) pattern is written in all N cells.
2. For each cell $C(j)$:
   2.1. Complement $C(j)$.
   2.2. Repeat the following sequence for $k \neq j$.

2.2.1. Read $C(j)$.

2.2.2. Read $C(k)$.

2.3. Restore the content $C(j)$.

This process looks like a sliding 1 is covering the array. As illustrated in Fig. 12.8a, step 1 requires N write operations. For each iteration of step 2, the target cell $C(j)$ is complemented twice, which is equivalent to two write operations. There are $2(n-1)$ read operations. Thus the toal number of operations is $N + N(2 + 2(N-1)] = N + N(2N+1)$. Since the test will be repeated with the array initialized to 1, the total test time is $T = 2(2N^2 + N)$. A variation of this sequence that omits step 2.2.2, known as the walking 0/1, is illustrated in Fig. 12.8b. Testing time is still $O(N^2)$. This test detects:

- Stuck-at faults, since each cell is read once when it has value 0 as well as value 1
- Transition faults on the cells, since each cell is switched from 1 to 0 and from 0 to 1
- Decoding circuits, as only one cell at a time contains logic value 1
- Some state coupling faults, since any two arbitrary cells under states 00, 01, and 10 are read

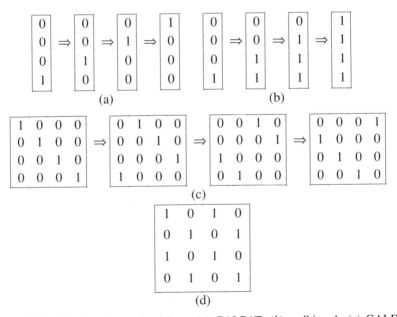

Figure 12.8 Functional test algorithms: (a) GALPAT; (b) walking 1; (c) GALDIA; (d) checker.

There are different test sequences that are based on GALPAT but are less complex than this algorithm. Although in the GALPAT only one cell has 1 at a time, all the cells in an entire row, column, or diagonal can have 1, and this configuration of 1's gallop through the RAM array. The most widely used is the GALDIA. First, the algorithm places 1's in all cells on the diagonal, that is, cells whose row and bit addresses are the same. On all other cells, 0's are placed. The cells are then read in an ascending order. The test is repeated with 0's in the diagonal. The test is illustrated in Fig. 12.8c for shifting 1's on the diagonal.

12.5.3   Algorithmic Test Sequence

The algorithmic test sequence (ATS) is one of the first test sets that were developed for RAMs [Knaizuk 1977]. The patterns are applied to three partitions of the memory array. Each partition corresponds to addresses Mod 3. Thus the cells of a partition did not occupy adjacent locations. The three partitions are Π_0, Π_1, and Π_2.

1. Write 0 in all cells of Π_1 and Π_2
2. Write 1 in all cells of Π_0
3. Read 1 from Π_1 if 0, then no faults, else, faulty
4. Write 1 in all cells of Π_1
5. Read 1 from all cells of Π_2 if 1, then no faults, else, faulty
6. Read 1 from Π_0 and Π_1 if 1, then no faults, else, faulty
7. Write 0 in, then read 0 from Π_0 if 0, then no faults, else, faulty
8. Write 1 in, then read 1 from Π_2 if 1, then no faults, else, faulty

This test sequence requires $4N$ operations. It detects SAF faults on the memory cells, decoders, and MDR and MAR registers. The algorithm has subsequently been improved and resulted in the MATS algorithm [Nair 1979] and an MATS+ [Abadir 1983]. These test sequences and other variations given in the next sections improve the complexity and permit the detection of more faults than with the GALPAT algorithm.

12.5.4   Marching Pattern Sequences

There is a class of test sequences called March tests [van de Goor 1998]. A special version of the walking 0/1, given in Section 12.5.1, minimizes the excessive reading operations. After initialization, each cell is read, then complemented. As the last cell is reached, the RAM array is filled with 1's. The process is repeated but in the reverse addressing order; that is, the process of reading and complementing starts with the last cell examined. This early example of a March test is due to [Cocking 1975] and was used to test memory in microprocessors [Barraclough 1976]. The algorithm requires only 10 read/write operations per cell. It is $O(N)$. The test sequences developed by [Nair 1978], March A and March B, have a complexity of $30N$. Improvements on this version were developed by [Suk 1981], [Marinescu 1982], and [van de Goor 1998].

The March C test, a solid 0 pattern, is first placed in the RAM. The contents are then read in ascending order. A check is made to assert that each cell C_j content is 0; then a 1 is written in the location. The process is repeated until 1's are written on all cells. The cells are then read in a descending order to verify that they contain 1's. The 1's are then changed to 0's. The same test is repeated but starting from a solid 1 pattern.

1. Starting from the lowest to highest address, $j = 1$ to N:
    1.1. Write 0 in all $C(j)$.
    1.2. Read $C(j)$, write 1 in $C(j)$.
    1.3. Read $C(j)$, write 0 in $C(j)$.
    1.4. Write 0.
2. Starting from the highest to the lowest address, $j = N$ down to 1:
    2.1. Read $C(j)$, write 1 in $C(j)$.
    2.2. Read $C(j)$, write 0 in $C(j)$.
    2.3. Read $C(j)$.

This sequence requires 11 read/write operations, and therefore testing time is $T = 11N$. It detects faults on decoders, transition faults (since it verifies that the cells can undergo transitions from 1 to 0, and vice versa), and some coupling faults.

There are several variations of this algorithm that either improve or reduce the timing sequence, or detect more faults. For example, the March C test reduces the operations to 10 per cell by removing step 2.3 while detecting the same faults [van de Goor 1998]. March C+ adds a read operation after each write operation in the march C−. Testing time is then increased to $14N$. However, it detects, in addition to the faults detected by March C, stuck-open faults and some timing faults. It is also possible to eliminate the last two steps of March C−, reducing the test complexity to $10N$ while detecting the same faults.

12.5.5 Checkerboard Test

This test places each cell in a state different from its immediate neighboring cells, as depicted in Fig. 12.8d. The bits are thus partitioned into two sets; those in state 0 form partition Π_0 and those in state 1 form partition Π_1. The algorithm writes and reads one set, then the other. The test is then repeated from the alternative logic values. This test detects and locates stuck-at faults and shorts between adjacent cells assuming that the decoding subcircuit is fully tested. Algorithmically, it can be expressed as:

TABLE 12.1   Algorithms and Their Complexity

Algorithm	Complexity	1	2	3	4	5	6	7	8
		Detectable Faults[a]							
MSCAN	$4N$		×						
GALPAT	$2(2N^2+N)$	×	×	×	×		×		
Marching 0/1	$2(N^2+N)$		×	×	×				
GALDIA	$(2N^{1/2}+4)N+5N^{1/2}$		×	×	×			×	
GALCOL	$3N^{3/2}+6$		×	×					×
March A	$30N$	×	×	×	×				
March B	$16N$	×	×	×	×				
March C	$11N$	×	×	×	×				
March C–	$10N$	×	×	×	×				
March C+	$14N$	×	×	×	×	×			
Checker	$N+32N\log_2 N$		×	×	×				

[a]1, Address faults; 2, SAF; 3, transition faults; 4, coupling faults; 5, timing faults; 6, faulty address transitions between each cell and every other; 7, faulty address transitions between each cell and every other; 8, faulty address transition between each cell and the cell row.

1. Write 0(1) in cells of Π_0 (Π_1).
2. Read all cells.
3. Write 1(0) in cells of Π_0 (Π_1).
4. Read all cells.

The test detects (1) all stuck-at faults; (2) data retention faults; and (3) 50% of transition faults. Notice that address decoder faults and state coupling faults are not covered. The algorithm requires $4N$ read/write operations. Their complexity and effectiveness in fault detection are summarized in Table 12.1, where fault models 1, 2, 3, ... refer to address faults, SAF on the cells, TF, CF, stuck-open, and timing respectively. No one single of these algorithms is capable of detecting all fault models. A more comprehensive set of algorithms can be found in [Sharma 1997 and van der Goor 1998].

12.6   MEMORY BIST

Memory ICs have many features that make them easily testable: regular memory array structure, shallow sequential depth, and accessible I/O. For embedded memories, discussed in a later section, the controllability and observability are reduced. Another characteristic of memory testing is the excessive length of the tests. For example, the simple sliding diagonal algorithm requires hours for a 1-Mbit chip. The GALPAT requires a testing time of order $O\ (N^2)$. Although this takes only a few seconds for 1 kbit, it may take days for a 1-Mbit chip. This complexity of algorithm is compounded by the fact that usually, external testing is not run at speed. Moreover, with increasing size of memories, the volume of test data is becoming too large for efficient handling through automatic test equipment (ATE). An attractive solution to this bottleneck problem is the use of built-in self-testing (BIST).

In the Early 1980s, a built-in testing scheme consisted of building address and data generators and an output comparator on the chip and to leave the flexibility of defining the patterns and timings to outside the chip. The test patterns are controlled from test pins and run at speed. In this fashion, few extra pins are needed. Using transitional LFSRs for address and data generation requires only a few modifications of the RAM circuitry but may not guarantee full fault coverage, due to aliasing, as explained in Chapter 11. The aliasing may be reduced with special added test structure [Zorian 1990]. However, it is possible to make use of the regularity or RAM architecture and the simple configuration of the test patterns to obtain higher fault coverage. For example, to implement the ATS algorithm, we need to address Mod 3 [Bardell 1988]. In this case we can configure the MAR to function at testing time as a Mod 3 counter. The data applied for writing the MDR, is also configured to generate 0's or 1's according to the test routine requirement. A controller is then needed to coordinate these operations—configuring the registers for test mode and applying the patterns according to the algorithm.

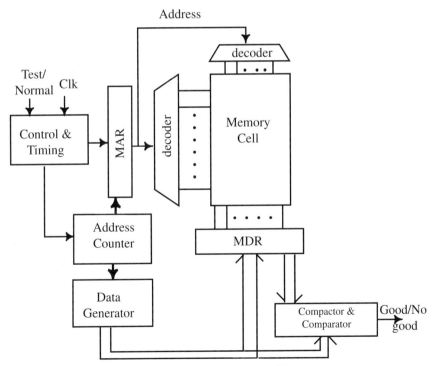

Figure 12.9 Memory BIST.

TABLE 12.2   Applying ATS in BIST Configuration

State	R/W	Data	Partition	ATS Step
1	W	0	Π_1	1
2	W	0	Π_2	1
3	W	1	Π_0	2
4	R	0	Π_1	3
5	W	1	Π_1	4
6	R	1	Π_2	5
7	R	1	Π_0	6
8	R	1	Π_1	6
9	W	0	Π_0	7
10	R	0	Π_0	7
11	W	0	Π_2	8
12	R	0	Π_2	8

Figure 12.9 shows the block diagram for the RAM configuration with BIST. In testing mode, the controller will be in one of the states listed in Table 12.2 for the application of the ATS sequence given in Section 12.5.3. If several RAMs are placed on the same chip, they may share the BIST circuitry. The controller would also need to be modified to select the chip [Nadeau-Dostie 1990]. BIST has also been demonstrated to be successful with DRAMs [Ohsawa 1987], FIFO [Zorian 1994], ROMs [Zorian 1992], and other types of memory.

12.7   MEMORY DIAGNOSIS AND REPAIRS

It has been a practice for stand-alone memory to include spare rows and columns of cells to replace possible faulty ones. Typically, a fuse-blow process using external laser equipment is required in order to swap faulty rows or columns with the spare ones. These repairs are also needed for embedded memories because ICs embed large and dense memories storing from 256 Kb to 256 Mb. It is necessary to determine the faulty cells in the embedded memory arrays and to reconfigure them. However, the amount of test data to be transferred in and out of the chip is too large. To minimize the high bandwidth interaction with the outside world, built-in self test and redundancy may be used, hence only the data that include information about the repairs are relayed to the tester. A more advanced solution is to include online the test and repairs resources for the embedded memories. In this case the embedded memory test goes beyond fault detection to include failed-bit diagnosis, redundancy analysis, and self-repair.

REFERENCES

Abadir, M. S., and H. K. Reghabati (1983), Functional testing of semiconductor random access memories, *Comput. Surv.*, Vol. 15, No. 3, pp. 175–198.

Bardell, P. H., and W. H. McAnney (1988), Built-in test for RAMs, *IEEE Des. Test Comput.*, Vol. 5, No. 8, pp. 29–37.

Barraclough, W., A. C. L. Chiang, and W. Sohl (1976), Technique for testing the micro-computer, *Proc. IEEE*, Vol. 64, No. 6, pp. 943–950.

Brown, J. R. (1972), Pattern sensitivity in MOS memories, *Digest Symporium on Testing to Integrate Semiconductor Memories into Computer Mainframes*, pp. 33–46.

Cocking, J. (1975), RAM test patterns and test strategy, *Digest of Papers Semiconductor Test Symposium*, pp. 1–8.

Hawkins, C., and J. M. Soden (1985), Electrical characteristics and testing considerations for gate oxide shorts in CMOS ICs, *Proc. IEEE International Test Conference*, pp. 544–555.

Hayes, J. P. (1975), Detection of pattern sensitive faults in random access memories, *IEEE Trans. Comput.*, Vol. C-24, No. 1, pp. 150–157.

Knaizuk, J., Jr., and C. R. P. Hartmann (1977), An algorithm for testing random access memories, *IEEE Trans. Comput.*, Vol. C-26, No. 4, pp. 414–416.

Marinescu, M. (1982), Simple efficient algorithms for functional TAM testing, *Proc. IEEE International Test Conference*, pp. 236–239.

Meershoek, R., et al. (1990), Functional and I_{DDQ} testing in static RAMs, *Proc. IEEE Int. Test Conference*, pp. 929–937.

Millman, S. D., and E. J. McCluskey (1990), Diagnosing CMOS bridging faults with stuck-at fault dictionaries, *Proc. IEEE International Conference*, pp. 860–870.

Nadeau-Dostie, B., A. Silburt, and V. K. Agrawal (1990), A serial interfacing technique for external and built-in self-testing of embedded memories, *IEEE Des. Test Comput.*, Vol. 7, No. 2, pp. 56–64.

Nair, R., S. M. Thatte, and J. A. Abraham (1978), Efficient algorithms for testing semiconductor random-access memories, *IEEE Trans. Comput.*, Vol. C-27, No. 6, pp. 672–676.

Nair, R. (1979), Efficient algorithms for testing semiconductor random-access memories, *IEEE Trans. Comput.*, Vol. C-28, NO. 6, pp. 672–676.

Ohsawa, T., et al. (1987), A 60-ns 4-Mbit CMOS DRAM with built-in self-test function, *IEEE J. Solid State Circuits*, Vol. 23, No. 10, pp. 663–668.

Poulton, J. (1997), An embedded DRAM for CMOS ASICs, *Proc. 17th Conference on Advanced Research in VLSI*, pp. 288-302.

Rideout, V. L. (1989), One-device cells for dynamic random-access memories: a tutorial, *IEEE Trans. Electron Devices*, Vol. ED-26, No. 6, pp. 839–851.

Saluja, K. K., and K. Kinoshita (1985), Testing for coupled cells in random access memories, *Proc. IEEE International Test Conference*, pp. 439–451.

Sharma, A. K. (1997), *Semiconductor Memories: Technology, Testing, and Reliability*, IEEE Press, Piscataway, NJ.

Suk, D. S., and S. M. Reddy (1980) Test procedures for a class of pattern sensitive faults in semiconductor random access memories, *IEEE Trans. Comput.*, Vol. C-29, No. 6, pp. 419–429.

Suk, D. S., and S. M. Reddy (1981), A March test for functional faults in semiconductor random access memories, *IEEE Trans. Comput.*, Vol. C-30, No. 12, pp. 982–985.

Tsuruda, T., et al. (1997), High speed/high-bandwidth design methodologies of on-chip DRAM cores multimedia system LSI's, *IEEE J. Solid State Circuits*, Vol. 32, No. 3, pp. 477-482.

van de Goor, A. J., and C. A. Verruijt (1990), An overview of deterministic functional RAM chip testing, *ACM Comput. Surv.*, Vol. 22, No. 1, pp. 5–33.

van de Goor, A. J., et al. (1991), Locating bridging faults in memory arrays, *Proc. IEEE International Test Conference*, pp. 685–694.

van de Goor, A. J. (1998), *Testing Semiconductor Memories: Theory and Practice*, Comptex Publishing, Gouda, The Netherlands.

Watanabe, S., et al. (1995), A novel circuit technology with surrounding gate transistors for ultra high density DRAMs, *IEEE J. Solid Stae Circuits*, Vol. 30, No. 9, pp. 960–970.

Yang, Z., and S. Mourad (1999), Deep submicron on-chip crosstalk," *Proc. IEEE International Conference on Measurement and Technology*, pp. 1788–1793.

Zorian, Y. (1990), A structured approach to macrocell testing using BIST, *Proc. IEEE Custom Integrated Circuit Conference*, pp. 28.3.1–28.3.4.

Zorian, Y. and A. Ivanov (1992), An effective BIST scheme for ROM's, *IEEE Trans. Comput.*, Vol. C-41, No. 5, pp. 646–653.

Zorian, Y. (1999), Testing the monster chip, *IEEE Spectrum*, Vol. 36, No. 7, pp. 54–60.

PROBLEMS

12.1. List various fault models used in memory testing. Consider a 3×3 SRAM and develop test patterns to detect each of these faults.

12.2. Consider one cell of SRAM implemented as shown in Fig. 12.3*b*. Use the test patterns generated in Problem 12.1 for stuck-at faults on the cell and determine which SAR of the circuit are detected by the test.

12.3. Find the complexity of the following algorithm:
For $j = 1$ to N;

       Write 0 in $C(j)$

       Read cell $C(j)$

End

For $j = 1$ to $N/2$

       Write 1 in $C(j)$

       Read $C(j)$

End For

For $j = 1$ to N;

Read cell $C(j)$
End For
For $j = N/2$ to 1
      Write 0 in $C(j)$
      Read $C(j)$
End For

12.4. Does the algorithm of Problem 12.3 detect decoder faults? Explain.

12.5. Develop the test sequence for a marching 1/0 algorithm for a 32-bit RAM implemented in an 8 × 4 array.

12.6. Compute the test time required to apply **(a)** a March test, then **(b)** a GAL-PAT test to a 32k RAM given that it takes 150 ns to apply each pattern.

13

TESTING FPGAs AND MICROPROCESSORS

13.1 INTRODUCTION

This chapter is devoted to two widely used types of specialized circuits: field-programmable gate arrays (FPGAs) and microprocessors. Two of the most attractive attributes of these circuits are their availability off the shelf as IC and embedded cores and their flexibility to customize to many applications. In addition, they come in different technologies and a wide range of architectures. For example, some FPGAs are reprogrammable and may serve in particular applications that need different reconfigurations on the fly, such as emulators, and one-time programmable FPGAs satisfy dedicated applications. Similarly, microprocessors may have general-purpose RISC or dedicated CISC architectures in order to serve a variety of data processing, from a simple controller to a sophisticated image-processing engine. FPGAs and microprocessors nowadays are often used as embedded cores in a system on chips. In the beginning, FPGAs have been used as prototyping devices instead of wire wrapping small and medium-sized ICs (SSI and MSI). In certain applications, designs are first implemented in FPGAs. Then, as they mature, they are migrated to mask-programmable gate arrays (MPGAs).

The architecture of FPGAs is quite regular. The testing of FPGAs depends on their type of programmaing. For RAM-based FPGAs, it is possible to reprogram them in different configurations and test the blocks as well as the interconnects. One-time programmable FPGAs are not as easily tested for any possible application. In this chapter we concentrate on reprogrammable FPGAs and, in particular, RAM-based FPGAs. Testing these devices requires configuring them in specific designs and then applying test patterns. Selecting the appropriate test configurations and the test sequence has to be optimized for minimal testing time and low volume of testing data.

The microprocessor, on the other hand, has a more complex and less regular architecture. It consists of only a few components that are quite diverse: registers and functional units (the datapath), and an FSM (the controller). In addition, they have complex associated RAMs and data architectures. Present technology processors are characterized by high integration and high speed. They are produced in high volume and are cost-sensitive ICs. Testing microprocessors does not require them to be configured. Several models have been formulated to facilitate their testing. Descriptions of FPGAs and their testing are covered in Sections 13.2 to 13.4; microprocessors and their testing are the topics of Sections 13.5 to 13.7.

13.2   FIELD-PROGRAMMABLE GATE ARRAYS

An FPGA consists of several uncommitted logic blocks in which the design is to be encoded. Each logic block consists of several universal gates. A large portion of die area is for programmable routing. The connectivity between any two blocks is programmd via different types of devices, such as SRAM, EEPROM, or antifuse [Trimberger 1994]. The architecture of the FPGAs depends on the fashion by which the blocks are connected. They may be in one of the forms shown in Fig. 13.1 and listed below:

- Islands in a matrix with horizontal and vertical channels (Lucent, Quicklogic, and Xilinx devices)
- Rows separated by routing channels such as these in MPGAs (Actel devices)
- Blocks of AND/OR logic arrays (Altera devices)

The advantages of FPGAs are (1) rapid turnaround, (2) availability of parts off the shelf, (3) larger gate counts and more design flexibility than in SSI and MSI chips, (4) low risk, and (5) reprogrammability (for some FPGAs). Their limitations include: (1) circuit delays depends on the performance of the tools used in design implementation; (2) the delay parameters can be extracted only after placement and routing, typically a time-consuming process; (3) mapping the design on FPGA architecture requires sophisticated tools; and (4) FPGAs are less dense and slower than traditional gate arrays.

13.2.1   Architecture

The principal components of an FPGA are (1) an array of uncommited logic blocks (LBs) in which the design is to be encoded, (2) an interconnect network, and (3) programmable elements. A logic block comprises a number of universal blocks or a group of gates; these are gates that can be used to implement

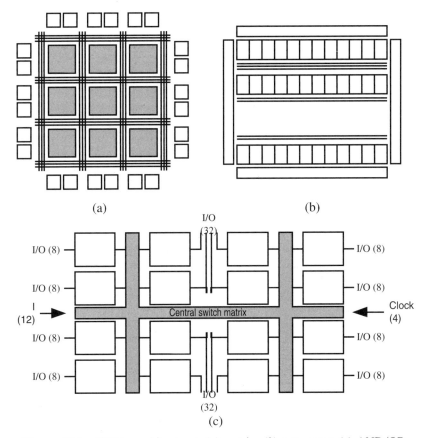

Figure 13.1 FPGAs architectures: (*a*) matrix; (*b*) gate array; (*c*) AND/OR.

any combinational function. Examples of these gates are NAND or NOR gates, combinations of AND/OR/NOT gates or of XOR/AND gates, mutliplexers, and RAMs.

The logic blocks may be organized in the form of a matrix, a gate array, or a configuration of AND/OR blocks. The three types are illustrated in Fig. 13.1. Examples of matrix architectures are XACT 3000 and 4000 by [Xilinx 1993] and Pasic by [Quicklogic 1993]. In the first example, the LB is a small RAM, also referred to as a lookup table (LUT). The second example is a mix of multiplexers and primitive gates. The two examples are illustrated in Fig. 13.2*a* and *b*, respectively. In the matrix architecture, the blocks are interconnected along horizontal and vertical channels. ACT1 and ACT2 are examples of the gate array type [Actel 1991]. Each block consists of a set of multiplexers arranged as shown in Fig. 13.2*c*. Interconnections between the block are in the channels separating the rows of LBs. The EPM5128 is a typical AND/OR block architecture [Altera 1991].

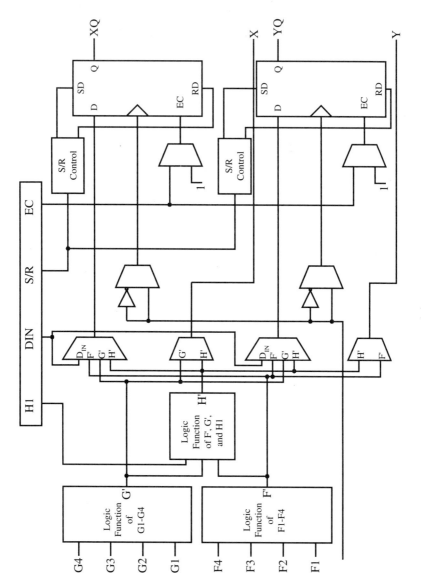

(a)

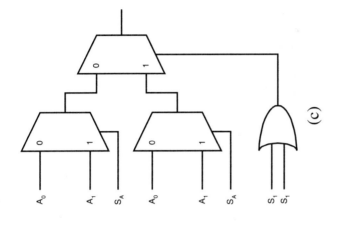

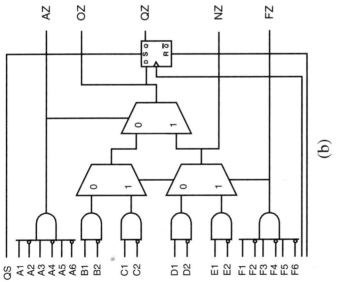

Figure 13.2 Logic blocks: (*a*) Xilinx 4000; (*b*) Quicklogic; (*c*) Actel.

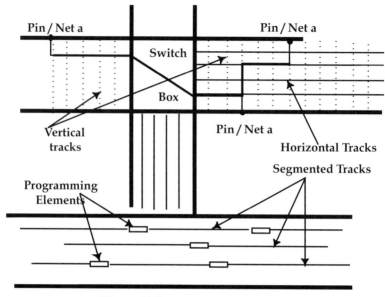

Figure 13.3   Interconnect resources.

Figure 13.3 shows general interconnect structure and resources in an FPGA. The routing in the channels is along segmented wires that may be joined through the programmable interconnecting points (PIPs). The programming elements are described in the next section. The change of routing from a horizontal to a vertical channel, or vice versa, is through switch boxes.

13.2.2   Programmability

The programming device is an important part of the FPGAs. The main types used are EEPROM, SRAM, and antifuse. All devices are illustrted in Fig. 13.4. Lookup table–based FPGAs use programmable devices in both the logic block and the interconnect structure. For other types of LBs, the programmable device is used only for the interconnections. Finally, for AND/OR blocks, the programmable device, which is primarily an EEPROM type, is used in a fashion similar to its use in PALs and PLDs and is shown in Fig. 13.4a.

The metal segments in RAM-based FPGAs are connected via pass transistors that is on or off depending on the state of the SRAM cell attached to its gate. This is illustrated in Fig. 13.4b. The antifuse is used only for programming the interconnect structure. An antifuse is a nonvolatile, two-terminal element. It occupies the area of a via and connects two metal segments when programmed. Two types of antifuse are shown in Fig. 13.4c. The Plice, Actel's antifuse, consists of a dielectric layer between two conducting materials, polysilicon and diffusion. When the appropriate voltage is applied across the dielectric, the two

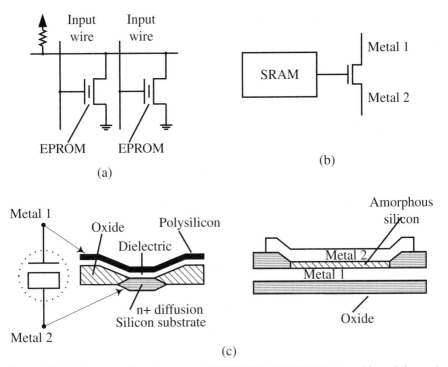

Figure 13.4 Programming elements: (*a*) EEPROM; (*b*) SRAM; (*c*) antifuse: Plice and ViaLink.

metal segments are connected throug a very low resistance, about 500 Ω. The off resistance of the antifuse is higher than 100 kΩ [Hamdy 1988]. Quicklogic's ViaLink is a metal-to-metal antifuse separated by a thin layer of amorphous silicon. It has a smaller resistance than the Plice, on the order of 50 Ω.

The size of the RAM cell is large compared to the antifuse. In addition to size, the other two most important characteristics of programmable devices are the volatility and the electrical characteristics: resistance and capacitance. Table 13.1 compares the three important elements.

TABLE 13.1 Comparison of Programming Elements

Feature	SRAM	EPROM	Antifuse
Programming method	Shift register	UV erasure	Breakdown
Area	Very large	Large	Small
Resistance	≈ 2 kΩ	≈ 1.8 kΩ	≈ 50 or 500 Ω
Capacitance	≈ 50 fF	≈ 15 fF	≈ 5 fF

13.3 TESTABILITY OF FPGAs

Similar to any other IC, FPGAs are tested upon manufacturing. However, testing FPGAs is completely different from that of other ICs or embedded cores. Testing traditional VLSI requires test pattern generation that is applicable only to a specific IC. In contrast, testing FPGAs of a certain type is the same regardless of their use. Therefore, it is reasonable to invest in testing FPGAs since, in addition to increasing yield, testing cost is quickly amortized over the large production volume.

13.3.1 Defects and Faults

In addition to failures common to conventional ICs, FPGAs are subject to unique failure modes that are identified as programmability failures [Chan 1994]. Thee types of failures may be due to a defective chip or malfunction of the device programmer. A noncalibrated programmer may apply too little or too much voltage for incorrect periods of time and cause an antifuse to remain intact or connected with higher resistance. This might cause missing and extra connections between the segments of the routing wires. In addition, they may introduce longer delays in the interconnect wires.

Each FPGA is also prone to specific physical defects, depending on the underlying architecture and programming technology. For instance, in EPROM-based technology, erasure of programmed connections can occur with exposure to light. The logic cell array using volatile static memory makes it prone to problems caused by noise, radiation, and power dropouts. Independent of the technology, failure modes in the interconnects between metal wire segments may result in extra or missing connections. Depending on the position of the extra or missing connection, the failure results in different possible fault models. Examples of mapping these types of failures into known fault models are listed in Table 4.2. Other causes for interconnect failures may result from incorrect or corrupted programming data files or the use of the wrong device type. These types of failures can result in logic block functional failures as well as interconnect failures.

13.3.2 Approaches to Testing FPGAs

The approach to testing FPGAs varies with programmability type. One-time programmable devices usually include testing circuitry that is used for manufacturing testing. The vendor provides both hardware and software testing resources for the user to verify postprogramming functionality [Actel 1991]. In the case of reprogrammable FPGAs, it is possible to reconfigure the device such that all programmable points are tested. However, reconfiguration is time consuming and contributes to the complexity of testing. Furthermore, these FPGAs have an extremely large number of PIPs and relatively limited I/Os. This makes testing of these devices complex and challenging. In the next section we concentrate on testing RAM-based, reprogrammable FPGAs. We refer to them simply as FPGAs, without identifying them as RAM-based.

13.4 TESTING RAM-BASED FPGAs

Although RAM-based FPGAs came on the market in 1989 [Xilinx 1989], it is only recently that interest in their testing has peaked. Testing FPGAs consists of testing (1) the LUT, the RAM of the LB; (2) the associated logic, mostly multiplexers, and flip-flops; and (3) the interconnect structure and resources. In the remainder of this section we take a functional approach to testing the above-mentioned components and the requirements for testing them. In addition, we describe I_{DDQ} testing, BIST application, and diagnosis testing. As we will see, the regular structure of the FPGA array will be exploited to expedite testing. The array is configured into the one- and two-dimensional ILAs that we dicussed in Chapter 8.

13.4.1 Functional Testing

Testing an FPGA requires programming in several configurations, called *testing configurations* (TCs). Changing configurations implies reprogramming costs. This imposes a requirement to determine the minimum number of test configurations that will cover all the faults of the structural fault model. Associated with each TC is a sequence of *test patterns* (TSs). The objective of the TS is the detection of a set of faults. The *test procedure* (TP), is one or a succession of (TC,TS): TP = {TC,TS)1, (TC,Ts)2, ...} [Inoue 1997]. A fault that is redundant in one configuration may be detected by another. If a fault is redundant for all TCs, it is undetectable. Thus the fault coverage depends on both the configurations and the test sequences applied [Renovell 1997b].

We consider here the three components of the LB: the RAM, the multiplexers, and the flip-flops. Figure 13.5 is a model for the LB that was detailed in Fig. 13.2a. The model distinguishes between the MUXs, the LUT/RAM, the control, and the flip-flops [Huang 1997]. The configuration registers, designated as "Conf." in the figure, are connected in a long stream of serially entered bits.

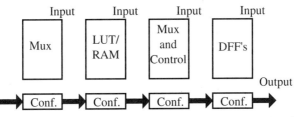

Figure 13.5 Testing the logic block [Huang 1996 © IEEE. Reprinted by permission].

13.4.1.1 Testing the LUT/RAM. We describe the approach followed by [Renovell 1999]. Accessing the RAM of an LB requires a number of input pins and an output pin as illustrated in Fig. 13.6*a*. The test applied to this RAM may be any of the algorithms used for memory testing, and a wealth of these are given in Chapter 12. To test the RAM of an LB, only one TC is needed and the TS is short, as we show next. The XACT 4000 LB has two 16-bit RAMs. The marching $1/0$ test consists of $14N$, where N is the number of bits in the RAM. The test length is thus 224 patterns. This test will detect SAF, TF, and addressing faults (see Chapter 12). Testing one module at a time requires a long testing period. Also, the limited number of I/Os make it impossible to test them concurrently. It is possible, however, to organize the RAMs of all modules in a single one-dimensional array as illustrated in Fig. 13.6*b*. The output of one module is connected to the input of the following one. This organization overcomes the I/O limitation, but it does make the controllability and observability of the modules more difficult unless a TC is used for each LB [Huang 1997]. It is also possible to have the output of one module connected to its flip-flop [Renovell 1999]. The read/write of the RAM and the block of the flip-flops are coordinated such that shifting a value from the primary input to the primary output is through the flip-flops and the modules. This approach is referred to as a *pseudo shift register.*

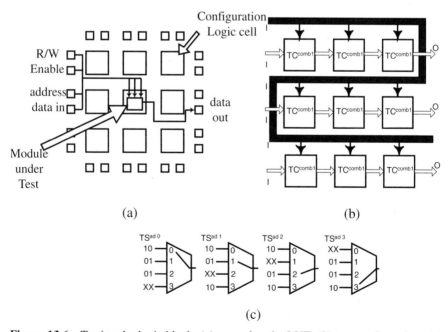

Figure 13.6 Testing the logic block: (*a*) accessing the LUT; (*b*) test configuration; (*c*) MUX configurations [Renovell 1996 © IEEE. Reprinted with permission].

13.4.1.2 Testing the Logic Block. To test the multiplexers functionally, each data line is selected and verified to pass a 0 or a 1. Selecting one data line is a TC. For a 4-to-1 MUX, four configurations are needed, and the test sequence consists of two vectors, as illustrated in Fig. 13.6c. Since the logic block consists of the RAMs and several multiplexers, similar test configurations of the set of MUXs can be arranged while applying the appropriate test sequences. For the XACT 4000 series, it has been found that eight TCs are sufficient to test the LB fully [Renovell 1999].

As in the case of the RAM above, the concurrency in testing the LBs is restricted by the number of I/Os. It is appropriate therefore, to join all the LBs in one array, as illustrated in Fig. 13.6b. All the blocks are first configured with the same TC, and then the test sequence is applied. To expedite the test application, each row of N blocks is receiving the TS and the observation is through the last block in the row. In such an arrangement, the TS to each block is received from the block to its left and the output is observed through the block to its right. For this testing, we then have to guarantee that the test sequence is actually passed to each embedded LB, and the response to the TS is eventually observed at the primary outputs. The test configurations include also testing the flip-flops that are connected as one shift register. For an LB having two flip-flops, FF1 and FF2, such as the XACT devices, all FF1's form one shift register and all FF2's form another shift register.

13.4.1.3 Testing the Interconnect Structure. The interconnect structure consists of vertical and horizontal tracks and switchboxes that allow connectivities between the two types of channels. The one used in XACT 3000, shown in Fig. 13.7a, allows 20 different configurations. A switch matrix is a programmable connecting element receiving n lines on each side. Some pins in the matrx cannot be connected. These are NC pins. Pins that can be connected are C pins.

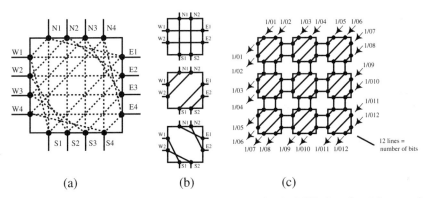

(a)   (b)   (c)

Figure 13.7 Testing the interconnects [Renovell 1998 © IEEE. Reprinted by permission].

Connection of C pins is governed by the programmed configuration. The failure modes of the interconnecting lines are short and open. The short may be with ground or V_{dd}. Therefore, stuck-at faults are also included. A general fault model is derived. Permanent connection (PC) is a short on any pair of NC pins or a bridge of any pair of C pins. Permanent disconnection (PD) is an open on any pair of C pins. To test these faults, a lower bound of three configurations was found sufficient for 100% test configuration coverage [Renovell 1997a]. The three TCs are shown in Fig. 13.7b for a simplified model of the swtich box. They are independent of the array size. Application of the first configuration to the entire array of switch boxes is shown in Fig. 13.7c. The number of vectors required to fully test any one of these three configurations is $\log_2(2km + 2)$, assuming that the switch box has k pins and that there are m boxes along the diagonal. This formula is based on previous results for an n-bit bus [Kautz 1974, Feng 1995]. The TS has to be repeated for each TC [Renovell 1998b]. Routing resource testing was also investigated by [Huang 1996b].

13.4.2 I_{DDQ} Testing

In Chapter 7 it was established that I_{DDQ} testing is an important supplement to voltage testing. In addition to detecting traditional fault models, it has the advantage of revealing defects that are not represented by these faults. For FPGAs, besides defects that occur in traditional ICs, there are programmability failures that result in open and bridging faults. I_{DDQ}, testing can help uncover these faults. Since observability is guaranteed in I_{DDQ} testing, focus is on controllability. Thus, configuring the FPGAs for testing can be less demanding. On the other hand, since current measurement is usually a slow process, it is important to strive for as few TCs as possible. This issue is particularly important because it usually takes more time to configure an FPGA than to measure the current.

I_{DDQ} can be used to test bridging faults (BFs) in all components of an FPGA, the logic block, the I/O blocks, and the routing resources. A hierarchical approach for I_{DDQ} testing was developed for FPGAs [Zhao 1998]. This is similar to that proposed by the leakage fault model test pattern generation described in Chapter 7 [Mao 1990]. A test library was built for each type of module in the LB, at the circuit or functional level. The test library consisted of different sets of TCs and TSs to test all internal BFs. External BFs are tested using the same test libraries with some additional vectors. The two-dimensional structure of the FPGA helps in applying the TCs and TSs to all the LBs simultaneously. All LBs in a row (or a column) receive the stimulus directly from the primary inputs and, of course, there is no need to propagate the results to a primary output.

13.4.3   BIST

Any FPGA includes in its resources a number of flip-flops that have been used effectively in testing the circuit, as we mentioned in Section 13.4.1. These flip-

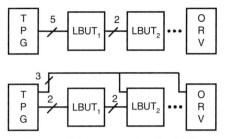

Figure 13.8 ILA testing for BIST.

flops may also be used for BIST application at no extra hardware cost. Some of the flip-flops can be configured for test pattern generation (TPG) and others for output response verification (ORV). This approach is similar to that used in BILBO and circular BIST, as described in Chapter 11. In the case of FPGAs, in addition to configuring the TPG and ORV, the array itself has to be configured before executing the test. If the FPGA includes IEEE Standard 1149.1, a RUNBIST instruction can be used for testing (see Chapter 10). BIST can be used to test the LBs, the I/O blocks, and the routing resources. The LB may be tested as a RAM using the memory BIST approach, or it may be configured as a logic function.

There are different ways of configuring the FPGA for BIST. At one extreme it is possible to consider one LB at a time, configure it, and apply the test. The testing complexity is $O(N^2)$ for an $N \times N$ array. However, it is possible to take advantage of the regularity of the array and organize it in a one- to two-dimensional iterative array, as explained in Section 13.4.1. However, this cannot be done without special arrangement of the LBs because there are fewer outputs than inputs in each LB. Propagating the test from one LB to the next in the ILA cannot easily be accomplished. To overcome this limitation, it is possible to supply the missing signals from TPG directly to each reconfigured LB as shown in Fig. 13.8. However, this will cause routing congestion. Instead, it is possible to pair the LBs in such a way that one is placed under test (LBUT) and the other is a helper to provide the missing signals to the next LBUT in the array [Stroud 1996]. Figure 13.9 shows this arrangement. If the output of the LBUT and its helper are still fewer than the inputs of the LB, the missing signals are provided from the TPG directly to make the array C-testable. The test can be applied simultaneously to all paired arrays. The testing time is then $O(N)$. Once the test is completed, the LBUTs and the helpers exchange roles and the test is applied again. Finally, the LBs used for TPG and ORV are tested in the same fashion. The organization of the FPGA for the method described is illustrated in Fig. 13.10 [Stroud 1996]. The FPGA is, therefore, tested with three configurations. As any C-testable array, the technique is applicable to any size array. Each LB is tested exhaustively and the results verified by comparing the results of two consecutive rows. BIST was used effectively to test the interconnect resources [Stroud 1998] and for diagnosis [Stroud 1997].

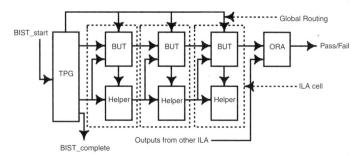

Figure 13.9   ILA testing for BIST test configuration [Stroud 1997 © IEEE. Reprinted by permission].

13.4.4   Diagnosis Testing

FPGA fault diagnosis is as important as testing. It applies to either programmed or unprogrammed FPGAs. Fault diagnosis of an unprogrammed FPGA leads to identification and isolation of a faulty part prior to programming, enabling designers to implement logic function using only fault-free parts. A universal fault diagnosis approach was introduced by [Inoue 1998]. The fault model includes stuck-at, incorrect-access, nonaccess, and multiple-access faults of lookup tables in LBs; functional faults of multiplexers; and the D flip-flops in LBs [Inoue 1997]. The assumption is that several of these faults may occur simultaneously in a CLB, and the number of LBs including these faults in a FPGA is at most one [Inoue 1998]. The complexity of this diagnosis test is $O(N^2 n \log n)$.

The approach of this testing is based on the concept of scalability testing discussed in Section 8.6.1. Consider a LB_j that is configured with TC. The response R_j to the application of test TS to this LB is used as part of the TS for the following LB_{j+1} under the same TC. The concept is illustrated in Fig. 13.11a. Based on these results, we can then test a sequence of LBs with a small

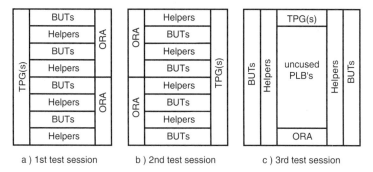

Figure 13.10   ILA testing for BIST, configuration for C-testability [Stroud 1996 © IEEE. Reprinted by permission].

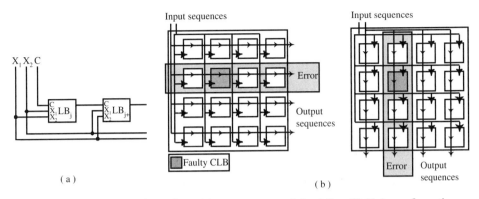

Figure 13.11 Diagnosis testing: (*a*) arrangement of the LBs; (*b*) ILA configuration [Inoue 1998 © IEEE. Reprinted with permission].

number of I/Os. The problem, however, is that it is not possible to identify the erroneous LB. To identify one LB and expedite the test application, the TS is applied simultaneously to all rows (or columns) of LBs. In this arrangement of the LBs, which is shown in Fig. 13.11*b*, it is possible to identify the faulty block as the coincidence of a row and a column. A C-diagnosable approach, which is independent of the size of the array dimension N, was also devised, and its complexity is $O(n \log n)$, where n is the size of the LUT. FPGAs diagnosis was also investigated by [Feng 1995] and [Liu 1995].

13.5 MICROPROCESSORS

It is superfluous to emphasize the indispensable role microprocessors have in present society. Their reliability is crucial to proper functioning of almost every aspect of this society. Since the introduction of the Intel 4004, microprocessors have proliferated in size, performance, and complexity. In addition, in some applications, instead of just customizing a generic microprocessor to the applications, special-purpose processors have been developed, such as digital signal processing (DSP). Although one can consider them as another electronic product and as such can test them with techniques already developed, microprocessors have a whole host of unique problems. For example, microprocessor systems differ from other digital circuits in that they include a software program that can be a source of errors in both the actual program and the memory where the program is stored. Thus, for microprocessors, one needs to pay attention not only to the associated logic, glue logic, but also to the problems associated with the execution of instructions. In this case, as in the case of memory testing, functional faults are used rather than structural faults. That is, the functional faults will cover the effects of structural faults. For example, a March test detects stuck-on or stuck-open faults in the RAM cells and the decoding circuits conditions.

Testing modern microprocessors presents a real challenge, for the following reasons [Needham 1998]:

- They have diverse and complex architectures.
- They have embedded memories, other storage devices, and FPGAs.
- They use deep-submicron devices, which bring an entirely new set of problems.
- They run at very high speed, which requires accurate equipment.

In the remaining sections of this chapter we first examine the models for the generic structure of a microprocessor and then how to use the model for validating the design. Subsequently, using case studies, we review how DFT approaches are used to test actual microprocessors.

13.5.1 Microprocessor Models

Circuit models are useful in the development of test algorithms as well as for simulation. At its most abstract level, the microprocessor may be depicted as shown in Fig. 13.12. It consists of a processor and an attached memory system. In turn, the processor is decomposed into a data path and a control unit. The data path includes functional units and registers; that is, circuits that operate on the data, shifting, adding, or multiplying. The control unit is an FSM with its associated registers. It interprets the instructions to manipulate the datapath.

A microprocessor may also be represented by interaction between registers $R = \{R_1, R_2, R_3, \ldots\}$ according to an instruction set $I = \{I_1, I_2, I_3, \ldots\}$. Both sets are used to construct a system graph, the S-graph $G(R, I)$, in which the vertices represent the registers and the edges represent the instructions [Thatte 1980]. An example of an S-graph is shown in Fig. 13.13. In addition to the registers, two of the nodes, *In* and *Out*, represent the outside world to the processor. They may be viewed as buses that connect the registers to memory locations and

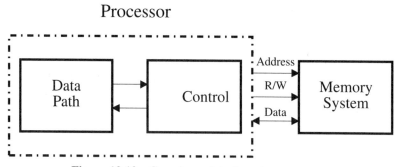

Figure 13.12   Generic model of a microprocessor.

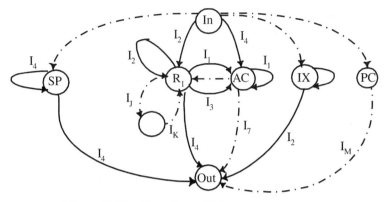

Figure 13.13 S-graph model for a microprocessor.

peripheral devices. Vertices may thus be connected by several edges since the registers may be used by several instructions. Also, an instruction may result in more than one edge. Only a few registers are shown: the program counter (PC), the accumulator (AC), the index register (IX), the stack pointer (SP), and one data register (R1). The three instructions in Table 13.2 will be used to illustrate the description of the graph. An instruction may be represented by several edges. For example, I_1 consists of an edge from R1 to A and from A to itself. I2 is represented by three edges. I3 is another instruction requiring the same transfer form R1 to A will be represented by another edge, from the first register to the second.

The more complex the set of instructions, the more complex is the graph. The graph might become too complex, thereby losing its effectiveness as a model. By drawing a relationship among the registers and among the instructions, it is possible to simplify the model [Brahme 1984]. The instructions are broken in to microcode. Several instructions might require the same microinstruction.

TABLE 13.2   Example Instructions of S-Graph in Fig. 13.13

	Instruction	Operations	Edges
I1	ADD A, R1	$A \leftarrow$ R1	R1 $\rightarrow A$
			$A \rightarrow A$
I2	ADD R1, (IX)	R1 $\leftarrow$ R1 + (IX)	IX $\rightarrow$ Out
			IN $\rightarrow$ R1
			R1 $\rightarrow$ R1
I3	MOV A, R1	$A \leftarrow$ R1	R1 $\rightarrow A$
I4	PUSH R1	SP $\leftarrow$ R1	SP $\rightarrow$ Out
		SP $\leftarrow$ SP + 1	R1 $\rightarrow$ Out
			SP $\rightarrow$ SP

To contain the explosion of microprocessor states due to the increasing complexity of the instructions and of the datapath size, an abstraction is extracted directly from the behavioral level to model the processor. This model encapsulates the control flow as well as the effect of the datapath on the control in a form of a FSM. An example of this is the extracted control flow machine model (ECFM) [Moundanos 1998].

Another model for a bit-slice microprocessor was also developed [Sridhar 1981]. A basic device, U (also called a *cell* or *slice*), performing a set of operations on n-bit operands is known to be *bit sliced* if a system that performs the same set of operations on k-bit operands, where $K = N \cdot n$ can be built by interconnecting N copies of U in a normal way. In a *bit-sliced* microprocessor, the cell, U, performs the functions of the ALU and the register file or scratchpad RAM of a computer. In past microprocessor designs, the use of bit slicing introduced structural simplicity and normality in the connections between IC chips at the printed circuit board level [AMD 1976]. Bit slicing has also been used within IC chips to achieve simplicity and compactness in many VLSI designs.

The models described in this section are used for testing microprocessors. However, before addressing this topic, we examine briefly the validation of these processors.

13.5.2   Microprocessor Validation

As pointed out in Chapters 1 and 5, simulation remains to be the most feasible verification means for VLSI design. However, with increased complexity, emulation can be on the order of six times faster than simulation, and this is why emulators are becoming more feasible for microprocessor verification. Emulators were used for Pentium and Motorola 68060 microprocessors. The actual time required for verifying real-time operation of the latter processor was six weeks; simulation of the same system would have taken 13 years [Kumar 1995]. Emulators facilitate true concurrency since the traditional software simulation is run in sequential fashion. The emulator is constructed at the gate level obtained for synthesis of an RTL model. Emulators built out of reprogrammable FPGAs are becoming common.

As the complexity of microprocessors increases, it is more realistic to use a more abstract model such as the ECFM, which was described briefly in Section 13.5.1. The higher abstraction model has the advantage of being more general and can cover a wider range of processors. The benefit of the model is that it can also be linked to the structural level to guide in test pattern generation, as described in the next section and illustrated in Fig. 13.14.

13.6   TESTING MICROPROCESSORS

In the early years of microprocessors, ad hoc testing techniques were used. For example, they were tested by exercising some of their functionality. That is,

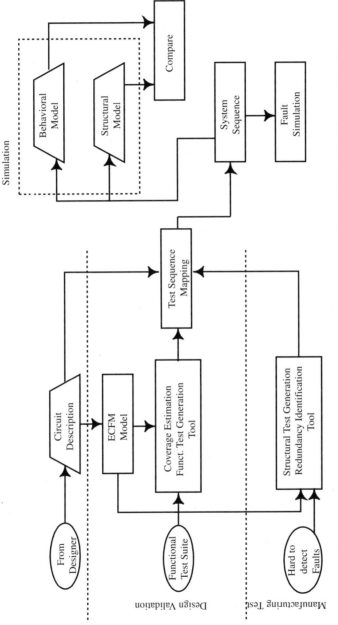

Figure 13.14 Unified verification and testing model [Moudanos 1998 © IEEE. Reprinted by permission].

337

every instruction is exercised for a selective set of operands. Thus the approach used for one microprocessor architecture could not be applied easily to another architecture. Also, no fault model was used, and hence it was difficult to assess the effectiveness of the test. On the other hand, testing microprocessors using traditional fault models requires extremely long test sets and is time consuming. Toward easing these difficulties, a functional fault model was developed and used successfully. One such approach proposed to test the CPU according to the following methodology [Robach 1976]. The system is divided into *macroblocks*, which in turn are subdivided into *microblocks*. Testing is carried out at the microlevel. The information is then passed to the macroblocks for fault detection and diagnosis. The emphasis in testing microprocessors has then shifted to instruction set verification.

13.6.1   Instruction Set Verification

The methodology aimed at testing the control unit under normal sequencing of the instructions was proposed by [Robach 1978]. The model used an FSM to represent the control commands and assumes that the results of these commands on the operands are directly observable. However, the operands are not directly available at the function units, and the results are not directly observable.

With the advent of the bit-sliced processor, a testing approach more suitable for this architecture has been developed based on the bit-slice model described above [Sridhar 1981]. The main idea was to develop the test for one slice, then arrange the slices in an iterative array, and develop a C-testable strategy for testing the array. The test patterns are generated by injecting faults in the datapath and the controller, for example, modifying the truth table of the combinatorial parts or the controller's state table of the controller. On the basis of this model, test patterns are generated to verify the different components of the datapath. The advantage of this method is that the efforts are concentrated on one slice and then generalized for the entire processor. However, the controller is not tested adequately except indirectly as the functional components are tested.

Toward a more generic approach to testing microprocessors, the S-graph model described above is used to verify the instruction set systematically [Thatte 1980]. The problem with the model was the increased complexity of the graph, as mentioned in Section 13.5.1. An improvement over this approach was realized with a closer examination of the instructions. They can be broken into microcodes [Annaratone 1982]. Examples of these microcodes are listed in the last column of Table 13.2. Generation of the test patterns does not require the details of the instructions but operates on the microcode [Brahme 1984]. The consequences of executing the instructions determine which circuit components are to be tested by which instructions. Accordingly, the following microprocessor component can be tested: register decoding, instruction decoding and sequencing, data transfer functions and data storage, and other functional units, such as the ALU. By relating a sequence of instructions with every one of these components, it is possible to generate the appropriate test patterns: the necessary data to decode the instructions and the data to place in the register, if necessary. For more details of test pattern generation using the model described above, refer to [Brahme 1984, Miczo 1986, Abramovici 1990, Rajsuman 1992].

Although this approach is a step in the right direction, as the instruction set proliferates, the model used becomes unmanageable. For this, a unified framework for multilevel and test generation was developed based on the ECFM mentioned for microprocessor modeling in Section 13.5.1 [Moundanos 1998]. Techniques used in formal verification are used to generate test to exercise the control transitions. Test generation is performed on the ECFM model. This sequence may have to be augmented by traditional ATPG techniques. The framework is outlined in part in Fig. 13.14. A similar unified principle was cited as the key to the Pentium Pro's testing success [Carbine 1998].

13.6.2   Testing the Datapath

The datapath consists of a set of functional modules of different types: arithmetic functional modules (adders, multipliers, etc) and logic functional modules (logical operations, comparators, shifters). All modules interact among each other and with embedded memories using registers. To optimize on the use of the functional modules and the registers, a set of multiplexers are used to facilitate the flow. The different interactions are performed under the control unit. A model for the datapath as described here is shown in Fig. 13.15.

Such tightly structured architecture causes several testability problems due to low controllability and observability of the embedded modules. There are known schemes to test the functional unit independent of each other. Some of the modules are modeled as iterative logic arrays (ILAs), some as one-dimensional (adders) and others as two-dimensional (array multipliers). In Chapter 8

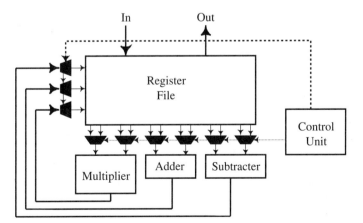

Figure 13.15   Datapath model [Gizopoulos 1996 © IEEE. Reprinted by permission].

we addressed the testability of ILAs. As we have learned, an exhaustive test set is applied on the basic cell of the ILA. The test guarantees the detection of all combinational faults in the unit and the faults are then propagated through subsequent units in the array. This approach is illustrated in Fig. 8.13 for an n-bit adder consisting of a succession of full adder. The exhaustive test is applied to the first full adder (FA) and the response from the carry c is applied to the second FA. For maximum concurrency testing (MCT), the patterns on a_2 and b_2 are arranged to apply an exhaustive test on this second adder. As shown in the figure, the test set for the adder is independent of the number of bits, the datapath size. Another example, the array multiplier, was shown to be C-testable after a slight modification to the basic cell [Shen 1984]. Only 16 patterns were sufficient to test any word-size multiplier. The concept was also extended to the Booth multiplier.

However, the main problem with datapath testability is the inaccessibility of the embedded module. BIST has been shown to be effective in solving this problem since it reduces the interaction with the outside world and utilized existing registers for pattern generation and response compaction. An effective BIST scheme for datapath would be capable to performing testing at speed with high fault coverage and reasonable overhead. A generic BIST approach will result in a long test set whose effectiveness is to be assessed with a fault simulator. Also, it depends on the functional unit under test and the width of the datapath. A more efficient BIST would rely on the C-testability of the functional modules, as described above. Some of the registers of the datapath are configured to generate the sets. Other registers can be configured as a SSA or a MISR. In addition, other types of compaction may be used, a count-based compaction [Ivanov 1992] or rotate carry addition [Rajski 1993]. The latter scheme was used successfully for BIST application to multipliers [Gizopoulos 1995].

The discussion above concentrated on the functional modules; however, the registers and multiplexers in the path of the tested modules are also tested fully. Those registers or multiplexers that were not exercised should be tested by extra patterns. BIST is also used for other circuitry, such as the embedded ROMs [Zorian 1992] and the RAMs [Zorian 1994]. Use of BIST for datapath is very popular and has also been reported to be successful with special-purpose processors [Pichon 1996]. In this example, due to the large number of ROMs and RAMs, the test pattern for BIST application was software generated and stored in a ROM.

13.7   DFT FEATURES IN MODERN MICROPROCESSORS

In the discussion above we have given the foundation of testing microprocessors based on generic architecture. The first known commercial processor that led in incorporating testability structures was the Motorola [Kuban 1984]. In the 386 processors, Intel introduced testing structure in their products [Gelsinger 1989]. At present, most DFT constructs and techniques are commonly utilized to test microprocessors. As a conclusion of this chapter, the testability features of some leading processor companies microprocessors are given next. This presentation is useful since it puts in perspective the different concepts leaned so far in the book. In addition, it will give a sense of the real-life handling of testability issues for large and complex systems.

The information in the remainder of this chapter is abstracted from a collection of papers on microprocessor testing that were presented at the International Test Conference in 1997 and 1998 [ITC 1997, 1998]. The highlights of these presentations were also published in two issues of *Design and Test of Computers* magazine [D&T 1997, 1998]. Testing covered the UltraSparc [Levitt 1997], Alpha 21164 [Bhavsar 1997, Stolicny 1998], Intel Pentium Pro [Carbine 1998], AMD-K6 [Fetherston 1998], IBM S/390 [Foote 1998], HP PA8500 [Brauch 1998], and M69060 [Kumar 1997].

13.7.1   Testing Sun Microsystems Processors

As a complex project, the design of sun processors has numerous goals that are not compatible. The achievement of high performance coupled with reduced chip area conflicts with the need for a design that is easy to debug, to test, and to manufacture. The UltraSparc incorporates the following DFT constructs:

- Full-scan design for all units except one that features partial scan using one scan clock
- IEEE Standard 1149.1, which includes a special memory test mode
- All built-in devices accessible via the IEEE Standard 1149 test port
- Chips tested using standard testers

The approach to the design of the UltraSparc combined structural techniques that added 3.5% to the die area and a fully custom design in the precharged logic and memory arrays. These arrays used 44% of the total transistor count. Test pattern generation was accomplished through commercial tools and the percentage fault coverage was in the high 90s. Table 13.3 shows how the DFT structures were deployed in testing, debugging, and manufacturing.

13.7.2   Testing Digital's Alpha 21164

The Alpha 21164 processor combines structured and ad hoc DFT solutions [Bhavsar 1997]. One of the features that complicated the design was the number of embedded RAMs that included extra cells to be used in memory repair. For this, a mix of hardware and software BIST was adopted. Attention was given, in particular, to the instruction cache that was designed with BIST and built-in self-repair (BISR). The RAM was organized as several columns of $N \times 1$ arrays stacked side by side.

Another improtant feature was the observability LFSRs (OBL), which served

TABLE 13.3   Taxonomy of DFT Uses

Feature	Test	Debug	Manufacturing
Scan	×	×	×
IEEE 1149.1	×	×	×
SRAM test mode	×	×	×
I_{DDQ}	×		×
Clock control	×	×	×
Observation bus	×	×	×
Process control monitors	×		×

Source: Data from [Levitt 1997 © IEEE. Reprinted with permission].

as MISR and scan cells to observe during testing and at speed, a single snapshot of data for chip debugging. The chip has 27 OBLs that observe 550 internal nodes. They are organized in the three scan chains to cover different geographic areas on the chip. The OBLs were configured to shorten the scan chain whenever needed. The cost of DFT is summarized in Table 13.4.

In addition, the chip also has a unique testability feature, the *test port*, which is provided solely for interfacing with the testability functions, including customer usable ones. Its 13 dedicated pins support three interface modes: normal, manufacturing, and debugging. Normal mode operation supports customer-usable test features. Two pins select the test mode. The boundary-scan port complies with IEEE Standard 1149.1 except for two deviations that actually enhance value to the end user:

- The optional reset pin has an internal pull down instead of pull up, as required by the standard.

TABLE 13.4   Alpha 21164 Cost Parameter Summary

Characteristic	Percent of Cost
Die area	2.2
Number of transistors	
Percent of total	0.5
Percent of core logic	3
Number of chip pins	
Percent of all pins	2.6
Percent of signal pins	4.4
Design effort	5
Performance decrease	0

Source: Data from [Bhavser 1997 © IEEE. Reprinted with permission].

- Two differential-oscillator input pins do not have any associated boundary-scan cells, which allows a meaningful test of the oscillator input pins.

13.7.3 Testing the Intel Pentium Pro

This high-performance Intel architecture microprocessor was designed for desktop, workstation, and server applications [Carbine 1998]. Its unique business requirement of meeting very high production, performance, and test quality goals simultaneously strongly influenced its design-for-test (DFT) direction. A set of constraints limits their design teams' ability to use DFT and test generation techniques (full or partial scan and scan-based BIST). For example, an increase of the die by 15% would have cost Intel a new fab! Nevertheless, they are able to optimize the design for low die area, high performance, low power dissipation, high-quality test, and low test cost.

Intel was guided by customer satisfaction for performance and testability goals that were conflicting in optimization of the design as illustrated in Fig. 13.17. The intent was to follow the four important DFT principles:

1. Have zero performance impact
2. Have minimal die area impact
3. Have multiuse features wherever possible
4. Be designed in from the start, that is, coded and validated in the register-transfer-level (RTL) model

The last principle was the key to the Pentium Pro's success. Intel coded the DFT features into the RTL model, verified their schematics using schematic

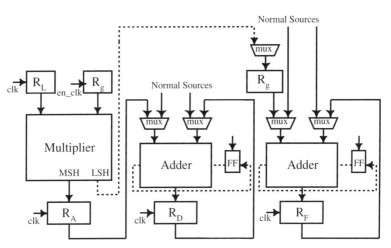

Figure 13.16 Testing the datapath.

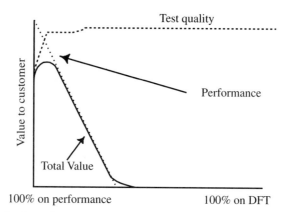

Figure 13.17 Intel: performance versus testability [Carbine 1999 © IEEE. Reprinted by permission].

formal verification tools, and verified their functionality through RTL simulation.

Intel's processor includes the following DFT features:

- *JTAG test access port (TAP).* The Pentium Pro processor TAP is a compliant implementation of IEEE Standard 1149.1, which includes seven public TAP instructions.

- *Scan-out.* Scan-out is a patented observation-only, scanlike technique that provides observability of internal nodes in the design. It works in two modes; a *snapshot mode* takes a single sample of the key processor state and shifts it out through the JTAG port while the processor is running, and a *signature mode* that enables continuous sampling and signature compression of the same key state elements in the design. Both modes can be used while the processor is running normally at full clock frequency without affecting its operation.

- $I_{DDQ}mode.$ This testing mode allows a private TAP instruction to be used to disable all devices that draw static current.

13.7.4 Testing AMD's K6

The heart of AMD's K6 processor is a reduced-instruction-set computer (RISC) core known as the enhaned RISC86 microarchitecture [Fetherston 1998]. From the onset, the designer team decided to make the AMD's K6 fully scannable, and therefore all internal storage elements are scannable and I/O cells are equipped with scan to support boundary scan and IEEE Standard 1149.1 compatibility. The internal storage element is a shared master, rising edge-triggered scan flip-flop. In normal operation, it operates as a simple flip-flop. In scan operation, the clock, which is embedded and controlled by a PLL, is held low and data are sanned using nonoverlapping shift clock pulses. AMD uses a commercial ATPG tool to create scan test patterns for stuck-at faults in their processor. By using a gate-level-fault simulation tool, the K6 contains just over 1 million simulator-primitive elements. The full-scan implementation makes combinational ATPG possible, thus providing a realistic opportunity to push the stuck-at-fault coverage goal for this large design to 100%.

The K6 also incorporates BIST into its DFT process. Each RAM array in the K6 processor has its own BIST controller. At power-on, the processor initializes the BIST controllers to a safe reset state. To start the testing, a single scan operation sets the StartBIST register in each BIST controller. BIST executes simultaneously on all arrays for the precomputed number of clock cycles that ensures completion for the largest array. Therefore, BIST execution time depends on the size of the largest array. For I_{DDQ} testing a special procedure was followed to counteract the effect of high subthreshold current.

13.7.5 Testing IBM S/390

The DFT framework of the 500-MHz CMOS S/390 microprocessor uses a wide variety of tests and techniques to ensure the highest reliability of components within a system [Stolicny 1998]. Some of the same test patterns applied in chip manufacturing are applied again at the system test level. Scan-based design is the fundamental requirement for all these techniques; that is, all system latches are scannable and configurable into one or more scan chains. The chip uses full scan and follows most level-sensitive scan design (LSSD) rules. It uses both an edge-triggered latch and a race-free latch clocked by a single-system clock. The chip contains a logic BIST (LBIST) engine to allow logic testing with minimal tester support. The LBIST places random patterns into the scan chain and compresses the results into a signature register. Register arrays are tested through the scan chain, while nonregister memories are tested with a programmable RAM BIST (RAMBIST). During logic test, RAMs become transparent and data are allowed to pass directly from RAM inputs to RAM outputs, making test generation simpler and faster. The overall objective of the S/390 chip test is 99% static stuck-at-fault coverage and 90% dynamic test coverage. Dynamic test coverage is the ratio of the total number of transition faults detected to the total number of transition faults in the design.

LBIST patterns are the first to be fault simulated during logic test generation and detect the majority of the logic fauls. On the S/390 chip set, test coverage is 95% static coverage with 256,000 LBIST patterns. The basic operation consists of loading the scan chains from the pseudo-random pattern generators (PRPGs) with scan clocks, applying system clocks to perform the actual logic test, then unloading the scan chains into the signature registers. As multiple patterns are applied, the load and unload operations combine into a single scan.

The RAMBIST engine is programmed more like a small microprocessor than a finite state machine. It contains a small microcode array initialized through the scan chain with the intended program/test application. Once initialized, the

RAMBIST decodes instructions from its microcode array and forms patterns to apply to the memory under test. Application of memory tests involves the scan initialization of the RAMBIST engines and microcode arrays, a series of clocks long enough to complete the test, and a scan-out of the resulting compression registers. Smaller memories have their results compressed into MISRs. With typical RAMBIST techniques, the set of tests applied is hard coded into the RAMBIST finite state machine. With programmable RAMBIST, unique memory tests can be accomplished at the chip, module, and system levels.

13.7.6   Testing Hewlett-Packard's PA8500

The PA8500 is a 0.25-μm fabrication, superscalar processor [Brauch 1998]. The small geometry provides a significant increase in speed and allows the large first-level cache to be placed entirely on a chip as the CPU. On-chip memories communicate directly with the CPU, eliminating the need to connect memory I/Os to external pins. One option for testing on-chip memories is to take the additional steps needed to port the memory I/Os and use a direct-access test (DAT) technique. However, a large two- or four-word-accessible cache requires many I/O pins (or pads). In addition, the cache must receive a large amount of data at high speeds to achieve acceptable fault coverage. To achieve a fast and thorough production test, the cache test hardware's first requirement is the ability to perform March tests. In general, March tests are an effective way to find several kinds of functional faults. The second requirement is the ability to apply arbitrary patterns as part of more general characterization tests.

Hewlett-Packard must assure that their hardware can support tests such as Walking Ones/Walking Zeros and GALPAT. These tests can write a value to a single memory cell and then perform reads on the entire array. The written value is then "walked" through every cell in the memory. Address sequencing is a fundamental part of most memory tests, so it is important to provide a great deal of flexibility in this area. When combined with pseudorandom patterns, pseudorandom address sequences can detect faults that deterministic tests might miss. The final required test function is the ability to create a bitmap of the memory. Generating bitmaps means determining the location of one or more failing cells. In addition to helping with electrical characterization and debugging, a bitmap can be a powerful tool when used to isolate and document possible process problems.

REFERENCES

Abramovici, M., M. M. Breuer, and A. D. Friedman (1990), *Digital Systems Testing and Testable Design*, IEEE Press, Piscataway, NJ.

Abramovici, M., and C. Stroud (199), Using ILA BIST for FPGAs, *Proc. IEEE International Test Conference* pp. 68–75.

Actel (1991), *ACT Family Field-Programmable Gate Array Databook*, Actel Corporation, Santa Clara, CA.

Altera (1992), *Applications Handbook*, Altera Corporation, San Jose, CA

AMD (1976), *A 2900 Bipolar Microprocessor Family*, AMD, Sunnyvale, CA.

Annaratone, M. A., and M. G. Sami (1982), An approach to functional testing of microprocessors, *Digest of Papers 12th Annual International Symposium on Fault-Tolerant Computing*, June, pp. 158–164.

Bhavsar, D. K.,a nd J. Edmondson (1997), Alpha 21164 testability strategy, *IEEE Des. Test Comput.*, Vol. 14, No. 1, pp. 25–33.

Brahme, D., and J. A. Abraham (1984), Functional testing of microprocessors, *IEEE Trans. Comput.*, Vol. C-33, NO. 6, pp. 475–485.

Brauch, J., and J. Fleischman (1998), Design of cache test hardware on the HP PA8500, *IEEE Des. Test Comput.*, Vol. 15, NO. 3, pp. 58–63.

Carbine, A. (1998), Pentium Pro Processor design for test and debug, *IEEE Des. Test Comput.*, Vol. 15, No. 3, pp. 77–82.

Chan, P. K., and S. Mourad (1994), *Digital Design Using Field Programmable Gate Arrays*, Prentice Hall, Upper Saddle River, NJ.

D&T (1997), *IEEE Des. Test Comput.*, Vol. 14, No. 1, pp. 8–41.

D&T (1998), *IEEE Des. Test Comput.*, Vol. 15, No. 3, pp. 56–104.

Feng, C., W. K. Huang, and F. Lombardi (1995), A new diagnosis approach for short faults in the interconnections, *Proc. IEEE Symposium on Fault-Tolerant Computing*, pp. 331–338.

Fetherston, R. (1998), Testability features of the AMD-K6 microprocessor, *IEEE Des. Test Comput.*, Vol. 15, No. 3, pp. 64–69.

Foote, T. D., et al. (1998), Testing the 500-MHz IBM S/390 microprocessor, *IEEE Des. Test Comput.*, Vol. x, No., x, pp. 389–395.

Gelsinger, P., S. Iyengar, J. Krauskopt, and J. Nadir (1989), Computer aided design and built in self test on the i486 CPU, *Proc. International Conference on Computer Design*, p. 199.

Gizopoulos, D., A. Paschalis, and Y. Zorian (1995), An effective BIST scheme for carrysave and carry-propagate array multipliers, *Proc. 4th IEEE Asian Test Symposium*, pp. 286–292.

Gizopoulos, D., A. Paschalis, and Y. Zorian (1996), An effective BIST scheme for datapaths, *Proc. IEEE International Test Conference*, pp. 76–85.

Hamdy et al. (1988), Dielectric based antifuse for logic and memory IC, *Proc. International Electron Device Meeting*, pp. 786–788.

Huang, W. K., X. T. Chen, and F. Lombardi (1996), On the diagnosis of programmable interconnect systems: theory and application, *Proc. 14th IEEE VLSI Test Symposium*, pp. 204–209.

Huang, W. K., F. J. Meyer, and F. Lombardi (1997), Testing memory modules in SRAM-based configurable FPGAs, *Proc. IEEE International Workshop on Memory Technology*, pp. 204–209.

Inoue, T., S. Miyazaki, and H. Fujiwara (1997), *Universal Fault Diagnosis for Lookup Table FPGAs*, Technical Report NAIST-IS-TR97020, Graduate School of Information Science, Nara Institute of Science and Technology, (*http://isw3.aist-nara.ac.jp/IS/TechReport/report/97020.ps*).

Inoue, T., S. Miyazaki, and H. Fujiwara (1998), Universal fault diagnosis for lookup talbe FPGAs, *IEEE Des. Test Comput.*, Vol. 15, No. 1, pp. 39–44.

ITC (1997), *Proc. IEEE International Test Conference*.

ITC (1998), *Proc. IEEE International Test Conference*.

Ivanov, A., and Y. Zorian (1992), Count-based BIST compaction schemes and aliasing probability computation, *IEEE Trans. Comput.-Aided Des.*, Vol. 11, No. 6, pp. 768–777.

Kautz, W. H. (1974), Testing for faults in wiring networks, *IEEE Trans. Comput.*, Vol. C-23, No. 4, pp. 358–363.

Kuban, J. R., and W. C. Bruce (1984), Self-testing the Motorola MC6804P2, *IEEE Des. Test Comput.*, Vol. 1, NO. 5, pp. 33–41.

Kumar, J. (1995), Emulation verification of the M68060 for concurrent verification, *Proc. International Conference on Computer Design*, pp. 150–158.

Kumar, J. (1997), Prototyping the M68060 for concurrent verification, *IEEE Des. Test Comput.*, Vol. 14, No. 1, pp. 34–41.

Levitt, M. E. (1997), Designing UltraSparc for testability, *IEEE Des. Test Comput.*, Vol. 14, No. 1, pp. 10–17.

Liu, T., F. Lombardi, and J. Salinas (1995), Diagnosis of interconnects and FPICs using a structured walking-1 approach, *Proc. Fourteenth IEEE VLSI Test Symposium*, pp. 256–261.

Mao, W., et al. (1990), Quietest: a quiescent current testing methodology for detecting leakage faults, *Proc. IEEE International Test Conference*, pp. 280–283.

Moundanos, D., J. A. Abraham, and Y. V. Hoskote (1998), Abstraction techniques for validation coverage analysis and test generation, *IEEE Trans. Comput.*, Vol. C-47, No. 1, pp. 2–14.

Needham, W. (1998), Microprocessors testing today, *IEEE Des. Test Comput.*, Vol. 15, No. 3, pp. 56–57.

Pichon, F. (1996), Testability features for a submicron voice-coder ASIC, *Proc. IEEE International Test Conference*, pp. 377–385.

Quicklogic (1993), *Very High-Speed Programmable ASIC*, Quicklogic, Santa Clara, CA.

Rajski, J., and J. Tyszer (1993), Test responses compaction in accumulators with rotate carry adders, *IEEE Trans. Comput.-Aided Des. Integrated Circuits Syst.*, Vol. 12, No. 4, pp. 531–539.

Rajsuman, R. (1992), *Digital Hardware Testing: Transistor-Level Fault Modeling and Testing*, Artech House, Norwood, MA.

Renovell, M., J. Figueras, and Y. Zorian (1997a), Test of RAM-based FPGA: methodology and application to the interconnect, *Proc. Fifthteenth IEEE VLSI Test Symposium*, pp. 203–237.

Renovell, M., et al. (1997b), Test pattern and test configuration generation methodology for the logic of RAM-based FPGA, *Proc. IEEE Asian Test Symposium*, pp. 254–259.

Renovell, M., et al. (1998a), SRAM-based FPGA: testing the LUT/RAM modules, *Proc. IEEE International Test Conference*, pp. 1102–1111.

Renovell, M., et al. (1998b), Testing the interconnect of RAM-based FPGAs, *IEEE Des. Test Comput.*, Vol. 15, No. 1, pp. 45–50.

Renovell, M., et al. (1999), SRAM-based FPGA: testing the embedded RAM modules, *J. Electron. Test. Theory Appl.*, Vol. 14, No. 1/2, pp. 159–167.

Robach, C., and G. Saucier (1978), Dynamic testing of control units, *IEEE Trans. Comput.*, Vol. C-27, No. 7, pp. 617–623.

Shen, J. P., and F. J. Ferguson (1984), The design of easily testable VLSI array multipliers, *IEEE Trans. Comput.*, Vol. C-33, No. 6, pp. 554–560.

Sridhar, T., and J. P. Hayes (1981), A functional approach to testing bit-sliced microprocessors, *IEEE Trans. Comput.*, Vol. C-30, No. 8, pp. 563–571.

Stolicny, C. (1998), Alpha 21164 manufacturing test development and coverage analysis, *IEEE Des. Test Comput.*, Vol. 15, No. 3, pp. 98–104.

Stroud, C., et al. (1996), Using ILA testing for BIST in FPGAs, *Proc. IEEE International Test Conference*, pp. 68–75.

Stroud, C., et al. (1997), BIST-based diagnostics of FPGA logic blocks, *Proc. IEEE International Test Conference*, pp. 539–547.

Stroud, C., et al. (1998), Built-in self test of FPGA interconnect, *Proc. IEEE Internationals Test Conference*, pp. 404–411.

Thatte, S. M., and J. A. Abraham (1980), Test generation for microprocessors, *IEEE Trans. Comput.*, Vol. C-29, No. 6, pp. 429–441.

Trimberger, S. M. (ed.) (1994), *Field-Programmable Gate Array Technology*, Kluwer Academic, Norwell, MA.

Xilinx (1989), *The Programmable Logic Data Book*, Xilinx, Inc., San Jose, CA.

Zhao, L., D. M. H. Walker, and F. Lombardi (1998), Detection of bridging faults in logic resources of configurable FPGAs using IDDQ, *Proc. IEEE International Test Conference*, pp. 1037–1046.

Zorian, Y., and A. Ivanov (1992), An effective BIST scheme for ROMs, *IEEE Trans. Comput.*, Vol. C-41, No. 5, pp. 646–653.

Zorian, Y., A. J. Van De Goor, and I. Schanstra (1994), An effective BIST scheme for ring-address type FIFO's, *Proc. IEEE International Test Conference*, pp. 378–387.

PROBLEMS

13.1. Discuss the suitability of using a functional fault model approach to testing microprocessors and FPGAs. State the benefits and the drawbacks.

13.2. In a paper on testing an 8-bit microprocessor, [Thatte 1980] reported the results of simulating a sample SSFs. The fault coverage obtained was 96%. Do you consider this coverage acceptable? Why or why not? Using the same methodology, is it possible to reach such fault coverage for modern microprocessors?

13.3. In a RAM-based FPGA, one RAM cell connecting two metal segments is defective and the two metal segments cannot be connected. What are the consequences of such a failure? Consider different scenarios for the location of the defect in the logic circuit.

13.4. Consider the testing of the LUT of a Xilinx FPGA. In Fig. 13.6b, a configuration was recommended to connect the LBs in a one-dimensional array and shift the results through the LB's flip-flops. Use the schematic of Fig. 13.3a to detail the connection of two consecutive LBs.

PART V

ADVANCED TOPICS

14

SYNTHESIS FOR TESTABILITY

14.1 INTRODUCTION

Earlier chapters have established some important facts about VLSI testing and the need for embedded test (ET) constructs to facilitate testing. As the technology feature sizes continue to decrease and circuits continue to get larger, testing is becoming more and more difficult. Its cost keeps rising and constitutes a large fraciton of the total product cost unless ET is incorporated. If ET is viewed as an add-on feature after the completion of the design, it is very likely to be dropped because of urgency to meet time-to-market deadlines or because it results in changes to the area and performance specifications. To assure proper embedded testing while delivering products on time, it is better to consider the testing strategy early in the design cycle.

The chip area used to have a strong impact on the cost. Area minimization resulted in an unprofitable increase in the design time. As the process technology improved and wafers size increased, area became a less important parameter than performance and power. Moreover, no matter how fast or low powered they are, ICs have to be easily testable. The emergence of design automation tools, in particular, synthesis, helped optimize design for time and productivity of random logic blocks. Hence optimizing testability should be part of synthesis and a parameter of equal importance to performance and power in the trade-off process. Including DFT in synthesis is what we refer to in this chapter as *synthesis for testability* (SFT).

Nowadays, SFT from RTL to gate level is a reality. Testimonies to this are the DFT tools that insert embedded testing constructs on these levels of abstraction [ITC 1994, D&T 1995]. Attempts to include high-level synthesis (HLS), which is commonly referred to as *behavioral synthesis*, in commercial products are also noticeable.

To discuss synthesis for testability, we need knowledge in three main areas: (1) circuit modeling, (2) synthesis concepts, and (3) testability analysis. Each oen of these areas has been covered in one or more chapters of this book. In Chapter 3 we discussed different circuit representations that we will use throughout this chapter. Synthesis, which we presented initially in Chapter 4, changes the design model from one level of abstraction to another closer to the final mask. Finally, testability analysis, as used here, refers to several issues that we specify in the next sections.

14.2   TESTABILITY CONCERNS

Testing a circuit often involves the generation of patterns that detect the faults. We have realized that test pattern generation is a NP-complete problem. In addition, there are many logic constructs that make the circuit difficult to test. For example, redundancy causes undetectable faults. Also, long feedback loops in sequential circuits increase the sequential depth and make ATPG more complex than it is for combinational circuits. To simplify test pattern generation and application for random logic blocks, different DFT approaches are used: addition of controllability and observability test points, exploitation of regularity of some circuits, partitioning of a large circuit, use of pseudorandom test patterns, insertion of scan-path or boundary-scan and enable self-test.

All of these approaches were presented in Chapters 8 to 13 as remedies to alleviate the difficulties encountered in testing. They were not presented in the context of the design cycle. Here we focus on how and when in the synthesis processes (1) constraints are placed to obtain easily testable circuits and (2) DFT constructs are incorporated. We concentrate on high-level synthesis. There are several advantages for addressing the testability problem on such a high level. First, the representation is closer to human reasoning than a multitude of logic gates that are a few folds removed from the circuit function. Also, any change in the design that can be made at this level has less impact on the time-to-market than had the change been done at the logic level. Although the intent is to concentrate on the high-level integration of testing to design, some test synthesis on the RTL and gate level is also discussed.

14.3   SYNTHESIS REVISITED

According to the taxonomy introduced in Chapter 3 and illustrated by Fig. 14.1, synthesis is the process that transforms the design from one level of abstraction to another that is closer to the physical circuit level in the physical domain. The highest level of design abstraction in the chart is the algorithmic code in the behavioral domain, point *A* in the figure. To idealize the design into an IC, we

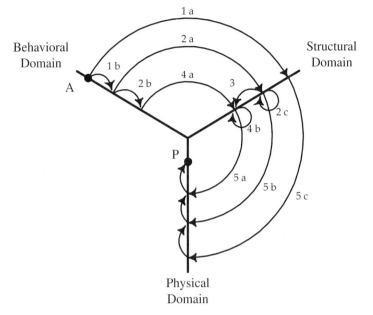

Figure 14.1 Design representations: Y-chart.

need to map the algorithmic representation to its final form before fabrication, the mask level in the physical domain, point P. The path from A to P consists of several transformations from one level to another in the same domain and from one domain to another at the same level, and optimization at any point along the path [Gajski 1983, Walker 1985].

The transitions shown in Fig. 14.1 illustrate the steps involved in the various synthesis processes [Michel 1992]. They were presented first in Chapter 4. Step 1a translates the different operations into concrete subsystems, processors, and so on. Step 1b results in an RTL-level description that relates the component from step 1a. The RTL synthesis consists of transitions 2a, 2b, and 2c. The first two steps transform the design to the gate level. Step 2c refines the gate-level design by optimizing on some attribute, number of gates or interconnect wires. Step 3 is the technology mapping. Steps 4a and 4b are for the logic synthesis and optimizaiton. Any or all of steps 5a, 5b, and 5c represents the physical synthesis. For a design based solely on standard cells, step 5a is sufficient. However, a system on a chip that includes hard cores would also involve step 5c.

14.4 HIGH-LEVEL SYNTHESIS

High-level synthesis is not a new technology. In the early 1970s, it was investigated at universities such as Carnegie–Mellon [Barbacci 1973], and the number of systems increased in the 1980s. Also, HLS was adopted by industry: for example, the Yorktown Silicon Compiler [Compasano 1988]. At present, HLS synthesis is gaining acceptance by industry for several reasons. Among them are shorter time-to-market, fewer errors, increased designer's productivity, and self-documenting design process [McFarland 1988]. HLS usually starts with a design described in algorithmic form, which is written in one of the programming languages discussed in Chapter 3: VHDL and Verilog HDL, or hardware C. They are all procedural languages in the sense that they describe operations on data by assignment statements that are organized in a combination of the following structures: sequence, If–Then–Else, and terminated repetitive iterations. In addition to being simulatable, the design has to be synthesizable. The design is next transformed from behavioral to RTL level through the following tasks:

- *Compilation:* creation of data and control graphs
- *Scheduling:* assignment of the operations to time steps
- *Allocation:* determination of the type and number of resources
- *Binding:* binding of the actual resources to the various operations

These tasks may be viewed in the context of the design taxonomy of Fig. 14.1. A possible interpretation of these tasks is shown in Fig. 14.2. Scheduling maps the algorithmic description (point *A*) into RTL level in the behavioral domain. Specific tasks are assigned to the time steps. The allocation maps the operations in the behavioral level to specific hardware blocks in the structural domain. Finally, binding transforms from the structural to the physical domain. Notice that neither the allocation nor the binding has to be on one level only. The structural parts may be cells, blocks, or cores.

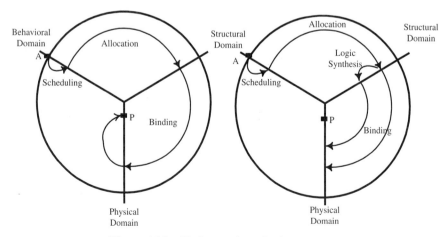

Figure 14.2   Y-chart and synthesis processes.

$$c = a + b$$
$$g = c * d$$
$$h = e - f$$
$$z = g + h$$

Behavioral Description

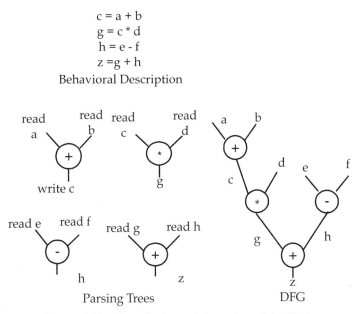

Parsing Trees   DFG

Figure 14.3 Compilation and formation of the DFG.

Scheduling and allocation are interrelated tasks, they are not independent, and neither has to be done in a particular order. Scheduling assumes some allocated resources, and allocation starts with a schedule. The result of the synthesis is a RTL-level structure that can then be synthesized to the logic gates. A comprehensive survey of HLS can be found in [Compasano 1990], [VLSI 1993, Gajski 1994, and Lee 1997].

14.4.1 Model Compilation

The HDL model is compiled into internal representation as illustrated in Fig. 14.3. The behavioral code is parsed in several trees, which are then assembled in a data flow graph (DFG). In addition, a control flow graph (CFG) is also constructed. In these directed graphs, the nodes represent the different operations of the model, and the edges represent the data on which the operation is performed, or the control flow. Figure 14.4*a* shows an example for a simple behavioral code of a square-root calculation problem [McFarland 1988]. The DFG, which is a one-to-one mapping of the textual sequence, is shown in Fig. 14.4*b*. The CFG is shown in Fig. 14.4*c*. At present there is a preference for a combined control data flow graph (CDFG). A version of such a graph is shown in Fig. 14.4*d*. Scheduling and allocation algorithms operate on the graphs to transform the HDL model into a structural model.

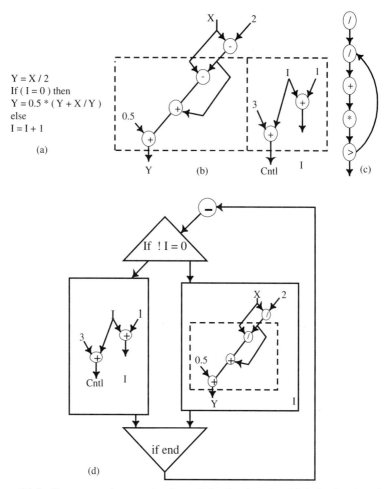

$Y = X / 2$
If (I = 0) then
$Y = 0.5 * (Y + X / Y)$
else
$I = I + 1$

(a)

Figure 14.4 Representation graphs for a behavioral code: (*a*) behavioral code; (*b*) DFG; (*c*) CFG; (*d*) one representation of the CDFG.

14.4.2 Transformations

Unlike software programming languages, a hardware behavioral description can be mapped by any one of several architectural structures. These structures constitute the *design space*. In exploring this design space, transformations are used to map from one structure to another. There are *compilerlike transformations* such as loop unrolling, and *optimizing transformations*. For example, it is possible to use a shift operation to substitute for a multiplication or division by a power of 2. Using transformations it is also possible to modify the height of the tree representing the assignment $z = a + b + c + d + e + f$. As illustrated in Fig. 14.5, the height is reduced from seven to three.

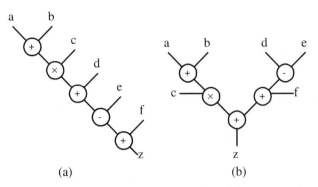

Figure 14.5 Transformations: reduction of the tree depth.

14.4.3 Scheduling

Scheduling is the process according to which the various operations of the DFG are performed relative to each other. In other words, it is assignment of tasks to the different control steps according to data dependency. The input to a scheduling algorithm is the CDFG, and the output is the temporal ordering of the operations. The ordering is performed for target architecture under some constraints for a particular goal. The goals are usually to optimize on one or more of the design attributes: area, performance, power, or testability. The target architecture may, for example, be a pipelined or busing structure. The outcome of scheduling is not unique. The schedule for the square-root problem may be one of the two presented in Fig. 14.6. The dependency of the operations performed on the data (X and Y) can only be scheduled as shown in Fig. 14.4a. The loop's index and control schedule may start in the first time slot or in the second, but no later, unless there is a good reason to delay the process. the first schedule shown in Fig. 14.6b illustrates what is known as the

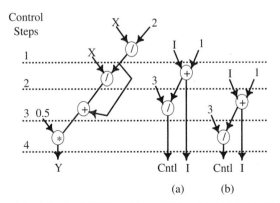

Figure 14.6 Schedule for SQRT problem of Fig. 14.4: (a) ASAP; (b) ALAP.

as soon as possible (ASAP) scheme, while the second is the *as late as possible* (ALAP) scheme. The advantage of the first case is that it is possible to use one adder and to share it in processing the data and the loop variable. The two additions are mutually exclusive. The second schedule does not allow such sharing. However, it is possible to implement the design with an ALU rather than with individual adders and multipliers. This is illustrated in Section 14.4.4.

Another example is shown in Fig. 14.7 for the behavior description of solving a differential equation [Walker 1995]. Three schedules are given. Those in Fig. 14.7a and b are ASAP and ALAP schedules. The overall number of time steps is the same, but the second may require fewer resources. For example, since the multiplication $3 * y$ is deferred to control unit 2, one of the multipliers used in the first control step can then perform this operation. However, these algorithms are generally used to assess the distribution of operations type in a control step. The multiplication operation $u*dz$ can be performed in time step 1, 2, or 3. This is expressed as the *mobility* of this operation. The higher the mobility, the more flexibility exists to have this operation share resources with other operations. Other more efficient algorithms can be found in the synthesis literature.

If the goal of scheduling is to minimize resources, the schedule shown in Fig. 14.7c is more appropriate. However, the reduction in resources resulted in increasing the design time. In addition, sharing resources require multiplexing on these resources. There are two drawbacks to this sharing: an increase in power and formation of loops to be formed in the hardware.

14.4.4 Allocation and Binding

The allocation task determines the type and quantity of resources ot be used in the circuit. These resources consist of functional units such as adders, memory elements, and buses as well as clocking system. The goal of allocation is to make appropriate trade-offs between the design's attributes. For example, if two addition operations are performed in the same time step, there is a need for two different adders. If the operations are at different time steps, that is, they are mutually exclusive, it is possible to make them share the same adder. To perform the rquired trade-off, say, for example, between area and performance, the area and delays have to be calculated. Use is made of a library of cells that are carefully characterized. It is rare to find a library now that does not include a scan flip-flop or an LFSR. Although in the past most allocation trade-offs concentrated on area, time, and power, nowadays there are several studies on allocation for testability [Avra 1991].

The next task to be performed is binding. This task assigns the operations within each clock cycle to available hardware. Resources can be shared by various operations. Operations that are performed at different time steps can share resources. From the schedule in Fig. 14.6, the two division operations are mutually exclusive and one divider can be shared. It is also possible to use one ALU to do all the operations, but first we need to determine the number of registers to use. For this it is important to determine, after scheduling, the lifetime of the

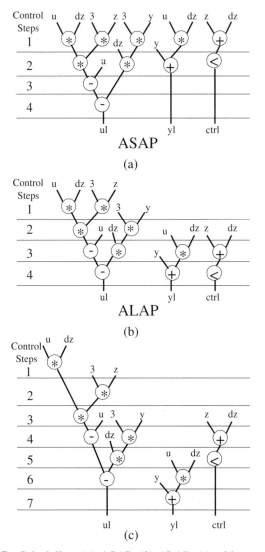

Figure 14.7 Scheduling: (*a*) ASAP; (*b*) ALAP; (*c*) with one multiplier.

operands and the variables. The *lifetime* diagram shown in Fig. 14.8 is derived from the scheduled DFG (SDFG) for the square-root problem and shown in the same figure. Each of the variables in the SDFG is represented by a line segment whose start and end represent the beginning (birth time) and end (death time) of the variable life. For example, the lifetime of X consists of time steps 1 and 2. The lifetime of Y consists of steps 2, 3, and 5. Both variable lifetimes overlap in time step 2 and it is not possible to use the same register for both variables. For similar reasons, Y and g cannot share a register. However,

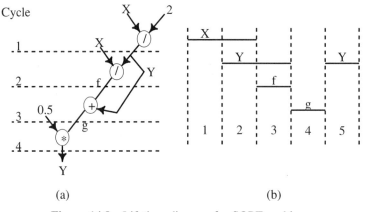

Figure 14.8   Lifetime diagram for SQRT problem.

X, g, and f can share the same register. Two registers, R1 and R2, are then sufficient to implement the design. According to this allocation of functional modules (the ALU) and of registers, the design is implemented as shown in Fig. 14.9. In this implementation, we substituted (transformed) the first division by 2 by and a multiplication by 0.5, and we added two multiplexers to facilitate sharing the resources. The control lines of the MUXs, S1 and S2, and the enable lines of the two registers are timed to perform the various operations. From the lifetime chart, it is also possible to bind register R2 to f in addition to y. In this case the implementation in Fig. 14.9 has to be modified. Another example of register allocation is shown in Fig. 14.10. In this example we have

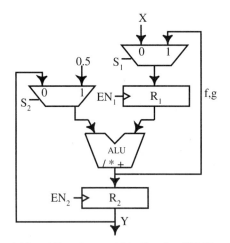

Figure 14.9   Allocation and binding for SQRT problem.

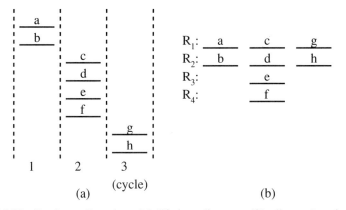

Figure 14.10 Register allocation: (*a*) lifetime diagram: (*b*) allocated registers [Lee 1997 © IEEE. Reprinted by permission].

many possibilities to allocate and bind the registers. In general, the operation is NP-complete, but heuristics are used to perform it. This problem is similar to channel routing, which involves the assignment nets to the various tracks of a channel. The left-edge algorithm (LEA) that has been used successfully in channel routing is applicable to allocate the minimal number of registers [Kurdahi 1987]. According to this algorithm, the lifetime diagram of the variables is organized by their birth time, indicated by the left end of the line segments in increasing order of time steps. We start with the leftmost segment, which has the longest continuous lifetime, variable *a*, and assign it to register R1. The same register can be assigned the next variable whose birth time is at, or after, the *death time* of variable *a*. This can be any of the variables *e*, *d*, *e*, or *f*. Variable *c* is selected. Next we select another variable whose birth time is at the end of *c*'s life. Register R1 is thus assigned to *a*, *c*, and *g*. We proceed in the same fashion and use another register R2 for *b*, *d*, and *h*. Variables *e* and *f* share the same lifetime and are each assigned to different registers: R3 and R4, respectively. In all, four registers are needed. The solution is not unique, but the number of register is minimal.

14.5 TEST SYNTHESIS METHODOLOGIES

One of the main difficulties of SFT at the behavioral level is the lack of a fault model that is easily mappable to a defect on the physical level [Wagner 1996]. Also, while HLS is a top-down process, testing has been considered a bottom-up process. Considering testability only during lower-level synthesis can lead to significantly increased overhead and reduced test quality [Dey 1998]. However, the enormous efforts that have been put in DFT techniques resulted in some design rules and constructs that can be used during high-level design.

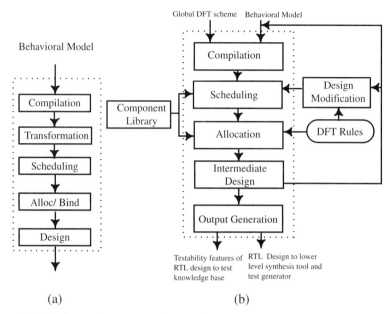

Figure 14.11 Comparing (*a*) HLS and (*b*) HLS for test [Varma 1993 © IEEE. Reprinted with permission].

The term *test synthesis* means many things to many people. It is used whenever the synthesis has something to do with testing. On one extreme this may be the substitution of flip-flops by their scan counterparts, in a completed design, for the purpose of test generation. This is considered as the two-pass approach to test synthesis [Aitken 1995]. On the other hand, test synthesis refers also to including testability among the design attributes used to make trade-offs during the scheduling and allocation processes, and in design rule checking.

Unlike area and delays, testability cannot be measured with one parameter. In addition, the optimization on area and performance is accomplished by iterating on the scheduling, allocation, and binding processes. This iterative process is not sufficient to optimize the testability of the design. More operations are required [Varma 1993]. Figure 14.11 outline these operations and compare traditional HLS and HLS for test.

In the remainder of this section we describe implementation of certain DFT techniques. Many of these topics have been discussed in special issues of [JETTA 1998, D&T 1995], among others.

14.5.1 Partitioning

Partitioning a large design has several reasons. First, synthesis tools still have difficulty handling complex designs. Second, performance may improve due to shorter intrconnection wires and the possibility of using architectural concurrency [McFarland 1983, Langnese 1989]. Third, and most relevant to this chapter, testing is simplified [Gu 1995]. Circuit partitioning is formulated as a graph-partitioning problem with constraints. It is a known to be NP-complete [Garey 1979], and therefore heuristics are used in partitioning algorithms. For example, it is possible to minimize redundancy [De 1993], to make the components resulting from the partition fan-out-free [Macii 1994]. Partitioning may also be performed to break feedback loops, insert test points at the worst testability lines, and make the circuit random pattern resistant-free [Gu 1995]. The approach taken in this work is first to partition for a goal, then synthesize the entire design instead of each partition separately, since in this fashion, the synthesis tool can make some of the physical components shared by the partitions. Afterwared, since the boundaries of the partitions are known, it is possible to generate patterns for each partition separately. The partitioning is guided by testability measures that are presented in Chapters 1 and 8.

14.5.2 Controllability and Observability

In datapath, traditionally, register allocation aims to minimize the number of registers used in a design as described in Section 14.4.4. However, this might make some of the variables not easily controllable and observable (C&O). The C&O of the registers can be increased using the following constraints on allocation [Lee 1992a]: (1) to assign the variables in such a way as to maximize the number of I/O registers connected to the primary I/O, and (2) to minimize the sequential depth from an input register to an output register. In addition, when two variables cannot share a register because of operating on both in the same control period, the CDFG is rescheduled such that an intermediate variable does not compete with another over an I/O register [Lee 1992b]. This is illustrated here by the example we used in Fig. 14.10. Let us assume that in this example, variables a and b are primary inputs, and variables g and h are primary outputs. The other variables represent intermediate results. The allocation shown in Figs. 14.10b and 14.12b makes registers R3 and R4 not directly controllable or observable. If we use a right-edge instead of the left-edge algorithm that we used for register allocation in Section 14.4.4, we can obtain the configuration shown in Fig. 14.12c. With this allocation, R3 and R4 become observable.

Testability measures have been used to guide test pattern generation, partitioning for test, and test point insertion, among other testing activities. Several techniques have been proposed to define testability measures on the RTL and high-level description of a design [Gu 1994, Hsu 1996]. The later work defines some controllability and observability measures starting from CDFG of the design. These measures are then used to insert test points to facilitate testing. We present the formulation of these measures and illustrate them using the behavioral code and its CDFG, shown in Fig. 14.13a and b. In evaluating the controllability, it is important to distinguish between a node that represents the results of an operation (rectangle in the figure) and one that represents a branch of a condition (diamond). A node is controllable if its precedents are

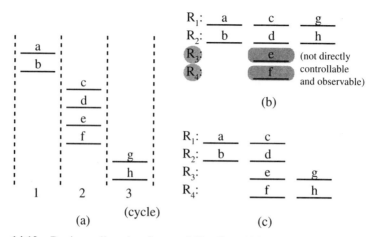

Figure 14.12 Register allocation for testability [Lee 1997 © IEEE. Reprinted by permission].

controllable by the primary inputs C clock cycles prior to performing the operation; otherwise, the node is not controllable. Similarly, a branch is controllable if the condition is controllable by the primary inputs C clock cycles before the branching is executed. Since no clock is associated with the behavioral description, the values of C are measured in a number of register transfer statements. Thus

$$C \text{ of an operation} = C_{max} \text{ of operands} + 1$$

The observability of a node is measured as the number of clock cycles required to observe the results at a primary output. Branches of conditions have no observability since this can be done by observing the results of operations they control.

$$O(\text{operand}_j) = \text{Max}\{O(\text{destination} + 1), C_{max}(\text{operands}_{k \neq j}) + 1\}$$

In the case of multiple branches as illustrated in Fig. 14.13c, the minimal controllability or observability numbers are used. After calculating the testability measures for the entire CDFG, hard-to-control nodes and hard-to-observe nodes are then identified (highest C and O values). Suitable solutions can be applied to improve their controllability and observability.

14.5.3 Feedback Loops

It has been demonstrated that sequential circuits with long feedback loops and large sequential depth increase the complexity of test pattern generation [Miczo

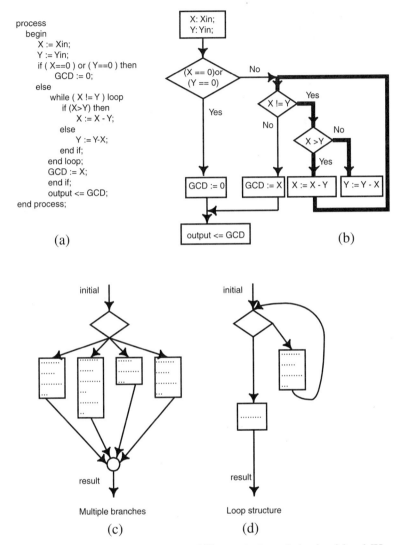

```
process
   begin
      X := Xin;
      Y := Yin;
      if ( X==0 ) or ( Y==0 ) then
            GCD := 0;
      else
         while ( X != Y ) loop
            if (X>Y) then
                  X := X - Y;
            else
                  Y := Y-X;
            end if;
         end loop;
         GCD := X;
      end if;
         output <= GCD;
end process;
```

(a)

(b)

Multiple branches   Loop structure

(c)   (d)

Figure 14.13   Example to illustrate testability analysis on behavioral level [Hsu 1996 © IEEE. Reprinted by permission].

1983]. Some of these loops cannot be avoided, but it is preferable to minimize them and to shorten their sequential depth. Loops in a CDFG can be broken by selecting the minimal feedback vertex set (MFVS) in the same manner as we described in Chapter 9 using the S-graph to select the scannable flip-flops. Even if no loops exist in the CDFG, they may be produced by hardware allocation and binding aimed to sharing resources. Therefore, it is possible to avoid these loops with appropriate transformations or rescheduling and binding. Next we explain

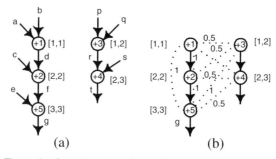

Figure 14.14 Example for allocation loop formation [Potkonjak 1995a © IEEE. Reprinted by permission].

how these unintended loops are formed and how resynthesis can eliminate them using the work by [Potkonjak 1995a].

Figure 14.14a shows two scheduled datapath segments. The numbers in brackets give the ASAP and ALAP control steps for each path. From this DFG a data dependency and compatibility graph (DDCG) is formed as shown in Fig. 14.14b. The solid lines indicate *dependency* of the data; the dashed lines indicate the *compatibility*. The compatibility means that the two operations can be done with the same hardware (e.g. an adder). If no sharing is allowed, no loops will be formed. However, if we allocate only two adders, A1 and A2, loop formation depends on selectin of the compatible operations to share the same adder. Some of the scheduling and the binding may result in one of the following configurations:

1. A1: +1, +3, +5, A2: +2, +4
2. A1: +1, +4; A2: +2, +3, +5
3. A1: +1, +2, +5; A2: +3, +4

The first two assignments will cause looping. However, in the third binding assignment each path has a dedicated adder and no sharing between the operations of the two paths. Thus, only self-loops are possible. The results of the first and last assignments are shown in Fig. 14.15. If looping cannot be avoided, the DDCG can be used to break loops in the same fashion described above in this section—finding the MFVS.

14.5.4 Scan Path

Insertion of scan path in a design seems to be a straightforward process that does not actually have to be done on a high-level design presentation. However, this two-pass design approach will affect adversely the circuit performance since the placement of the scannable flip-flops optimal for high performance. This is particularly a problem in deep submicron technology circuit where the inter-

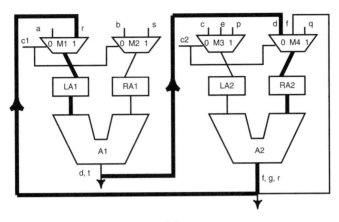

(a)

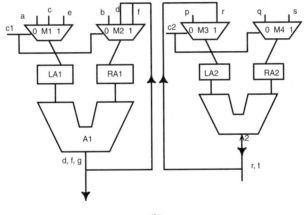

(b)

Figure 14.15   Hardware configurations for the flow in Fig. 14.14: (*a*) loop forming; (*b*) self-loop only [Potkonjak 1995a © IEEE. Reprinted with permission].

connect will contribute significantly to the delays in the circuit. Performance is the main reason for considering scan from the onset of the design process—this is the one-pass approach to the design.

Including the knowledge about the scan storage units and their need for special signals, scan enable (SE), scan in (SI), and scan out (SO) are needed for design rule checking and for mapping to the gate level. More important, this knowledge will help in the formation of multiple scanchains. For example, if different clocks control different parts of the design, a separate chain can be configured for each clock.

The other reason why scan path should be considered during high-level synthesis is for ordering the scan flip-flops for minimal test application time and minimal interconnect. As we discussed it in Section 9.11.1, rearranging the order in which the flip-flops were scanned can significantly decrease test application. If done on the gate level, any rearrangement might require major changes and affect the design time adversely.

There is a definite advantage to considering partial scan path in HLS. When using the S-graph to select the MFVS set of nodes to be scanned, the complexity is much reduced on the high level compared to the gate level. As we have seen in the preceding section, appropriate scheduling and allocations reduce the feedback loops and consequently the cardinality of the MFVS. In addition, it has been shown that further reduction can be idealized using high-level transformations [Potkonjak 1995b].

14.5.5 BIST Insertion

BIST is one of the most popular DFT structures used in today's systems. It is often used in conjunction with scan path and boundary scan. We discussed these topics in Chapters 9 to 11. Figure 14.16 shows the main features of BIST that need to be considered during synthesis.

We may view BIST simply as appending LFSRs to a design as a PRPG and a MISR. However, there are other issues to be addressed, as depicted in Fig. 14.16, which illustrates the main BIST features that are of concern to designers. These features include development of a controller for BIST execution and selection of optimal test points to facilitate the detection of RPR faults as discussed in Chapters 8 and 11. In addition, BIST is generally used in conjunction with scan path, and we have just explained why scan constructs need to be done

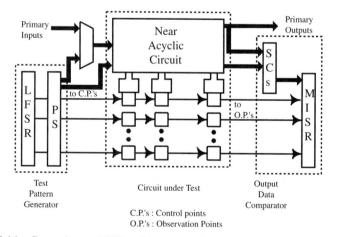

Figure 14.16 General scan-BIST architecture [Lin 1993 © IEEE. Reprinted with permission].

during HLS. With register-intensive designs, it is possible using BIST to deploy registers from the circuit to be used as PRPG and MISR.

Instead of modifying an existing design, as it used to be done in the past, the purpose of test synthesis is to select the appropriate registers during the design itself. The merit of the early decision about the register during scheduling and allocation is to optimize the design, area, and performance while taking testability into account. This type of synthesis is thus more elaborate than just adding the registers [Avra 1991, 1993; Harmanani 1993; Harris 1994; Papachristou 1998].

14.5.5.1 Test Point Insertion. For test point insertion, a certain metric needs to be used to determine optimal locations for these points. To avoid time-consuming fault simulation, testability measures can be used. On the gate level some techniques were proposed for test point insertion in BIST [Savaria 1991, Lin 1993]. It is, however, more appropriate to use some high-level TMs, such as those discussed in Section 14.5.2. Regardless of the method used to generate the TM, there must be a strategy to utilize the results. Since optimal selection of these points is an NP-complete problem, a heuristic must be used [Rao 1991]. It is worth noticing, however, that whenever an observation point is inserted, it affects an entire cone of the circuit for which it is a vertex. Usually, this point is handled as any primary output in the sense that it is connected to the test response compactor. The compactor will probably have fewer stages than the circuit has primary outputs. A space compaction scheme is thus required for all POs, including the observation points. Similarly, a control point will affect the controllability of the nodes influenced by this control point. But the observability of the nodes in the shaded area will also be affected.

14.5.5.2 BIST for Datapth. As reported in Chapter 11, in register-intensive design such as datapath, some of the registers feeding into a combinational block (CB) are configured as a PRPG, while others connected to the output of the same CB are configured as a MISR. No register serves both functions for the same CB. Sometimes, however, the same register is connected to the inputs and outputs of the same CB, as illustrated in Fig. 14.17. Registers with this

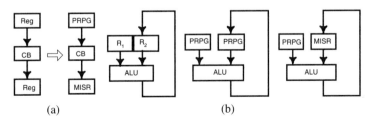

Figure 14.17 BIST allocation: (*a*) the configuration; (*b*) problems with self-adjacency [Avra 1991 © IEEE. Reprinted by permission].

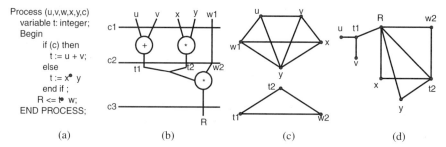

Figure 14.18   BIST allocation: (*a*) behavioral code; (*b*) constraint graph for life-sharing variable; (*c*) constraint graph for self-adjacency [Avra 1991 © IEEE. Reprinted by permission].

property are called *self-adjacent*. In this case, the register can be configured as PRPG and MISR concurrently for the same CP. The implementation of such an arrangement is costly in terms of area and delay penalties. Hence it is advantageous to minimize self-adjacent registers. This is done during scheduling and allocation. A problem is represented with a conflict graph [Avra 1991]. A vertex in this graph represents a register that stores a variable. An edge between two vertices indicates that the variables in the corresponding registers cannot be assigned the same register. That is, the two variables have an overlapping lifetime. An example of a scheduled data flow diagram and its register conflict graphs at each time steps is shown in Fig. 14.18. The conflict graphs in Fig. 14.18*c* are part of the synthesis process independent of testability issues. Notice that w is represented by two vertices, w_1 and w_2, at the consecutive time steps, c_1 and c_2. In Fig. 14.18*d*, the testability conflict graph is constructed such that an edge is added between vertices when one vertex represents an input to an operation and the other vertex stands for the output of this operation. Therefore, there is an edge between t_2 and x and y and similarly, an edge between R and x, y, t_1, t_2, and w_2. These are added constraints on register allocation and scheduling which guarantee that the input and output of a CB are not assigned to the same register. That is, no self-adjacent register will be synthesized.

Another BIST approach strives to test all CBs concurrently to minimize testing time and improve fault coverage, CBIST [Harris 1994]. Any two components in a circuit may not be testable concurrently if their testing paths share hardware. Figure 14.19 illustrates a dataflow graph after allocation and binding. The vertex inc1 can be bound to IN1 or IN2. In the first case, shown in Fig. 14.19*b*, both A1 and S1 will be observed through IN1. This causes a conflict. To avoid the conflict, inc1 is bound to IN2 as illustrated in Fig. 14.19*c*. During scheduling, the goal is to put clock boundaries between dataflow vertices with limited test path options. This will therefore increase their conflict probability.

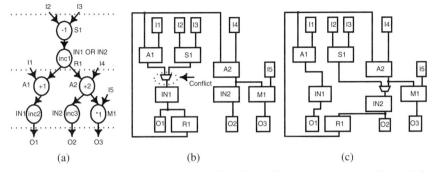

Figure 14.19 Concurrent BIST: (*a*) CDFG; (*b*) conflict, sharing paths; (*c*) resolving the conflict [Harris 1994 © IEEE. Reprinted by permission].

REFERENCES

Aitken, R. C. (1995), An overview of test synthesis tools, *IEEE Des. Test Comput.*, Vol. 12, No. 2, pp. 8–15.

Avra, L. (1991), Allocation and assignment in high-level synthesis for self-testable data paths, *Proc IEEE International Test Conference*, pp. 463–472.

Avra, L. (1993), Synthesizing for scan dependence in built-in self-testable designs, Ph.D. dissertation, Stanford University, Stanford, CA, July. Also *Proc. IEEE International Test Conference*, pp. 734–743.

Barbacci, M. (1973), Automatic exploration of the design space for register transfer (RT) system, Ph.D. dissertation, Department of Computer Science, Carnegie Mellon University, Pittsburgh, PA.

Compasano, R. (1988), Structural synthesis in the Yorktown silicon compiler, *Proc. VLSI 87: VLSI Design of Digital Systems*, C. Sequin (ed.), North-Holland, Vancouver, pp. 61–72.

Compasano, R. (1990), From behavior to structure: high-level synthesis, *IEEE Des. Test Comput.*, Vol. 7, NO. 1, pp. 8–19.

De, K., and P. Banerjee (1993), Logic partitioning and re-synthesis for testability, *Proc. IEEE International Test Conference*, pp. 906–915.

Dey, S., A. Raghunathan, and K. D. Wagner (1998), Design for testability techniques at the behavioral and register transfer levels, *J. Electron. Test. Theory Appl.*, Vol. 13, No. 2, pp. 79–91.

D&T (1995), *Des. Test Comput.*, Vol. 12, No. 2.

Gajski, D. D., and R. Kubar (1983), Introduction: new VLSI tools, *IEEE Comput.*, Vol. 6, No. 12, pp. 11–14.

Gajski, D. D., and L. Ramachandran (1994), Introduction to high-level synthesis, *IEEE Des. Test Comput.*, Vol. 11, No. 4, pp. 44–54.

Garey, M. R., and D. S. Johnson (1979), *Computers and Intractability: A Guide to the Theory of NP-Completeness*, W. H. Freeman, New York.

Gu, X., K. Kuchinski, and Z. Peng (1995), An efficient and economic partitioning approach for testability, *Proc. IEEE International Test Conference*, pp. 403–412.

Harmanani, H., and C. Papachristou (1993), An improved method for RTL synthesis with testability tradeoffs, *Proc. International Conferenceon Computer-Aided Design*, pp. 30–35.

Kuchinski, K., X. Gu, and Z. Peng (1994), Testability analysis and improvement from VHDL behavioral specifications, *Proc. EURO-DAC*, pp. xxx.

Harris, I. G., and A. Orailoglu (1994), SYNCBIST: synthesis for concurrent built-in self-testability, *Proc. Interational Conference on Computer Design*, pp. 101–104.

Hsu, F., E. M. Rudnick, and J. H. Patel (1996), Enhancing high-level control-flow for improved testability, *Proc. IEEE International Conference on Computer-Aided Design*, pp. 322–328.

ITC (1994), *Proc. IEEE International Test Conference*.

JETTA (1998), *J. Electron. Test. Theory Appl.*, Vol. 13, No. 2.

Kurdahi, F. J., and A. C. Parker (1987), REAK: a program for register allocation, *Proc. Design Automation Conference*, pp. 210–215.

Langnese, E. D., and D. E. Thomas (1989), Architectural partitioning for system level design, *Proc. 25th IEEE/ACM Design Automation Conference*, pp. 62–67.

Lee, T.-C., et al. (1992a) Behavioral synthesis of easy testability in data paths allocation, *Proc. International Conference on Computer Design*, pp. 612–615.

Lee, T-C., et al. (1992b), Behavioral synthesis of easy testability in data paths scheduling, *Proc. International Conference on Computer Design*, pp. 616–619.

Lee, T. C. (1997), *High-Level Test Synthesis of Digital VLSI Circuits*, Artech House, Norwood, MA.

Lin, C.-J., Y. Zorian, and S. Bhawmik (1993), PSBIST: a partial-scan based BIST scheme, *Proc. IEEE International Test Conference*, pp. 507–516.

Macii, E., and A. R. Meo (1994), A test generation program for sequential circuits, *J. Electron. Test.*, Vol. 5, No. 1, pp. 115–120.

McFarland, M. C. (1983), Computer-aided partitioning of behavioral hardware descriptions, *Proc. 20th IEEE/ACM Design Automation Conference*, pp. 472–478.

McFarland, M. C., A. C. Parker, and R. Camposano (1988), Tutorial on high-level synthesis, *Proc. 24th ACM/IEEE Design Automation Conference*, pp. 330–336.

Michel, P., U. Lauther, and P. Duzy (eds.) (1992), *The Synthesis Approach of Digital System Design*, Kluwer Academic, Norwell, MA.

Miczo, A. (1983), The sequential ATPG: a theoretical limit, *Proc. IEEE International Test Conference*, pp. 143–147.

Papachristou, C. A., M. Baklashov, and K. Lai (1998), High-level test synthesis of behavioral and structural designs, *J. Electron. Test. Theory Appl.*, Vol. 13, No. 2, pp. 167–188.

Potkonjak, M., S. Dey, and R. K. Roy (1995a), Behavioral synthesis of area-efficient testable designs using interaciton between hardware sharing and partial scan, *IEEE Trans. Comput.-Aided Des.*, Vol. 14, No. 9, pp. 1141–1154.

Potkonjak, M., S. Dey, and R. K. Roy (1995b), Considering testability at the behavioral level: use of transformations for partial scan cost minimization under timing and area constraints, *IEEE Trans. Comput.-Aided Des.*, Vol. 14, No. 5, pp. 531–546.

Rao, N., and S. Toida (1991), Computational complexity of test point insertion and decomposition, *Proc. International Conference on VLSI*, pp. 223–231.

Savaria, Y., et al. (1991), Automatic test point insertion for PR testing, *Proc. International Symposium on Circuits and Systems*, pp. 1960–1963.

Varma, K. K., P. Vishakantaiah, and J. A. Abraham (1993), Generation of testable designs from behavioral descriptions using high level synthesis tools, *Proc. VLSI Test Symposium*, pp. 124–130.

VLSI (1993), Special issue on high-level synthesis, *IEEE Trans. VLSI Syst.*, Vol. 1, No. 3, pp. 230–341.

Wagner, K. D., and S. Dey (1996), High-level synthesis for testability: a survey and perspective, *Proc. Design Automation Conference*, Las Vegas, NV, pp. 131–136.

Walker, R. A., and D. E. Thomas (1985), A model of design representation and synthesis, *Proc. 22nd Design Automation Conference*, pp. 453–459.

Walker, R. A., and S. Chaudhuri (1995), Introduction to the scheduling problem, *IEEE Des. Test Comput.*, Vol. 12, No. 2, pp. 60–69.

PROBLEMS

14.1. Develop **(a)** the data flow graph and **(b)** an RT architecture for the following behavioral description:

$$d = a * b$$
$$x = d + c$$

14.2. For the behavioral description; develop the data flow graph, then a schedule for the operations.

$$f = a + b$$
$$g = f + c$$
$$h = d + e$$
$$z = g + h$$

14.3. From the schedule of Problem 14.2, deduce the lifetime table of the variables and allocate registers and adders.

14.4. For the schedule of Problem 14.2 construct the constaint graph for the registers.

14.5. How can you assign the registers of Problem 14.2 such that they become easily testable?

15

TESTING SOCs

15.1 INTRODUCTION

The shift toward very deep submicron technology has encouraged IC designers to increase the complexity of their designs to the extent that an entire system is now implemented on a chip. This new design paradigm of system on a chip (SOC) has changed the approach to design and testing. To increase the design productivity and decrease time-to-market, reuse of previously designed modules is becoming common practice in SOC design. However, the reuse approach is not limited to in-house designs, but is extended to modules that have been designed by others as well. Such modules are referred to as *embedded cores*. This approach to design has encouraged the founding of several companies that specialize in providing embedded cores to service multiple customers. It is predicted that in the near future, cores, of which 40 to 60% will be from external sources [Zorian 1997], will populate 90% of a chip. Except for a very few, individual companies do not have the wide range of expertise that can match the spectrum of design types in demand today.

Core-based design that is justified by the need to decrease time-to-market has created a host of challenging problems for the design and testing community. First, there are legal issues regarding the intellectual property (IP) for the core provider and the user. Second, there are problems in integrating and verifying a mix of proprietary and external cores that is more involved than simply integrating ICs on a printed circuit board (PCB). The third issue, which is the basic concern of this book, is the need for an efficient testing strategy.

Before embarking on discussing the possible testing strategies and their corresponding design approaches, we first examine the various types of cores and core-based design approaches. The testing strategies presented here are taken from academic and industrial sources.

15.2 CLASSIFICATION OF CORES

A typical SOC configuration is shown in Fig. 15.1. It consists of several *cores* that are also referred to as *modules*, *blocks*, or *macros*. Often, these terms are used interchangeably. These cores may be DSP, RAM modules, or controllers. This same image of an SOC may be perceived as a PCB with the cores being the ICs mounted on it. It also resembles standard cells laid on the floor of an IC. In the latter cases, the blocks are of identical types. That is, they are all ICs in the PCB case or all standard cells in the IC case. For an SOC, they may be a mix of types, as we describe next.

Reusable cores are classified in three categories: hard, firm, and soft [Gupta 1997]. *Hard cores* are optimized for area and performance and they are mapped into a specific technology and possibly a specific foundry. They are provided as layout files that cannot be modified by the users. *Soft cores*, on the other extreme, may be available as HDL technology-independent files. From a design point of view, the layout of a soft core is flexible, although some guidelines may be necessary for good performance. This flexibility allows optimization to the desired levels of performance or area. *Firm cores* are usually provided as technology-dependent netlists using library cells whose size, aspect ratio, and pin location can be changed to meet the customer needs. Table 15.1 summarizes the attributes of reusable cores. From this table there clearly is a trade-off between design flexibility on one hand and predictability and hence time to market performance complexity on the other. Soft cores are easily embedded in a design. The ASIC designers have complete control over the implementation of this core, but it is the designer's job to optimize it for performance, area, test, or power.

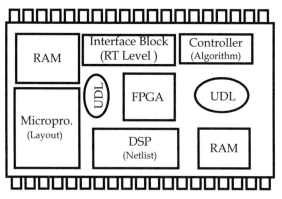

Figure 15.1 System on a chip.

TABLE 15.1 Categorizing Reusable Blocks

Type	Flexibility	Design Flow	Representation	Libraries	Process Technology	Portability
Soft	Very flexible, unpredictable	System design RTL design	Behavioral RTL	Not applicable	Independent	Unlimited
Firm	Flexible	Floor planning Placement	RTL, blocks, netlist	Reference Footprint, timing model	Generic	Library mapping
Hard	Inflexible, predictable	Routing verification	Polygon data	Process specific library and design rules	Fixed	Process mapping

Source: Data from [Hunt 1996 © IEEE. Reprinted with permission].

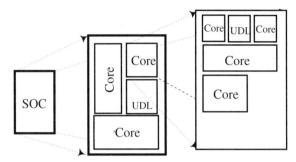

Figure 15.2   Hierarchical organization of cores.

Hard cores are very appropriate for time critical application, whereas soft cores are candidates for frequent customization. The relationship between flexibility and predictability is illustrated in Fig. 15.2. The cores can also be classified from a testing perspective. For example, there is typically no way to test a hard core unless the supplier provides a test set for this core whereas test set for the soft core need to be created if not provided by the core provider. This makes hard cores more demanding when developing a test strategy for the chip. For example, it would be difficult to transport through hard cores a test for an adjacent block that may be another core or a UDL component. In some special cases, the problem may be alleviated if the core includes a well described testability functions. We discuss this feature in Section 15.6.3.

15.3   DESIGN AND TEST FLOW

As we have advocated in the beginning of the book, an integrated design and test process is highly recommended. This approach cannot be more appropriate than it is for core-based systems. Conceptually, the system on a chip paradigm is analogous to integration of several ICs on a printed circuit board. However, there is a fundamental difference. Whereas in a PCB the different ICs have been designed, verified, fabricated, and tested independently from the board, fabrication and testing of an SOC are done only after integration of the different cores. This fact implies that even if the cores are accompanied by a test set, incorporation of the test sets is not that simple and must be considered while integrating the system. In other words, reuse of design does not translate to easy reuse of the test set. What makes this task even more difficult is that the system may include different cores that have different test strategies. Also, the core may cover a wide range of functions as well as a diverse range of technologies, and they may be described using different HDL languages, such as Verilog, VHDL, and Hardware C to GDSII.

The basic design flow, shown in Fig. 3.3, still applies to SOC design in the sense that the entire system needs to be entered, debugged, modified for testability, validated, and mapped to a technology; but all of this has to be done in an *integrated framework*. Before starting the design process, an overall strategy needs to be chartered to facilitate the integration. In this respect, the specification phase is enlarged and a test strategy is included. This move toward more design on the system level and less time on the logic level has been emphasized in the book from the onset and illustrated in Fig. 1.1.

The design must first be partitioned. Then decisions must be made on such questions as:

- Which partition can be instantiated by an existing core?

- Should a core be supplied by a vendor or done in-house?

- What type of core should be used?

- What is the integration process to facilitate verification and testing?

Because of the wide spectrum of core choices and the diversity of design approaches, SOC design requires a *metamethodology*. That is, a methodology that can streamline the demands of all other methodologies used to design and test the reusable blocks as well as their integration with user-defined logic. To optimize on the core-based design, an industry group deemed it necessary to establish a common set of specifications. This group, known as the Virtual Socket Interface Alliance (VSIA), was announced formally in September 1996. Its intent is to establish standards that facilitate communication between core creators and users, the SOC designers [IEEE 1999a].

An example of using multiple cores is the IBM-designed PowerPC product line, based on the PowerPC 40X chip series [Rincon 1997]. The PowerPC microcontroller consisted of a hard core and several soft cores. For timing critical components such as the CPU, a hard core was selected, while soft cores were used for peripheral functions such as the DMA controller, external bus interface unit (EBIU), timers, and serial port unit (SPU). The EBIU may be substituted by, say, a hard core from Rambus.

A change in simulation and synthesis processes is required for embedded cores due primarily to the need to protect the intellectual property of the core provider. Firm cores may be encrypted in such a manner as to respond to the simulator without being readable by humans. For synthesis, the core is instantiated in the design. In the case of a soft core, sometimes the parameters are scaled to meet the design constraints. To preserve the core performance, the vendor may include an environment option to prevent the synthesis program from changing some parts of the design. This will protect the core during optimization. However, the designer may remove such an option and make some changes in the design. A hard or a firm core is treated as a black box from the library and goes through the synthesis process untouched.

15.4 CORE TEST REQUIREMENTS

There are many issues that arise in attempting to test a core-based SOC: The system is too large, it has a heterogeneous complection since it integrates a wide range of core types, and the cores themselves may be core-based, as illustrated in Fig. 15.3. In traditional ASIC, the design consists of several components usually selected from a cell library. Based on fault models, test patterns are generated for the complete design. This may be done manually or automatically, but it is usually applied to the primary inputs of the design and its responses are observed at the primary outputs. For core-based circuits, it is possible to merge the soft cores with the UDL components and a test can then be generated by the core user for the merged design. Such an approach is not feasible with nonmergeable cores, such as hard cores. Cores might also come with different embedded test schemes that the core provider includes in the design, such as a BIST controller or scan path for internal test. Assume for a moment that internal test sets are available for each core. The next issue is to devise mechanisms and a strategy for test execution in the individual cores; that is, it is necessary to have test access to justify the test patterns at the inputs terminal of the core. In addition, it is important to propagate the test response from the output terminals as depicted in Fig. 15.4. In this paradigm, the patterns are supplied from the source of test data, and the response verification is considered as the sink of the test data. Another important requirement is testing the interconnections and the "glue" logic among the cores on the chip. In summary, the main test requirements for SOC testing are:

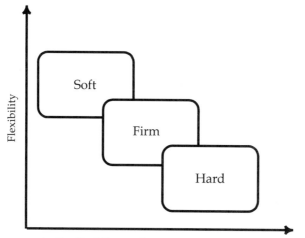

Figure 15.3 Trade-offs among types of cores [Hunt 1996 © IEEE. Reprinted with permission].

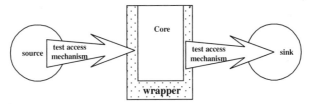

Figure 15.4 Test architecture.

- Internal test for embedded cores
- Access to the embedded core
- Isolation of one or more cores that will be tested simultaneously
- Facility to test the interconnections among the cores
- Facility to test all UDLs added logic to "glue" the cores

At the SOC level, to orchestrate application of the test sets to any core and the observation of the response, it is important to plan a strategy at the onset of the design. It is important to decide about the need for a *test controller* or *a test bus*. It would be too late to realize such needs after the SOC design is completed.

Recognizing the above requirements and the need for interoperability, a number of initiatives have begun to address the integration of testing with core-based chip design. Several organizations, including an IEEE Test Technology Technical Council (TTTC) [*www.computer.org/tab/tttc*], meet regularly to develop a standard test access for cores from various providers [Zorian 1997].

15.5 CONCEPTIONAL TEST ARCHITECTURE

One of the major challenges in the system chip realization process is the integration and coordination of the on-chip test and diagnosis capabilities. Compared to the conventional chips, the system chip test requirements are far more complex than the PCB assembly test, which, for instance, consists of interconnect and pin toggling tests. The system chip test is a single composite test. This test is comprised of the individual internal tests for each core, the UDL test, and the test of their interconnects. As discussed earlier, each individual core or UDL test may involve surrounding components. Certain peripheral constraints (e.g., safe mode, low power mode, bypass mode) are often required. This necessitates access and isolation modes. In addition to the test integration and interdependence issues, the system chip composite test requires adequate test scheduling. This is needed to meet a number of chip-level requirements, such as total test time, power dissipation, area overhead, and so on [Zorian 1993]. Also, test scheduling is necessary to run intracore and intercore tests in a certain order so as not to impact the initialization and final contents of individual cores. With the foregoing scheduling constraints, the schedule of the composite system chip test is created.

In addition to the foregoing differences between testing traditional chips and system chips, we have to note that system chips also have the typical testing challenges of the very deep submicron chips, such as defect/fault coverage, overall test cost, and time-to-market.

The generic test architecture for cores in SOCs consists of three structural elements, as depicted in Fig. 15.4.

1. *Test pattern source and sink.* The test pattern source generates the test stimuli for the embedded core, and the test pattern sink compares the resource(s) to the expected response(s).

2. *Test access mechanism.* The test access mechanism transports test patterns. It can be used for on-chip transport of test stimuli from a test pattern source to the core under test and for transport of test responses from the core under test to a test pattern sink.

3. *Core test wrapper.* The core test wrapper forms the interface between the embedded core and the environment. It connects the terminals of the embedded core to the rest of the IC and to the test access mechanism.

All three elements can be implemented in various ways, such that an entire palette emerges of possible approaches for testing embedded cores. In subsequent sections we review the various alternatives and classify current approaches.

15.5.1   Source and Sink of Test Data

The test pattern source generates the test stimuli for the embedded core. The test pattern sink compares the response(s) to the response(s) expected. Test pattern source as well as sink can be implemented either off-chip by external automatic test equipment (ATE), on-chip by built-in self-test (or embedded ATE), or as a combination of both, as we mentioned first in Chapter 1. Source and sink do not need to be of the same type; for example, the source of an embedded core can be implemented off-chip, while the sink of the same core is implemented on-chip. The choice for a certain type of source or sink is determined by (1) the type of circuitry in the core, (2) the type of predefined tests that come with the core, and (3) quality and cost considerations [Zorian 1998].

We distinguish three main types of circuitry used in system chips today: (1) logic, (2) memory, and (3) analog and mixed-signal. Simple cores consist of one circuitry type only; complex cores consist of multiple simple cores, possibly of different circuitry type. These three types of circuitry exhibit different defect behavior, and hence their tests are quite different in nature [Agrawal 1994]. This also requires different types of sources to generate the stimuli and sinks to compare the responses. ATE systems as well as BIST schemes for logic, memory, and analog are traditionally quite different. This diversity of circuit types may require a mix of off- and on-chip test applications.

The variety in types of core tests is much larger than the three circuitry types listed above. In addition, tests cannot be classified by the type of circuit they test, but also by the type of measurement they require (voltage vs. current), by the way they are generated (functional vs. structural), by the amount of core-internal adaptation they require (scan vs. test points), and so on. All these classification bases were discussed in one or more of the earlier chapters. For on-chip testing, the test patterns for digital logic core are typically generated by an LFSR dedicated to a core or shared with other cores.

15.5.2 Test Access Mechanism

The test access mechanism (TAM) takes care of on-chip test pattern transport. It can be used (1) to transport test stimuli from the test pattern source to the core under test and (2) to transport test responses from the core under test to the test pattern sink. The test access mechanism is by definition implemented on-chip. Although for one core the same type of test access mechanism is often used for both stimuli as well as response transportation, this is not required, and various combinations may coexist. The selection of the mechanism requires a trade-off between the bandwidth of test data transport, possible extra hardware, and test application time [Zorian 1998].

When implementing a core test access mechanism, we have the following options:

- A test access mechanism can either reuse existing functionality to transport test patterns or be formed by dedicated test access hardware.
- A test access mechanism can either go through other modules on the IC, including other cores, or pass around those other modules.
- One can either have an independent access mechanism per core or share an access mechanism with multiple cores.
- A test access mechanism can either be a plain signal transport medium or may contain intelligent test control functions.

The most straightforward TAM is the traditional direct access test bus shown in Fig. 15.5a. To access individual cores, which are competing on the IC pins, a multiplexing system needs to be added to connect the TAM to the IC pins [Immaneni 1990, Monzel 1997]. Another well-known mechanism is Boundary-Scan [Whetsel 1997]. A variation of this serial technique is shown in Fig. 15.5b [Touba 1997]. The TestRail access mechanism combines the strengths of the test parallel TAM and serial Boundary-Scan test. It is shown in Fig. 15.5c in the context of a core wrapper, TestShell [Marinissen 1998]. Use of these mechanisms is illustrated in Section 15.6 after wrappers have been described.

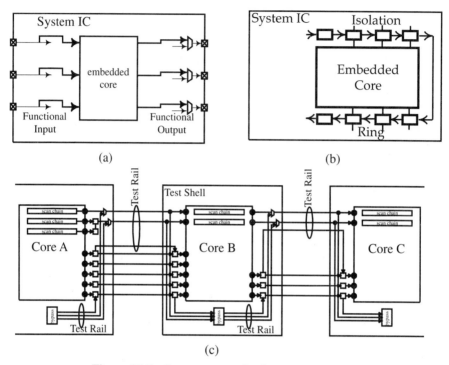

Figure 15.5   Test access mechanisms and wrappers.

15.5.3   Core Test Wrapper

The core wrapper is the interface between the embedded core and the rest of the chip. This wrapper should allow the following mandatory operational modes:

1. *Normal modes*, in which the wrapper is transparent to various interactions of the core with the other circuitry on the IC
2. *Internal core test mode*, in which the TAM is connected to the core for the transport of testing data
3. *External test mode*, in which the TAM supplies the test data to test UDLs and interconnections between cores on the SOC

Other optional modes may be used, such as a bypass mode. It is possible to combine several modes. The core wrapper connects the core I/Os to the test access mechanism. Whereas the I/Os of the core are determined by its functionality, the width of the access mechanism is a function of the bandwidth of the test sink and source and the amount of chip area available. There are probably more I/Os than the access mechanism can accommodate and some space compaction will be required in test application mode. The latter case would require a serial-to-parallel conversion of the core input signals and parallel-to-serial conversion of the output signal. This conversion function maybe implemented in the core test wrapper or the TAM.

Examples of core test wrappers are the *test collar* [Varma 1997] and the *TestShell* [Marinissen 1998]. The first example is illustrated in Fig. 15.5*b*, where each I/O of the core is attached to a flip-flop in a fashion similar to Boundary-Scan Cells (BSCs). In this example the TestShell is shown in Fig. 15.5*c*, where TestRail connects the I/Os of the various cores. The TestRail is the TAM and its width is three. The length of the TestShell varies according to the number of elements in each scan chain, but the TestRail is of constant width. For example, core A has three scan chains; one chain is connected directly to the TestRail, but the other two had to be multiplexes to the third wire in the TestRail.

15.6   TESTING STRATEGIES

In the rest of the chapter we describe several approaches that address one or more aspects of SOC's testability. The examples are intended to demonstrate use of the SOC test concepts and DFT techniques presented throughout the book. In this respect they serve the same purpose that the examples of micro-processors did in Chapter 13. The examples mostly use DFT techniques that have been proven successful in traditional IC design. They utilize a mix of the test architecture components described in Section 15.5. We categorize them by one of the access mechanisms discussed earlier and presented in Fig. 15.5.

15.6.1   Direct Access Test Scheme

The direct access that scheme (DATS) is symbolized by Fig. 15.5*a*. This is the approach taken to test ASIC design in which the Intel 80C186 and 80C51 are embedded as cores [Immaneni 1990]. Its intent is to make the core's inputs, out-puts, and the bidirectional ports accessible outside the chip by mapping them onto the chip's pins. In this fashion, any embedded core can be isolated, simu-lated, and tested independently of the rest of the chip. Test vectors can then be generated to check the interconnections between the various components of the chip. The scheme requires the modification of the I/O ports that are not primary I/Os of the chip to be modified as illustrated in Fig. 15.6. TMODE and TSEL are two added control pins to the chip. The circuit functions in normal mode whenever TMODE = 0. The mdoule under test, be it a core or a UDL, is in test mode when its TMODE = 1 and TSEL = 1. Meanwhile, all the other modules are not active in testing and their TSEL = 0. A change in a user-defined logic is as illustrated in Fig. 15.7, where only the embedded I/Os are modified while those that are connected directly to the pins of the chip are "untouched." The core also needs to follow the standard and, in addition, to be restructured to minimize the I/O modifications without jeopardizing its testing. We skip the modification made for these cores for the sake of showing how the scheme can be applied to the ASIC. In Fig. 15.8, two cores, block 1 and block 3, and a UDL,

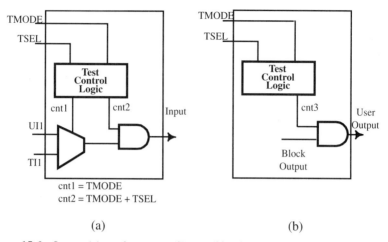

Figure 15.6 Input (*a*) and output (*b*) modifications [Immaneni 1990 © IEEE. Reprinted with permission].

block 2, are shown. TMODE at PIN3 is distributed to all components, while the multiplexing on the I/O pins is made only whenever necessary. The advantage of this approach is in its simplicity and in testing the core as if it is the only circuit on the IC. The drawbacks are the competition with other cores on the limited pins of the ICA. A large multiplexing circuitry is thus needed to accommodate all cores.

15.6.2   Use of Boundary-Scan

In Section 15.3 we likened the SOC to a PCB. As such, it is natural to apply the PCB testing approach, Boundary-Scan, to test the SOC. In Chapter 10 we presented the major features of boundary-scan, IEEE Standard 1149.1. It has the advantage of facilitating the testing of any chip from the edge of the board. It

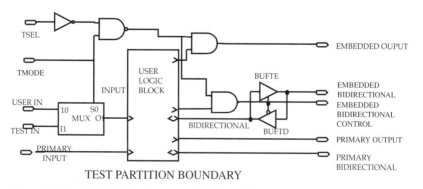

Figure 15.7   DATS UDL [Immaneni 1990 © IEEE. Reprinted with permission].

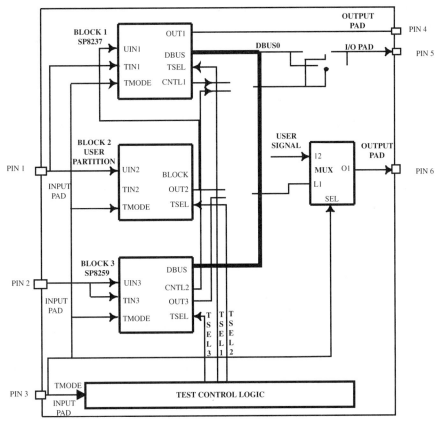

Figure 15.8   DATS implementation example [Immaneni 1990 © IEEE. Reprinted with permission].

has the flexibility to bypass a chip or integrate a DFT structure such as internal scan and BIST. Also, the availability of private instructions permits the customization of individual cores. A test access port (TAP) and its controller are used to orchestrate the testing scenario. By extending this testing technique to the SOC, one would extrapolate that each core on the chip will have a TAP and its controller. Cores originally designed with Boundary-Scan, include a TAP. When embedding them in a system, it is not advisable to remove the TAP since it might be used to regulate other DFT features. An extra TAP can be used to control cores that do not adopt IEEE Standard 1149.1.

An approach based on Boundary-Scan for testing SOC is illustrated in Fig. 15.9, and includes a mix of cores with a TAP controller (Taped) and non-Taped cores (NTC). TAP1 controls all NTC cores. The test data input (TDI) and test data output (TDO) are connected through all TAPs. This proposed arrangement does not comply with Standard 1149.1, which assumes one TAP per chip. It is important to make believe that all these TAPs are equivalent to one TAP.

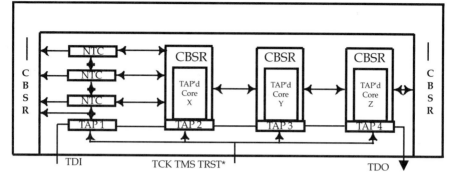

Figure 15.9 Example of IC containing four TAPs [Whetsel 1997 © IEEE. Reprinted with permission].

This is achieved by adding extra circuitry: the *TAP multiplexing link* (TML) and signals, as illustrated in Fig. 15.10.

The control signals, SELi and ENAi, are to facilitate communication between the TML and any TAPi. All TDO pins are connected to chip TDO through a multiplexing system. The TML enables and connects one or more TAPs to the accessed core via the IC's test pins. This arrangement allows testing of the core whose TAP is enabled. Similarly, the ENA signals can enable or disable the TAPs. These signals could also be included in the design of a TAP controller to force the TAP to go to or remain in the Run–Test–Idel state when disabled.

As depicted in Fig. 15.2 and described earlier in the chapter, any core could

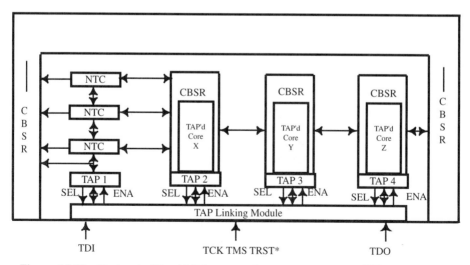

Figure 15.10 Design in Fig. 15.9 incorporating test access ports [Whetsel 1997 © IEEE. Reprinted with permission].

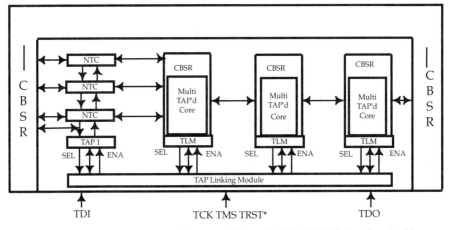

Figure 15.11 Reusable TLM architecture [Whetsel 1997 © IEEE. Reprinted with permission].

itself be composed of other cores. Cores including TML controllers are embedded in a chip at their turns, as illustrated in Fig. 15.11. This is the equivalent of Fig. 5.10 with each TAP being replaced by a TML, except for the TAP used for NTC cores. Thus the proposed testing approach can be done on a hierarchy of cores.

The main advantage of this approach is that it is based on a mature, already proven technology. The scheme has merit when short test sets are used, as in the case of testing the interconnects. However, due to its serial nature it has the disadvantage of a long test application time for long test sets.

A variation of the Boundary-Scan testing approach was recommended by [Touba 1997]. A flip-flop is placed at each I/O pin. All such flip-flops form an *isolation ring* around the core as a collar, as shown in Fig. 15.12a. This ring serves as a wrapper and provides full controllability and observability of the core as well as observability for the UDL driving it. For example, if a UDL has full scan, the test is applied through the scan chain and observed on isolation rings around the core, as illustrated in Fig. 15.12b. Also, the test to the core is supplied and the response is observed through the isolation ring. Instead of a full ring, it is possible to use only a partial ring and to modify the output space of the UDLs in such a way as to allow justification of the test vectors on the neighboring core [Pouya 1997]. Some of the inputs include flip-flops (IR) and some do not (NIR). The IR set is included in the ring and the others are not. To observe the outputs of the UDL, each output driving an NIR pin is XORed with an output driving an IR pin, and the resulting signal is fed to the partial ring. This amounts to the space compaction discussed in Chapter 12. As the test set is applied to the core through a UDL module, there is no guarantee that the patterns arriving at the core will detect all faults. Thus selection of input, IR, to be included in the ring has to be done with care to guarantee high

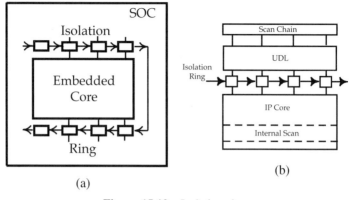

Figure 15.12   Isolation ring.

fault coverage. The selection scheme of the IRs depends on the itnerfacing modules.

15.6.3   Use of Scan Path

The intent of the scan-path approach is to make cores other than the one under
test transparent and to use them to transfer test data in and out of the chip.
This approach is based on using the scan chains of the cores as a TAM. It
assumes that all cores include scan path and will make a minimal change to
the I/O of the tested modules, cores, or UDLs. Such an assumption is realistic
since scan-path design is now practiced widely. Typically in the cases of soft
and firm cores, system designers include the scan path, whereas in hard cores
a scan path is already included. The scan-path chains of all the cores are then
used to justify (1) test vectors to an internal component and (2) test responses
to the chip's outputs.

The scan chains of the various modules of the system are used to regulate the
test data traffic in and out of the chip. For this each module can be made transparent by representing it with its scan chain only as shown in Fig. 15.13. We call
this reprsentation the *transparent model* of the module. A model is transparent
"in the sense that test data can be propagated through them without information
loss" [Murray 1996]. Also, the system as a whole may include feedback, and this
makes testing more difficult. Thus it is important to look at the circuit as a whole
and assure that no feedback is present in testing mode. After determining the scan
path for all cores, the S-graph for all UDL logic is formed and is merged with those
of the cores. The combined S-graph can then be searched for feedback loops and
the smallest number of them is open to make the entire system acyclic. This will
result in different scan chains through the chip that need to be connected in a *judi-
cial* way to allow the application of test patterns and observation of the responses
from any component on the chip.

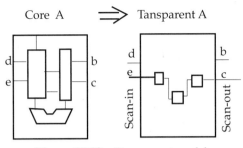

Figure 15.13 Transparent model.

The scan chains of the individual cores can be connected in one of the two principal schemes. In the first scheme, all scan-in (and scan-out) of the individual cores is connected to an IC pin through a multiplexer network. In the other scheme, all scan chains are connected in a daisychain [Aerts 1998]. Both schemes are illustrated in Fig. 15.14. The first scheme is actually equivalent to a direct access mechanism, and the cores will be tested one at a time. The test application time in the second scheme is as long as the total number of scan chains. To reduce this time, a bypass flip-flop is placed across each chain. The output of the scan chain and of the bypass flip-flop are multiplexed to control whether the core is working in bypass or scan mode. This arrangement gives various alternatives to testing the cores. For example, it is possible to test only one core and bypass the others, or vice versa, to bypass only one and test the others. Another alternative is to test all cores concurrently and as soon as one core runs out of patterns, put it in bypass mode [Aerts 1998].

Instead of using traditional scan testing as presented above, we can reduce the dependency on the external ATE bandwidth by using an on-chip test pattern generation as test data source and/or a space compaction of the cores' outputs signals in the data sink. The approach is illustrated with the example shown in Fig. 15.15. We will examine the testability of components B and U. We replace all the other modules by their transparent images, the scan chains. First, we describe the overall approach to testing the UDL and the core it drives, then we elaborate on details of performing the testing and the required extra hardware [Mourad 1999].

- *Testing U.* A test applied on U can be observable through the scan chain of B and then through C to the primary outputs of the chip.
- *Testing B.* The test for B is applied through U, then its response is shifted through C.

There are two main issues to consider next: observability of the test response at the outputs of U that are not connected to scan-in input of core B, and the justification of the test to core B at all of its inputs. Solving these problems is, however, dependent on the type of testing approach of these modules. We con-

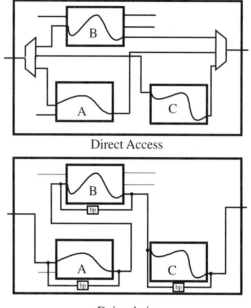

Direct Access

Daisychain

Figure 15.14   Scan path as test access mechanism.

sider two alternatives with different merits (advantages and disadvantages that we need to balance). The first approach uses space compaction of the outputs; the second is based on san-BIST.

15.6.3.1   Space Compaction.   Space compaction schemes have been discussed in Chapter 12 in conjunction with BIST testing. Often BIST involves more outputs than can be accommodated in the signature analyzer. For this, parity checking of some, if not all, of the output signals is used to replace the evaluation of each output. Such a compaction for UDL U is shown in Fig. 15.15a. The parity of the four outputs is connected to the scan-in input of core B through the 2-to-1 MUX. Sel = 0 indicates normal operation, and during testing, Sel = 1. If necessary, we can replace the outputs with parities of fewer outputs at a time, enlarging the MUX accordingly.

Next we describe how to supply the test patterns at the inputs of core B, which as no inputs reachable at the pins of the chip. A K-to-M weighted decoder is connected through multiplexers to the M inputs of core B as shown in Fig. 15.15b. Any input combination at the decoder control lines will result in a test pattern for B. Therefore, the signals to the control lines of the weighted decoder are, in effect, the patterns to the core under test (CUT). They are supplied through the scan chain. Typically, the number of control lines is, $K = \lceil \log_2(M) \rceil \ll M$. Consequently, this scheme decreases the external band-

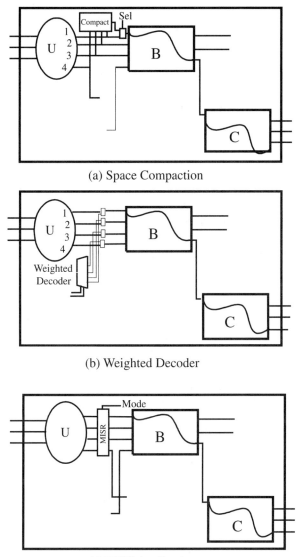

(a) Space Compaction

(b) Weighted Decoder

(c) BIST Application

Figure 15.15   Reducing the external bandwidth.

width by the ratio M/K. In addition, the overall area cost of this technique is much lower than that of a full collar wrapper [Greene 1999].

15.6.3.2   *Scan-BIST.* Let us assume this time that BIST is used to test the CUT and a multiple input signature register is used to compress the response to a PR test set. The output of the MISR can then be propagated to a primary output of the chip through the transparent images of the cores. Again, a MUX is used to select between normal operation and test, as illustrated in Fig. 15.15c. The same MISR can be reused to test other cores on the chip. A comprehensive BIST schedule and control method has been developed to deal with tradeoffs for BIST power dissipation, test time optimization, and BIST resource sharing [Zorian 1993]. This universal BIST scheduler is either customized to execute a predetermined BIST schedule for a manufacturing test or is programmed on the fly through the JTAG port to execute the test of individual cores during silicon debug and diagnosis.

15.7 TO EXPLORE FURTHER

In this chapter we have described a state-of-the-art SOC design and test. The entire area needs to be "explored further." However, we recommend that the reader follow up closely with the progress of the standards mentioned earlier: VSIA and P1500.

15.7.1 Virtual Socket Interface Alliance

The virtual socket interface (VSI) alliance consists mainly of representatives from the semiconductor industry companies [VSI 1998]. The organization's task is to identify and define interface standards for design reuse of virtual components. Seven development working groups (DWGs) have been formed to study and recommend standards for various aspects of the design and test integration, such as IP protection, implementation verification, on-chip bus, and manufacturing-related test. The scope of the latter DWG include [Zorian 1997]:

- Test access
- Test isolation
- Interconnect test
- Shadow logic test
- Test integration
- Test logic integration
- Failure identification

Among the expected deliverables by the DWGs are:

- Test data interchange specification:
- Guidelines for system chip integrators:
- Guidelines for care providers:
- Guidelines for mergeable IP providers:

Whenever possible, the VSI Alliance tries to endorse existing or emerging standards. In the scope of the manufacturing-related test DWG, these are IEEE P1500 [IEEE 1999*b*] and IEEE P1450, which is better known as STIL (standard test interface language) [Makson 1997, Wohl 1997, IEEE 1999*a*].

15.7.2 IEEE P1500 Standard

IEEE P1500 is a standard focusing on the critical aspects of ease of reuse and interpolarity with respect to testing of cores originating from distinct core providers. This is done by standardizing core test knowledge transfer and test access for embedded cores.

In September 1995 the Test Technology Technical Council (TTTC) of IEEE Computer Society initiated a Technical Activity Committee on embedded-core test to identify common needs in this domain. After several meetings in conjunction with major conferences, it concluded that there were indeed common needs and hence a design for standardization [Zorian 1997]. In June 1997 the IEEE Standard Board granted permission to start a standard activity; this was the official start of IEEE P1500. To improve efficiency of both core providers and core users, the IEEE P1500 standard facilitates the interpolarity with respect to testing when cores of various providers come together in one SOC, but it does not cover the core's internal test methods or DFT, or SOC test integration and optimizaiton. IEEE P1500 focuses on standardizing those areas that are at the interface between core provider and core user. Through a scalable standard, IEEE P1500 contributes to easy plug-and-play for testing, while maintaining the required flexibility to cope with different cores and system chips.

IEEE P1500 has active participation from leading experts in relevant industry segments, such as systems companies, electronic design automation (EDA) vendors, core providers, IC manufacturers, and ATE suppliers. Several task forces carry out the detailed technical work of P1500. The standard draft document is posted on the P1500 Web site [IEEE 1999b], where one can also find the minutes and presentation of working group meetings.

The two main elements of the IEEE P1500 standard are a language, called CORE Test Language (CTL), and a scalable core test architecture. The first is meant to standardize the core test knowledge transfer. CTL is based on another IEEE standard language, viz., IEEE Std., 1450.0, the Standard Test Interface Language (STIL) [IEEE 1999a], which is extended to accommodate specific core test constructs. IEEE P1500 only standardizes the test wrapper around the core and its interface to one or more TAMs. The test wrapper is comprised of the following basic components: a Wrapper, Boundary Register, a Wrapper Instruction Register, and a ByPass Register.

IEEE P1500 is currently in the development phase of the standard. The first version of the standard focuses on nonmerged digital logic and memory cores. It is planned to cover analog and mixed-signal cores, as well as the DFT guidelines for mergeable cores, in future expansions. Because the standardization work is not finalized at this point of time, we do not include its architectural details and the language specifics, but we recommend further exploration by either contacting the TTTC Web site (*http://:computer.org/tab/tttc*) or visiting the P1500 web site (*http//:grouper.ieee.org/groups/1500*).

REFERENCES

Aerts, J., and E. J. Marinissen (1998), Scan chain design for test time reduction in core-based ICs, *Proc. IEEE International Test Conference*, pp. 448–457.

Agrawal, V., et al. (1994), Built-in self-test for digital integrated circuits, *AT&T Tech. J.*, Vol. 73, No. 2, pp. 38.

Beenkar, F., B. Bennetts, and L. Thijseen (1995), Testability concepts for digital ICs: macro test approach, Vol. 3, in Kluwer Academic, Norwell, MA.

Gupta, R. K., and Y. Zorian (1997), Introduction to core-based system design, *IEEE Des. Test Comput.*, Vol. 14, No. 4, pp. 15–25.

Hunt, M., and J. A. Rowson (1996), Blocking in a system on a chip, *IEEE Spectrum*, Vol. 36 #11, pp. 35–41.

IEEE (1999a), P1450 Web site *http://grouper.ieee.org/groups/1450/*.

IEEE (1999b), P1500 Web site *http://grouper.ieee.org/groups/1500/*.

Immaneni, V., and S. Raman (1990), Direct access test scheme: design of block and core cells for embedded ASICs, *Proc. IEEE International Test Conference*, pp. 488–492.

Marinissen, E. J., et al. (1998), A structured and scalable mechanism for test access to embedded reusable cores, *Proc. IEEE International Test Conference*, pp. 448–457.

Matson, G. (1997), Structuring STIL FOR scan test generation based on STIL, *Proc. IEEE International Test Conference*, pp. xxx.

Monzel, J. (1997), Low cost testing of system-on-chip, *Proc. Digest of 1st IEEE International TECS*.

Murray, B. T., and J. P. Hayes (1996), Testing ICs: getting to the core of the problem, *IEEE Computer*, Vol. 29, No. 11, pp. 32–38.

Pouya, B., and N. A. Touba (1997), Modifying user-defined logic test access to embedded cores, *Proc. IEEE International Test Conference*, pp. 60–67.

Ricon, A. M., et al. (1997), Core design and system on a chip integration, *IEEE Des. Test Comput.*, Vol. 14, No. 4, pp. 26–35.

Touba, N. A., and B. Pouya (1997), Testing embedded cores using partial isolation rings, *Proc. 15th IEEE VLSI Test Symposium*, April, pp. 10–15. Also, *IEEE Des. Test Comput.*, Vol. 14, No. 4, pp. 52–58.

Varma, P., and S. Bhatia (1997), A structured test reuse methodology for core-based system chip, *Proc. IEEE International Test Conference*, pp. 294–302.

VSI (1998), VSI Alliance Web site *http://www.vsi.org/*.

Whetsel, L. (1997), An IEEE 1149.1 based test access architecture for ICs with embedded IP cores, *Proc. IEEE International Test Conference*, pp. 488–491.

Wohl, P., V. T. Williston, and J. Waicukauski (1997), A unified interface for scan test generation based on STIL, *Proc. IEEE International Test Conference*, pp. 1011–1019.

Zorian, Y. (1993), A distributed BIST control scheme for complex VLSI devices, *Proc. 11th IEEE VLSI Test Symposium*, pp. 6–11.

Zorian, Y. (1997), Test requirements for embedded core-based systems and IEEE P-1500, *Proc. IEEE International Test Conference*, pp. 191–199.

Zorian, Y., et al. (1998), Testing embedded-core based system chips, *Proc. IEEE International Test Conference*, pp. 135–149.

PROBLEMS

15.1. List the three main types of IP cores and compare the steps used in their design and testing.

15.2. You are to design an SOC that includes three hard cores, one soft core, and some logic that you defined. Use the format of Fig. 14.2 to represent the synthesis and optimization steps needed to complete the design.

15.3. The supply of the test patterns to the various cores on an SOC and the retrieval of the response to these patterns are two of the bottlenecks in testing these ICs. What would you recommend to alleviate this problem?

15.4. What test strategy would you suggest for a SOC that uses only hard cores and some user-defined logic?

15.5. SOC is likened to PCB. Is Boundary Scan appropriate to use for SOCs? List advantages and disadvantages.

15.6. What is meant by reuse in the context of SOC? How can test reuse facilitate testing of SOCs?

APPENDIX A

BIBLIOGRAPHY

I. BOOKS

Abramovici, M., *Digital Systems Testing and Testable Design*, Computer Science Press, New York, 1990.

Agrawal, V. D., and S. C. Seth, *Test Generation for VLSI Chops*, IEEE Computer Society Press, Los Alamitos, CA, 1988.

Amersekera, E. A., and D. S. Campbell, *Failure Mechanisms in Semiconductor Devices*, Wiley, New York, 1987.

Arazi, B. A., *Commonsense Approach to Error Correcting Codes*, MIT Press, Cambridge, MA, 1988.

Bakoglu, H. B., *Circuits, Interconnections and Packaging*, Addison-Wesley, Reading, MA, 1990.

Banerjee, P., *Parallel Algorithms for VLSI Computer-Aided Design*, Prentice Hall, Upper Saddle River, NJ, 1994.

Bardell, P. H., W. H. McAnney, and J. Savir, *Built-in Test for VLSI: Pseudorandom Techniques*, Wiley, New York, 1987.

Bateson, J., *In-Circuit Testing*, Van Nostand Reinhold, New York, 1985.

Beenker, F. P. M., R. G. Bennetts, and A. P. Thijssen, *Testability Concepts for Digital ICs: The Macro Test Approach*, Kluwer Academic, Norwell, MA, 1995.

Bennetts, R. G., *Introduction to Digital Board Testing*, Crane, Russak, New York, 1982.

Bennetts, R. G., *Design of Testable Logic Circuits*, Addison-Wesley, Reading, MA, 1984.

Bhattacharya, D., and J. P. Hayes, *Hierarchical Modeling for VLSI Circuit Testing*, Kluwer Academic, Norwell, MA, 1990.

Bleeker, H., P. ven den Eijnden, and F. de Jong, *Boundary-Scan Test: A Practical Approach*, Kluwer Academic, Norwell, MA, 1993.

Breuer, M. A. (ed.), *Design Automation of Digital Systems*, Prentice Hall, Upper Saddle River, NJ, 1972.

Breuer, M. A., and A. D. Friedman, *Diagnosis and Reliable Design of Digital Systems*, Computer Science Press, New York, 1976.

Chang, H. Y., E. G. Manning, and G. Metze, *Fault Diagnosis of Digital Systems*, Wiley-Interscience, New York, 1970.

Chakravarty, Sreejit, and Paul J. Thadikaran, *Introduction to IDDQ Testing*, Kluwer Academic, Norwell, MA, 1998.

Cortner, J. M., *Digital Test Engineering*, Wiley, New York, 1987.

Crouch, Alfred L., *Design for Test*, Prentice Hall, Upper Saddle River, NJ, 1999.

David, R., *Random Testing of Digital Circuits: Theory and Applications*, Marcel Dekker, Paris.

Davis, B., *The Economics of Automatic Testing*, McGraw-Hill, London, 1982.

Eichelberger, E. B., E. Lindbloom, J. Waicukauski, and T. W. Williams, *Structured Logic Testing*, Prentice Hall, Upper Saddle River, NJ, 1991.

Fee, W. G., *Tutorial: LSI Testing*, IEEE Computer Society Press, Los Alamitos, CA, C., 1978.

Feugate, R. J., and S. M. McIntyre, *Introduction to VLSI Testing*, Prentice Hall, Upper Saddle River, NJ, 1988.

Friedman, A. D., and P. R. Menon, *Fault Detection in Digital Circuits*, Prentice Hall, Upper Saddle River, NJ, 1971.

Fujiwara, H., *Logic Testing and Design for Testability*, MIT Press, Cambridge, MA, 1980.

Ghosh, A., S. Devadas, and A. R. Newton, *Sequential Logic Testing and Verification*, Kluwer Academic, Norwell, MA, 1992.

Hachtel, G. D., and F. Somenzi, *Logic Synthesis and Verification Algorithms*, Kluwer Academic, Norwell, MA, 1996.

Healy, J. T., *Automatic Testing and Evaluation of Digital Integrated Circuits*, Reston Publishing Company, Reston, VA, 1981.

Jensen, F., and N. E. Peterson, *Burn-in*, Wiley, Chichester, West Sussex, England, 1982.

Karpovsky, M. G. (ed.), *Spectral Techniques and Fault Detection*, Academic Press, San Diego, CA, 1983.

Kohavi, Z., *Switching and Finite Automata Theory*, 2nd ed., McGraw-HIll, New York, 1978.

Koren, I., *Defect and Fault Tolerance in VLSI Systems*, Plenum Press, New York, 1989.

Krstic, A., and K.-T. Cheng, *Delay Fault Testing for VLSI Circuits*, Kluwer Academic, Norwell, MA, 1998.

Lala, P. K., *Digital Testing and Testability*, Academic Press, San Diego, CA, 1997.

Lala, P. K., *Fault Tolerant and Fault Testable Hardware Design*, Prentice Hall International, London, 1985.

Lee, M. T.-C., *High-Level Test Synthesis of Digital VLSI Circuits*, Artech House, Norwood, MA, 1997.

Mahoney, M., *DSP-Based Testing of Analog and Mixed-Signal Circuits*, Tutorial, IEEE Computer Society Press, Los Alamitos, CA, 1987.

Maunder, C. M., and R. E. Tulloss, *The Test Access Port and Boundary Scan Architecture*, IEEE Computer Society Press, Los Alamitos, CA, 1990.

Michel Petra, U. Lauther, and P. Duzy, (eds.), *The Synthesis Approach to Digital System Design*, Kluwer Academic, Norwell, MA, 1992.

Miczo, A., *Digital Logic Testing and Simulation*, Harper & Row, New York, 1986.

Miller, D. M. (ed.), *Developments in Integrated Circuit Testing*, Academic Press, San Diego, CA, 1987.

Needham, W., *Designer's Guide to Testable ASIC Devices*, Van Nostrand Reinhold, New York, 1991.

Nicolaidis, M., Y. Zorian, and D. Pradhan, *On-Line Testing for VLSI*, Kluwer Academic, Norwell, MA, 1998.

Parker, Kenneth, P., *The Boundary-Scan Handbook Analog and Digital*, Kluwer Academic, Norwell, MA, 1998.

Parker, K. P., *Integrating Design and Test: Using CAE Tools for ATE Programming*, IEEE Computer Society Press, Los Alamitos, CA, 1987.

Pradhan, D. K. (ed.), *Fault-Tolerant Computing theory and Techniques*, Vols. I and II, Prentice-Hall, Upper Saddle River, NJ, 1986.

Pynn, C., *Strategies for Electronics Test*, McGraw-HIll, New York, 1986.

Rajski, Janusz, and Jerzy Tyszer, *Arithmetic Built-In Self-Test*, Prentice Hall, Upper Saddle River, NJ, 1998.

Rajsuman, R., *Iddq Testing for CMOS VLSI*, Artech House, Norwood, MA, 1995.

Rajsuman, R., *Digital Hardware Testing: Transistor-Level Fault Modeling and Testing*, Artech House, Norwood, MA, 1992.

Reghbati, H. K., *Tutorial: VLSI Testing and Validation Techniques*, IEEE Computer Society Press, Los Alamitos, CA, 1985.

Ronse, C., *Feedback Shift Registers*, Springer-Verlag, Berlin, 1984.

Roth, J. P., *Computer Logic, Testing, and Verification*, Computer Science Press, New York, 1980.

Russell, G., and I. L. Sayers, *Advanced Simultation and Test Methodologies for VLSI Design*, Van Nostrand Reinhold, New York, 1989.

Russell, G., et al., *CAD Test Pattern Generation*, Van Nostrand Reinhold, New York, 1990.

Sachdev, M., *Defects Oriented Testing for CMOS Analog and Digital Circuits*, Kluwer Academic, Norwell, MA, 1998.

Sharma, A. K., *Semiconductor Memories: Technology, Testing, and Reliability*, IEEE Press, Piscataway, NJ, 1997.

Singh, N., *An Artificial Intelligence Approach to Test Generation*, Kluwer Academic, Norwell, MA, 1987.

Sivaraman, M., and Strojwas, A. J., *A Unified Approach for Timing Verficiation and Delay Fault Testing*, Kluwer Academic, Norwell, MA, 1998.

Stevens, A. K., *Introduction to Component Testing*, Addison-Wesley, Reading, MA, 1986.

Stover, A. C., *ATE: Automatic Test Equipment*, McGraw-Hill, New York, 1984.

Timoc, C. C., *Selected Reprints on Logic Design for Testability*, IEEE Computer Society Press, Los Alamitos, CA, 1984.

Tsui, F. F., *LSI-VLSI Testability Design*, McGraw-Hill, New York, 1986.

van de Goor, A. J., *Testing Semiconductor Memories: Theory and Practice*, Comptex Publishing, Gouda, The Netherlands, 1998.

Wang, F. C., *Digital Circuit Testing*, Academic Press, San Diego, Ca, 1991.

Welkins, B. R., *Testing Digital Circuits: An Introduction*, Van Nostrand Reinhold, Wokingham, Berkshire, England, 1986.

Williams, T. W. (ed.), *VLSI Testing*, North-Holland, Amsterdam, 1986.

Yarmolik, V. N., *Fault Diagnosis of Digital Circuits*, Wiley, New York, 1990.

Yarmolik, V. N., and I. V. Kachan, *Self-Testing VLSI Design*, Elsevier, Amsterdam, 1993.

Zorian, Y., *Multi-Chip Module Test Strategies*, Kluwer Academic, Norwell, MA, 1998.

II. PERIODICALS

A. IEEE Publications*

IEEE Computer

IEEE Design and Test of Computers

IEEE Journal Solid State Circuits

IEEE Transactions on Circuits and Systems

IEEE Transactions on Computer-Aided Design

IEEE Transactions on Computers

IEEE Transactions on Electronics

IEEE Transactions on Reliability

B. British Publications

Electronics Letters

IEE Proceedings, Part E

IEE Proceedings, Part G

C. Other Publications

Bell System Technical Journal

Electronics Test

IBM Journal of Research and Development

International Journal of Electronic Testing, Theory and Applications (JETTA)

Test and Measurement World

VLSI System Design

III. CONFERENCES AND SYMPOSIA

ATS: *Asian Test Symposium*

CICC: Custom Integrated Circuits Conference

DAC: Design Automation Conference

DATE: Design, Automation and Test in Europe

DFTS: Defect and Fault-Tolerance Symposium

FTCS: Fault-Tolerant Computing Symposium

ICCAD: International Conference of Computer-Aided Design

ICCD: International Conference on Computer Design

IEEE Reliability Physics Symposium

ISCAS: International Symposium on Circuits and Systems

VTS: IEEE VLSI Test Symposium

IV. WEB SITES

IEEE Computer Society:
http://computer.org/
Home page

http://computer.org/conferences/
Conference

http://computer.org/conferences/calendar.htm
Conference Calendar

http://computer.org/cspress/catalog0.htm
Publications Catalog

http://computer.org/publications
Publications

http://computer.org/tab/tttc/
Test Technology Technical Council

Relevant Standards:
http://grouper.ieee.org/groups/1450
P1450 Web site

http://grouper.ieee.org/groups/1500
P1500 Web site

Association of Computing Machinery (ACM)
http://www.acm.org/dl/alpha_list.html

Benchmark Repository
http://www.cbl.ncsu.edu/www/

Institute of Electrical and Electronic Engineers (IEEE)
http://www.ieee.org/
IEEE National

http://www.ieee.org/conferences/tag/tag.html
IEEE Conferences

http://www.ieee.org/products/periodicals.html
IEEE Periodicals

http://www.ieee.org/usab
IEEE-USA

http://www.vsi.org/
Home of VSIA

APPENDIX B

TERMS AND ACRONYMS

ALAP	as late as possible
ALFSR	autonomous linear feedback shift register
ASAP	as soon as possible
ASIC	application-specific integrated circuits
ASM	algorithmic state machine
ATE	automatic test equipment
ATPG	automatic test pattern generator
ATS	algorithmic test sequence
BDD	binary decision diagram
BF	bridging fault
BIC	built-in current
BILBO	built-in logic blocks observer
BIST	built-in self-test bit lines
BL	bit line
BSC	boundary-scan cells
BSDL	boundary-scan description language
BSR	boundary-scan register
CAD	computer-aided design
CAM	content addressable memories
CDFG	control data flow graph
CF	coupling fault
CF_{ids}	idempotent fault
CF_{ins}	inversion coupling
CMOS	complementary MOS
CUT	circuit under test
DA	design automation

DAG	directed acyclic graph
DATS	direct access test scheme
DFG	data flow graph
DFT	design for testability
DIBL	drain-induced barrier lowering
digraph	directed graph
DOD	Department of Defense
DPM	defects per million
DR	data registers
DRAM	dynamic random access memory
DRC	design rule checker
DS	distinguishing sequence
DSP	digital signal processing
DWG	development working groups
EBIU	external bus interface unit
EDIF	electronic description interchange format
EEPROM	electrically erasable programmable read-only memories
EPROM	erasable programmable read-only memories
ESD	electrostatic discharge
ET	embedded test
FAN	fan-out-oriented test generation
FB	feedback
FF	fault-free
FPGA	field-programmable gate arrays
FSM	finite state machines
GDF	gate delay fault
GUI	graphic user interface
HDL	hardware description language
HLS	high-level synthesis
HS	homing sequence
I_{DDQ}	CMOS quiescent supply current
IEEE	Institute of Electrical and Electronic Engineers
ILA	iterative logic array
IP	intellectual property
IR	instruction register
ITG	iterative test generation
JTAG	joint test action group
LFSR	linear feedback shift register
LSB	least significant bit
LSSD	level-sensitive scan design
LVS	logic versus schematic
MAR	memory address register
MDR	memory data register
MFVS	minimal feedback vertex set
MISR	multiple input signature register

MPGA	mask-programmable gate array
MSA	multiple stuck-at fault
MST	minimal spanning tree
MTBM	mean time between metastability
MTC	maximum test concurrency
MTTF	mean time to failure
MUT	module under test
NFB	nonfeedback
NPSF	neighborhood pattern-sensitive fault
NS	next state
NTC	nontaped cores
OEM	original equipment manufacturer
OVI	Open Verilog International
P&R	placement and routing
PCB	printed circuit board
PDF	path delay fault
PI	primary input
PLA	programmable logic array
PLD	programmable logic device
PO	primary output
PODEM	path-oriented decision making
PR	pseudorandom
PROM	programmable read-only memory
PS	present state
PSA	parallel signature analysis
PSF	pattern-sensitive fault
RAM	random access memory
ROM	read-only memory
RPR	random pattern resistant
RTL	register transfer level
SA0	stuck at zero
SA1	stuck at one
SAF	stuck-at fault
SCF	state coupling faults
SE	scan enable
SEM	scanning electron microscope
SEU	single-event upset
SFT	synthesis for testability
SI	scan-in
SO	scan-out
SOC	system on a chip
SON	stuck-on fault
SOP	stuck-open fault
SPICE	simulation program, IC emphasis
SPU	serial port unit

SRAM	static random access memory
SRSG	shift register sequence generator
SS	synchronizing sequence
SSA	single signature analyzer, single stuck-at
SSN	simultaneous switching noise
STA	static timing analysis
STG	state transition graph
STSR	self-test SR registers
STUMPS	self-test using MISR and a parallel shift register sequence generator
TAP	test access port
TAM	test access mechanism
TCAD	technology CAD
TCK	test clock
TDI	test data input
TDO	test data output
TECS	testing embedded core-based system
TF	transition fault
TM	testability measure
TML	TAP multiplexing link
TMS	test mode select
TRST	test reset
TS	transition sequence
TTTC	Test Technology Technical Council
UDL	user-defined logic
UDP	user-defined primitive
UUT	unit under test
UVPROM	UV erasable programmable read-only memory
VHDL	very high speed integrated circuit hardware description language
VSI	virtual socket interface
VSIA	virtual socket interface alliance
WL	word line

INDEX